Fundamentals of Project Management

– Tools and Techniques –

kepublishing.com
rkepublishing.com

Fundamentals of Project Management

Rory Burke

ISBN: 978-0-9582733-6-7

Published: 2010
Reprinted: 2010

Email: rory@burkepublishing.com

Website: www.burkepublishing.com

Website: www.knowledgezone.net (Instructor's Manual)

Distributors: **UK:** Marston Book Services Limited, email: trade.orders@marston.co.uk

USA: Partners Book Distributing, email: saraspeigel@partners-east.com

South Africa: Blue Weaver Marketing, email: orders@blueweaver.co.za

Australia: Thames and Hudson, email: orders@thaust.com.au

Hong Kong: Publishers Associates Ltd (PAL), email: pal@netvigator.com

DTP: Sandra Burke

Cover Design: Simon Larkin

Sketches: Buddy Mendis, Michael Glasswell

Printer: Everbest, HK / China

Production notes: Page size (168 x 244 mm), Body Text (Minion Pro, 11 pt), Chapter Headings (Helvetica, bold, 60 pt), Subheadings (Helvetica, bold, 11 pt), Software InDesign, Photoshop, Illustrator, CorelDRAW, PC and Mac notebook computers.

ISBN: 978-0-9582733-6-7

Dedicated to Kath and Derek, who always have a plan!

'Help Derek, I'm taking off!'

'Hold on Kath I'm just checking the tent plan.'

Content

Foreword

Author's Note

Chapter 1	**Introduction to Project Management**	**18**
	1. What is a Project?	21
	2. Project Work vs Production Line Work	23
	3. Types of Projects	26
	4. What is Project Management?	29
	5. Types of Management	31
	5.1 Programme Management	31
	5.2 Portfolio Management	32
	5.3 General Management	32
	5.4 Technical Management	33
	6. Project Management Software	33
	7. Project Management Environment	33
	8. Role of the Project Manager	34
	9. Benefits of Using Project Management Techniques	35
Chapter 2	**History of Project Management**	**38**
	1. Project Management Techniques	39
	2. Program Evaluation and Review Techniques (PERT)	40
	3. Project Organisation Structures	42
	4. Project Management Triangle	43
	5. History of Project Management Computing	44
Chapter 3	**Project Management Standards**	**46**
Chapter 4	**Project Integration Management**	**52**
	1. Project Charter	54
	2. Project Reviews and Project Closeout Report	59
	3. Questionnaire	60
Chapter 5	**Project Management Process**	**62**
	1. Fayol's Management Process	65
	2. Eastonian Process	66
	3. Project Management Process	67
	4. Initiating Process	68
	5. Planning Process	69
	6. Executing, Monitoring and Controlling Process	70
	7. Closing Process	71
	8. Process Levels	73

Chapter 6	**Project Management Plan**	**74**
	1. What Is a Plan?	76
	2. Project Planning Steps	78
	3. Project Planning Documents	79
	4. Planning and Control Cycle	83
Chapter 7	**Project Lifecycle**	**84**
	1. Project Lifecycles	86
	2. Project Lifecycle (4 phases)	87
	3. Project Lifecycle (Generic)	88
	4. Level of Effort	90
	5. House Extension Project Lifecycle (4 phases)	92
	6. Phase-to-Phase Relationship (fast tracking)	94
	7. Lifecycle Methodology (Input, Process, Output)	95
	8. Product Lifecycle (8 Phases)	96
Chapter 8	**Feasibility Study**	**98**
	1. Stakeholders Analysis	101
	2. Define the Client's Needs	105
	3. Internal Project Constraints	107
	4. Internal Corporate Constraints	110
	5. External Constraints	111
	6. Evaluate Options and Alternatives	112
	7. Feasibility Study Summary Template	113
Chapter 9	**Scope Management**	**114**
	1. Scope Definition	117
	2. Scope Verification	119
	3. Scope Change Control	120
Chapter 10	**Work Breakdown Structure (WBS)**	**128**
	1. Checklists	131
	2. WBS Methods of Subdivision	132
	3. Product Breakdown Structure (PBS)	134
	4. WBS	136
	5. Work Packages	137
	6. Cost Breakdown Structure (CBS)	138
	7. Organization Breakdown Structure (OBS)	139
	8. Activity List	140
	9. Numbering System	141
	10. PBS, WBS, OBS, CBS Interfaces	142
	11. WBS Templates	143
	12. Spreadsheet Presentation	144
Chapter 11	**Time Management (Estimating Time)**	**146**
	1. WBS / Activity List	150
	2. Definition of an Activity	151
	3. Definition of an Event	152
	4. Activity Duration	153

Chapter 12	**Critical Path Method (CPM)**	**156**
	1. Network Diagram	159
	2. Logical Relationships	160
	3. How to Draw the Logical Relationships	161
	4. Activity Logic - Tabular Reports	162
	5. Critical Path Method Steps	163
	6. Forward Pass	164
	7. Backward Pass	167
	8. Activity Float	169
Chapter 13	**Gantt Chart**	**174**
	1. How to Draw a Gantt Chart	177
	2. Tabular Reports	178
	3. Activity Float	179
	4. Hammock Activities	180
	5. Events, Key dates, Milestones and Deadlines	181
	6. Rolling Horizon Gantt Chart	182
	7. Revised Gantt Chart	183
Chapter 14	**Procurement Schedule**	**186**
	1. Procurement Process	189
	2. Procurement Schedule	194
	3. Expediting	198
Chapter 15	**Resource Planning**	**200**
	1. How to Draw the Resource Histogram	203
	2. Resource Loading	205
	3. Resource Smoothing	206
Chapter 16	**Project Cost Management (Estimating Costs)**	**210**
	1. Estimating Costs	213
	2. Direct Costs	213
	3. Indirect Costs	214
	4. Fixed and Variable Costs	214
	5. Labour Costs	215
	6. Material Costs	217
	7. Unit Rates	218
	8. Assign Budgets	219
	9. Budget Format	220
	10. Cost Control	221
Chapter 17	**Project Cashflow**	**222**
	1. Project Cashflow Statement	225
	2. Cashflow Timing	227
	3. Cost Distribution	228
	4. How to Draw an Expense S Curve	230
	5. Benefits of Using a Project Cashflow Statement	232

Chapter 18	**Project Execution, Monitoring and Control**	**234**
	1. Planning and Control Spiral	236
	2. Project Control Cycle	238
	3. Baseline Plan	240
	4. Monitor Progress (Data Capture)	243
	5. How to Apply Project Control	248
	6. Reporting Frequency	248
Chapter 19	**Earned Value**	**250**
	1. Earned Value Terminology	251
	2. Earned Value Graph	253
	3. Earned Value Table	254
	4. Project Control	254
	5. Client's View of Earned Value.	255
	6. Earned Value Reporting	255
Chapter 20	**Project Risk Management**	**258**
	1. Risk Management Model	261
	2. Define Objectives	263
	3. Risk Identification	263
	4. Risk Quantification	266
	5. Risk Response	266
	6. Risk Control	268
Chapter 21	**Quality Management**	**270**
	1. Quality Definitions	273
	2. Quality Planning	275
	3. Quality Assurance	276
	4. Quality Control	277
	5. Quality Control Plan (QCP)	278
Chapter 22	**Project Communications Management**	**280**
	1. Communication Theory	283
	2. Communication Plan	286
	3. Communication Stakeholders	287
	4. Communication Content	288
	5. Communication Method	289
	6. Communication Timing	289
	7. Document Control	290
	8. Project Reporting	293
Chapter 23	**Project Meetings**	**296**
	1. How to Prepare a Meeting	298
	2. Handover Meeting	301
	3. Project Progress Meetings	303
	4. Brainstorming Workshops	304

Chapter 24	**Project Organization Structures**	**306**
	1. Functional Organization Structure	308
	2. Matrix Organization Structures	310
	3. Responsibility-Authority Gap	313
Chapter 25	**Human Resource Management (Project Teams) ...**	**316**
	1. Purpose of Project Teams	319
	2. Team Charter	320
	3. Team Development Phases	324
	4. Team Building Techniques	326
Chapter 26	**Project Management Office**	**328**
	1. Site Office	329
	2. Matrix Organization Structure	330
	3. Centre of Excellence	331
	4. Management-by-Projects	336
	5. Mobile Project Office	336
Chapter 27	**Managing Small Projects**	**338**
	1. Define Small Projects	339
	2. Managing Small Projects	341
	3. Product Lifecycle	342
	4. Small Business Management	344
Chapter 28	**Event Management**	**346**
	1. What is an Event?	347
	2. Project Management Techniques	348
Appendix 1 - Solutions		354
Booklist		366
Glossary		368
Index		

Abbreviations

The Fundamentals of Project Management includes the following abbreviations:

ABS – American Bureau of Standards

ACPM - Association of Construction Project Managers

AIPM - Australian Institute of Project Management

AoA – Activity-on-Arrow

AoN - Activity-on-Node

APM BoK – Association of Project Managers body of knowledge

AV – Actual Value

BAC – Budget at Completion

B2B – Business to Business

BoM – Bill of Materials

BS – British Standard

CBS – Cost Breakdown Structure

CEASA - Cost Engineering Association of South Africa

CEO – Chief Executive Officer

CND – Campaign for Nuclear Disarmament

CPM – Critical Path Method

CV – Cost Variance

DCF - Discounted Cashflow

DIN – Deutsches Institut fur Normung

DoD – Department of Defense

EAC – Estimate at Completion

EV – Earned Value

FRI - Forecast Rate of Invoicing

HRM – Human Resource Management

IPMA – International Project Management Association

ISO – International Organization for Standardization

JIT – Just-In-Time

LoE – Level of Effort

MBE - Management-by-Exception

MRP - Material Requirement Planning

MTBF – Mean Time Between Failures

NASA - National Aeronautics and Space Administration

NCR – Non Conformance Report

NQF – National Qualification Framework

OBS – Organization Breakdown Structure

PC – Percentage Complete

PC – Personal computer

PBS - Product Breakdown Structure

PERT – Program Evaluation and Review Technique

PDM - Precedence Diagram Method

PID - Project Initiation Document

PMBOK – Project Management Body of Knowledge

PMI – Project Management Institute

PMO – Project Management Office

PMP – Project Management Plan

PMP – Project Management Professional

PMSA - Project Management South Africa

PO – Project Office

PRINCE2 - PRojects IN Controlled Environments (second edition)

PV – Planned Value

QA – Quality Assurance

QC – Quality Control

QCP – Quality Control Plan

RAM – Responsibility Assignment Matrix

R&D – Research and Development

SAQA – South African Qualifications Authority

SOW – Scope of Work

SV – Schedule Variance

VO - Variation Order

TQM – Total Quality Management

WBS – Work Breakdown Structure

Project Management Series

Project Management Techniques
ISBN 978-0-9582733-4-3
Project Management BOK
Project Lifecycle
Feasibility Study
Planning and Control Cycle
Scope Management
Estimating Techniques
Risk Management
CPM
Gantt Charts
Procurement Management
Resource Management
Project Control
Earned Value
Communication / Meetings
OBS / Matrix

Project Management Leadership
ISBN 978-0-9582733-5-0
Leadership Styles
Team Creation
Team Development Phases
Team Dynamics
Delegation
Problem Solving
Decision Making
Creativity and Innovation
Negotiation
Conflict Resolution
Communication Theory
Team Meetings
OBS/Matrix
Outdoor Teambuilding

Project Management Series

Advanced Project Management
Implement Corporate Strategy
Project Methodology
Strategy Phase
Statement of Requirements
Business Case
Feasibility Study
Configuration Management
Design Phase
Planning Phase
Execution Phase
Commissioning Phase
Operation Phase
Project Finance
Issues Management
PMMM Maturity Models

Project Management Entrepreneur
ISBN 978-0-9582733-2-9

Entrepreneurs Toolkit
ISBN 978-0-9582391-4-1

Small Business Entrepreneur
ISBN 978-0-9582391-6-5

Foreword

If you are looking at this book you are probably looking for something that will introduce you to the world of managing projects. Perhaps you are an engineer who is working to deadlines and budgets, a team member who has to monitor and report project performance, or an entrepreneur starting a new venture. Whoever you are *Fundamentals of Project Management* will start your journey into a world of projects that will present some surprises, challenge some perceptions and lead you to some places you never knew existed!

Managing project work can be compared to juggling. Learning how to juggle needs an iterative, step-by-step process that involves trial-and-error, learning from your mistakes, getting it wrong and trying again. Of course, there is also *'getting it right'* and noticing what works and how to make it even better. In producing this book, Rory has applied that very technique. This book includes some new topics that have come from advances made in the project management profession. Particularly, updates to the Project Management Institute (PMI), Association of Project Management (APM) and the International Association of Project Management (IPMA) professional bodies' presentation of PM competence and knowledge. *Fundamentals of Project Management* provides those people new to project management, access to the latest thinking in the profession in a very accessible manner.

There is another side to managing projects, other than the techniques and tools presented here. Significantly, projects involve people. Indeed, without people most (if not all) projects would never be completed. Unfortunately, the opposite is also true - because of people, many projects are not completed! How we involve people in projects is critical to project success and significant in avoiding project failure.

This is why Rory and I have produced another book: *Project Management Leadership*, that provides helpful (and critical) information about how to involve and work with people on projects. It helps you to understand how to work with individual people and with teams of people; you will read how behaviours can affect project work and how this often changes over the life-cycle of a project. *Project Management Leadership* is easily connected with the content of *Fundamentals of Project Management* to provide a comprehensive toolkit for managing projects effectively.

You can work through the content of this book and learn all about the techniques and how to apply them. I am sorry to say it is likely this will not be sufficient to be a good project manager. If you want to be a juggler, could you claim to be able to juggle after reading a book? Of course not! To be able to juggle you need to be able to throw balls (for example) in the air and be able to catch them. The only way to do this is to practice throwing and catching (please don't try that just now, especially with this book), maybe starting with one item and progressing to more as your skill progresses.

It is the same with projects. Reading this book is just the start. You need to try out the techniques on your projects, possibly one at a time. Try out one technique that helps to solve an immediate problem, then try another. Over time you will collect a number of tools for your project management toolbox. A carpenter generally uses only one tool at a time. However, throwing things in the air one at a time does not make a juggler. The real skill of a project manager is getting the techniques to work together in concert, each one feeding information to the others until the project is a result of *'joined up'* organizing. This is not a *'mission impossible'* but the result of having a go and persistence in trying out the techniques until they work for you.

There is one last use of the juggling metaphor that I would like to leave with you. One of the key skills of the juggler is not about knowing where the balls ARE, but rather where they WILL BE. The hands of the juggler need to be ready to catch the ball at the end of its trajectory, not in mid air. The project management techniques have a similar prediction about where the project work is heading and helps to consider consequences of action.

So, like the juggler, you need to keep your balls in the air, think about where they are going to be and, most importantly, try not to drop one!

I wish you well in your journey into project management.

Steve Barron
Lancaster University
Director of the MSc in Project Management

'The project management juggler!'

Author's Note

Fundamentals of Project Management has been written to explain the fundaments of the project management planning and control techniques. The content of this book has been updated to include the new edition of the *Project Management Body of Knowledge* (fourth edition) [PMBOK 4ed] and the new edition of the *Association of Project Management Body of Knowledge* (fifth edition) [APM BoK 5ed].

This book includes worked examples and exercises to help explain how to calculate and apply the key project management principles, tools, techniques and processes.

It is **ideal for project managers** who are entering the field of project management and need a solid platform of project management techniques to manage small projects or sub-projects which have a limited scope of work, limited number of resources, a small budget and are of limited complexity.

This book is also ideal for **project team members** who need to understand the basic principles of project management so that they can support the project manager and carry out the project administration functions within the project office. As a team member this might involve gathering and processing project data, monitoring and reporting project progress, managing scope change control, administrating documentation control and expediting progress.

Learning Outcomes: The aim of this book is to help project managers and project team members acquire the **competency** and **knowledge** they need to calculate and process all the project management planning and control tools and techniques. It will also explain the content of the body of knowledge to give a clear understanding of its content, terminology and application.

Unit Standards: There are a number of countries offering national certificates in project management linked to their own unit standards. This book is mapped to SAQA 50080 (level 4) unit standards. A level 4 qualification recognises specialist learning and involves detailed analysis of a high level of information and knowledge in an area of work or study. Learning at this level is appropriate for people working in technical and professional jobs.

Body of Knowledge Mapping: A new feature of *Fundamentals of Project Management* is the body of knowledge mapping at the beginning of most chapters. The relevant knowledge area sections and unit standard outcomes are mapped to the chapters within the book. This should help lecturers in their course development and course approval.

Chapter Sequence: The sequences of the chapters are structured to strike a balance between the sequence used by the PMBOK 4ed knowledge areas and the modules used in project management courses.

Motivation for *Fundamentals of Project Management*:

1. To incorporate new content from the new PMBOK 4ed and the new APM BoK 5ed.
2. To align the chapters with the PMBOK headings and project courses modules.
3. To add more worked examples and exercises.
4. To start each chapter with a body of knowledge mapping to help link the chapters to the BoKs.

Chapter	Changes to *Introduction to Project Management*
Ch 1: Introduction to Project Management	Introduce the Project to Production Line Continuum. Expand section on Types of Projects.
Ch 2: History of Project Management	No changes
Ch 3: Project Management Standards	No changes
Ch 4: Project Integration	New chapter title to include an expanded Project Charter and the Project Closeout Report.
Ch 5: Project Management Process	New chapter title to explain how to manage the Project Management Processes (initiating, planning, execution, controlling and closing). Fayol's management process and the Eastonian process.
Ch 7: Project Management Plan	New chapter title, previously called Project Planning and Control Cycle.
Ch 5: Project Lifecycle	New Level of Effort example.
Ch 8: Scope Management	Expanded Scope Change Management and Configuration Management
Ch 9: Feasibility Study	Expanded Stakeholders' Analysis.
Ch 10: WBS	Expanded Methods of Subdivision, PBS, WBS, OBS and CBS, and include RAM / RACI diagrams.
Ch 11: Estimating Time	New chapter title to focus on Estimating Time.

Ch 12: CPM	Additional worked examples.
Ch 13: Gantt Chart	Additional worked example and exercises.
Ch 14: Procurement Schedule	Additional worked example and exercises.
Ch 15: Resource Planning	Additional worked example and exercises.
Ch 16: Estimating Costs	New chapter title to focus on Estimating Costs.
Ch 17: Project Cashflow	Additional worked example and exercises.
Ch 18: Project Execution, Monitoring and Control	New chapter title to explain the project progress measurement and control cycle.
Ch 19: Earned Value	Minor changes.
Ch 20: Risk Management	Minor changes
Ch 21: Quality Management	Expanded section on Quality Planning.
Ch 22: Communication Management	Expanded section on Document Control.
Ch 23: Project Meetings	Minor changes
Ch 24: Project Organization Structure	Minor changes
Ch 25: Project Teams	Additional new sections on the Team Charter, Team Development Phases, and Team Building Techniques.
Ch 26: Project Management Office	Minor changes
Ch 27: Managing a Small Project	Minor changes
Ch 28: Event Management	Minor changes
Glossary	Expanded

Lecturers: PowerPoint slides can be downloaded from: <www.knowledgezone.net>

Note: For clarity the project manager will always be referred to in the male gender, it is clearly understood that this equally applies to the female gender.

Acknowledgements: I wish to thank my international industry and education contacts who have supported me with this edition; with a special thank you to:

Australia: Rex Glencross Grant (UNE), and David Farewell (UNISA)

Canada: Sheryl Staub-French (UBC)

HK: Steve Rawlinson (HKU)

Ireland: Dermot Reidy (Dublin) and Bert Hamilton (Limerick)

NZ: John Tookey (Auckland)

SA: Mark Massyn (UCT), Libby Robb (Wits), Rolf Kuhnast (Damelin), Pieter Steyn (Cranefield), PD Rwelamila (UNISA), Les Labuschagne (UJ), and Gerrit Van Der Waldt (Potchefstroom)

UK: Miles Shepherd (IPMA), Neal Allen (Bath), Michael Haughton (Henley), John Power (PRINCE2), Geoff Reiss (Leeds), and Neil Alderman (Newcastle)

USA: John Colville (JC Consulting)

Foreword: Steve Barron for his inspirational foreword.

Proof Reading: Sandra Burke and Jan Hamon for their eagle eyes.

Sketches: Michael Glasswell for his creative sketches.

Cover: Simon Larkin for updating the cover.

Rory Burke
Bay of Islands

Interview at the BBC

1

Introduction to Project Management

Learning Outcomes

After reading this chapter, you should be able to:

- Define a project and compare it with business-as-usual
- Define project management and compare it with other types of management
- Outline the role of the project manager
- List the benefits of project management

Fundamentals of Project Management will explain how to use the latest project management tools and techniques to plan and control the administration of projects. This book is ideal for project managers and project team members who are new to the field of project management and have limited experience and knowledge of managing projects.

This book is structured in line with:

- The Project Management Body of Knowledge 4th edition (PMBOK 4ed)
- The Association of Project Managers Body of Knowledge 5th edition (APM BoK 5ed)
- The SAQA Unit Standards 50080 Certificate in Project Management (level 4).

It is designed to support short courses and project managers studying for their Project Management Professional (PMP) certification.

This book will benefit project team members who need to understand the special techniques of project management to support the project manager and help run the administration of the project management office (PMO). As a project team member this involves:

- Developing the project management plan
- Gathering and processing data
- Monitoring and reporting project progress
- Administrating scope change control, build-method, and configuration management
- Administration, communication, reporting and document control
- Expediting progress (particularly procurement).

Body of Knowledge Mapping

All the chapters in this book will start with a Body of Knowledge (BoK) mapping for the PMBOK 4ed, the APM BoK 5ed and the unit standards 50080 (level 4). The mapping will link the body of knowledge requirements to the chapters where the topics are discussed. The PMBOK 4ed, APM BoK 5ed and SA unit standards use different structures; therefore they will not always align perfectly with each chapter.

PMBOK 4ed: The *Introduction to Project Management* knowledge area includes the following:

PMBOK 4ed	Mapping
1.2: What is a project?	This chapter will define a project.
1.3: What is project management?	This chapter will define project management.
1.4: The relationship between project management, program management, and portfolio management.	This chapter will outline the relationship between project management, program management and portfolio management.
1.6: Role of the project manager.	This chapter will outline the role of the project manager.
1.7: Project management body of knowledge.	See *Project Management Standards* chapter.

These topics form the basis of the first module of most project management courses.

APM BoK 5ed: The *Project Management in Context* knowledge area includes the following:

APM BoK 5ed	Mapping
1.1: Project management	This chapter will define project management.
1.2: Programme management	This chapter will define programme management.
1.3: Portfolio management	This chapter will define portfolio management.
1.5: Project sponsor	See *Feasibility Study* chapter.
1.6: Project office	See *Project Management Office* chapter.

For this knowledge area the APM BoK 5ed lists similar requirements to the PMBOK 4ed.

Unit Standard 50080 (Level 4): The unit standard for the *Introduction to Project Management* knowledge area includes the following:

Unit 120372: *Explain fundamentals of project management*

Specific Outcomes	Mapping
120372/SO1: Explain the nature of a project.	This chapter will explain the nature of a project.
120372/SO2: Explain the nature and application of project management.	This chapter will explain the application of project management.
120372/SO3: Explain the types of structures that are found in a project environment.	See *Project Organization Structures* chapter.
120372/SO4: Explain the application of organization structures in a project environment.	See *Project Organization Structures* chapter.
120372/SO5: Explain the major processes and activities required to manage a project.	See *Project Management Process* chapter.

1. What is a Project?

The starting point for most project management courses is to define a 'project' and compare it with 'business-as-usual' or ongoing work. It is important to make this distinction because this difference underpins the whole purpose of using a project management approach and, therefore, the whole purpose of using project management tools and techniques.

This section will introduce the widely accepted definitions of a project and the features of a project, together with a number of examples of the different types of projects.

The PMBOK 4ed defines a **Project** as, *a temporary endeavour undertaken to create a unique product, service or result.*

The APM BoK 5ed defines a **Project** as, *a unique, transient endeavour undertaken to achieve a desired outcome.*

These two definitions are similar in content, where **temporary** and **transient** indicate the project has a finite duration, which means that a project must have a start and finish. And **unique** means each project creates a product (goods or services) which is different from the product (goods or services) created by previous projects.

Features of a Project: Considering the above definitions, the features of a project include the following:

Client and Project Sponsor	A project has a client and/or a project sponsor – they are responsible for making sure the project acquires tangible benefits for the company.
Stakeholders	A project has stakeholders – stakeholders are the people who are impacted upon by the project or, conversely, have an impact on the project. Stakeholders should be identified, and their needs and expectations analysed (see *Feasibility Study* chapter).
Project Lifecycle	A project has a lifecycle - this means the project passes through a number of phases. Typically these phases are; feasibility phase, design phase, execution phase and commissioning phase (see *Project Lifecycle* chapter).
Project Charter	A project has a project charter – this document officially initiates the project by setting out what the project is to achieve, and how it is to be managed. The project charter is owned by the project sponsor (see *Project Integration Management* chapter).
Project Manager	A project has a project manager who is appointed by the project sponsor and is acknowledged as the single point of responsibility. The project manager is responsible for delivering the project on time, within budget and to the required quality.
Non-Repetitive	A project is non-repetitive. They are one-offs - this means each project is different to previous projects. The difference could be in scope, or any of the other knowledge areas.

Scope of Work	A project has a **unique** scope of work (SOW) which defines what the project should achieve as well as the work that is not included (see *Scope Management* chapter).
Work Breakdown Structure	The Work Breakdown Structure (WBS) technique is used to subdivide the work into manageable work packages for better management and control (see *Work Breakdown Structure* chapter).
Schedule and Timeline	A project is **temporary** - this means that besides a start and a finish date for the project, there are also start and finish dates for all the activities, together with start dates for any milestones (see *Critical Path Method* chapter).
Procurement Schedule	A project has a procurement schedule – projects usually procure goods and services from outside the company. To do this the project team develops a procurement list and a procurement schedule (see *Procurement Schedule* chapter).
Resource Planning	A project uses resources (machines and people) to perform the work - the resource availability must mirror the resource loading otherwise there will have to be a trade-off. This is presented as a resource histogram (see *Resource Planning* chapter).
Budget	A project has a budget - this means an agreed amount of money is assigned to implement the project. Using a WBS the budget can be assigned at the work package level and even at the activity level. With tighter budgets comes tighter control (see *Project Cost Management* chapter).
Cashflow Statement	A project has a cashflow statement – a project cashflow statement can be produced by integrating the project schedule with the budget (see *Project Cashflow* chapter).
Risk	A project is confronted by risks – this means risks need to be identified, quantified and a response developed (see *Project Risk Management* chapter).
Quality	A project has a quality requirement – this means the required quality needs to be defined, planned and controlled (see *Quality Management* chapter).
Communication	Projects are run by good communication – this means the communication requirements need to be defined, planned and controlled (see *Project Communications Management* chapter).
Matrix OBS	A project has an Organization Breakdown Structure (OBS) – as project teams usually do not have their own work force many projects use workers from other departments operating through a matrix OBS (see *Project Organization Structures* chapter).
Project Team	A project has a project team – as projects are temporary a project team might also be temporary and brought together especially to manage the project. To centralise the project team, they typically operate out of a project office (PO).

The definitions of a project and the features of a project should help to establish the boundaries of where the project management approach can be effectively applied.

2. Project Work vs Production Line Work

One way to define project work is to compare it with production line work (also referred to as business-as-usual), where project work is temporary and unique, and production line work is ongoing and repetitive.

In practice, not every scope of work or job exactly fits the definition of a project or production line work, but has an element of both. This implies a continuum of possible situations. There are three project continuums this section will discuss, namely:

- Efficiency and productivity continuum
- Sales and marketability continuum
- Risk and uncertainty continuum.

Efficiency and Productivity Continuum: The efficiency and productivity continuum considers how to reduce the cost of making the project by increasing the efficiency and productivity. Although project work might be considered to be unique there is usually a certain amount of work/jobs which are similar. If these can be grouped together this should increase efficiency and productivity and, therefore, reduce manufacturing costs.

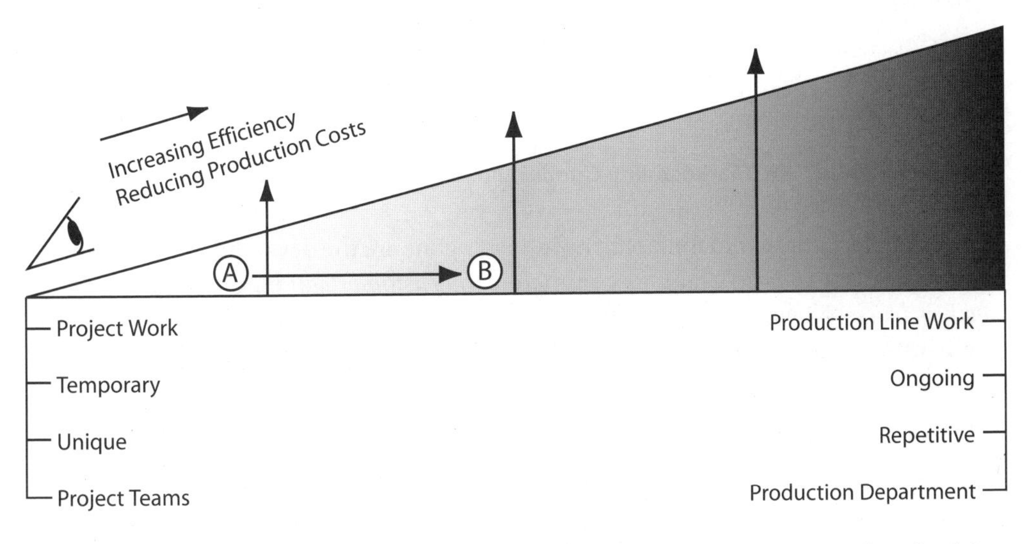

Figure 1.1: Efficiency and Productivity Continuum – shows how efficiency and productivity increase as the work moves from a one-off project to production line work

For example, building a ship is usually considered to be a project because it is temporary (it has a start and finish) and it is unique (the design of each ship is different). However, while the ship is being built there are many jobs which are repetitive, for example, cutting the frames, welding the frames, fitting the pipe work, and painting the steelwork. Therefore, it makes economic sense to perform all these jobs as a production run. Figure 1.1 shows how this essentially moves the project along the continuum from project work [A] towards production line work [B].

Sales and Marketability Continuum: If production line work is more cost effective than project work, why make one-offs? This is where the sales and marketing function considers how to make the project more marketable (appealing, attractive, desirable) to increase the sales figures. In a commercial environment, increased sales figures usually relate to achieving competitive advantage over the competition. Besides reducing the retail price, competitive advantage can also be achieved by adding more features and making the project/product more unique.

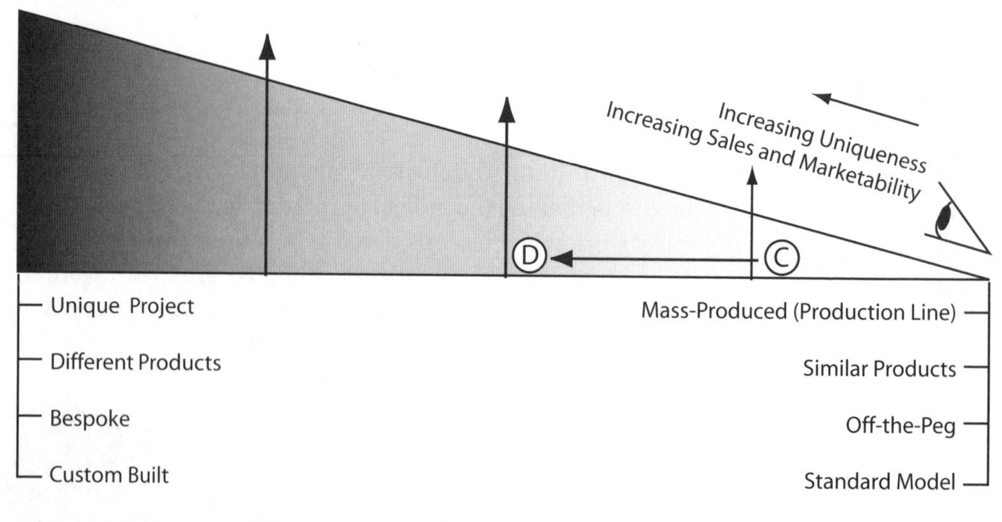

Figure 1.2: Sales and Marketability Continuum

For example, in the car manufacturing industry gone are the days when a car salesman can repeat Henry Ford's famous comment, '*The customer can have any color as long as it is black.*' Now it is essential to give the customers what they want to meet their demands. Consider the following:

- A range of colours
- A range of engine sizes
- A choice of leather or cloth seat covers
- The option of a sunroof and spoilers
- The choice of tinted or clear windows
- A range of audio and video facilities
- A range of exhausts
- A choice of mag wheels and low profile radials.

In figure 1.2, by offering this range of features the car manufacturer is essentially moving along the continuum from [C] to [D].

Risk and Uncertainty Continuum: The risk and uncertainty function considers how to reduce the risk of the project not achieving its objectives, or expressed the other way round, how to ensure the project meets its objectives.

As the level of risk is related to the amount of change, one would, therefore, assume that production line risk is relatively lower than project work risk. This is because production line work is similar on a day-to-day basis, whereas project work is continually changing.

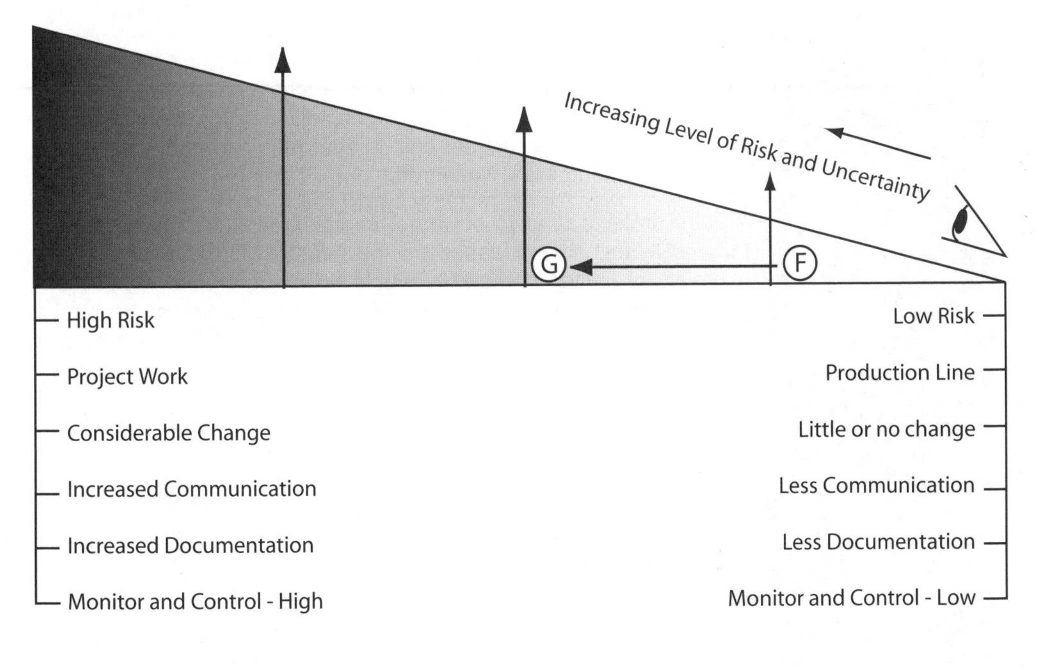

Figure 1.3: Risk and Uncertainty Continuum

For example, consider the car production line again. If each car is made exactly the same then the amount of information required is much less than if every car has a different specification. As cars become more unique this, effectively, moves the management process along the continuum from [F] to [G].

The project management response to the higher level of risk is a higher level of management. This is achieved through effective communication, controlled documentation, detailed scope management (build-method and configuration), scheduling, resource planning, procurement planning, project cashflow planning and quality management.

The above highlights that project managers should manage a project's work content to best suit the client's needs, the project's needs and the company's needs. This means their management approach will be dynamically positioned along the continuum between project work and production work.

Project management is, therefore, a balance between the cost of effective project management, the cost of mistakes, and the competitive advantage of making a unique project.

3. Types of Projects

Historically, project management techniques were developed for large capital projects within the construction industry, defence procurement and petrochemical type projects. But, in recent years, most proactive industries, commercial businesses and government departments have re-structured some of their work as projects.

The following introduces different types of projects that project managers new to project management might be involved with:

Construction Projects	A construction project of limited complexity would be; the construction of a small building, a sub-project within the construction of a bridge, or a sub-project within the construction of a power station. Construction projects are typically unique, built in situ, requiring temporary site offices and a temporary workforce.
Defence Projects	Military projects tend to be large, complex capital projects using innovative technology based on the latest R&D. An example of a project of limited complexity would be a sub-project within the procurement of a weapon system. Many of the planning and control techniques used today were initially developed for military projects.  ***'This new model definitely has increased fire power!'***
Petrochemical	Petrochemical projects include the search for oil and gas (particularly offshore), and also the large scale refining of fuels. These major projects are often in remote and dangerous locations, making logistics and communications a challenge. Petrochemical installations are typically large with high volume flows to gain an economy of scale to reduce unit costs. Safety standards are a big concern when handling such highly inflammable products. An example of a project of limited complexity would be a sub-project within the design and construction of a large petrochemical facility.
IT Projects	IT projects range from large capital projects, such as, computerizing the medical records of everyone in the UK, to designing a small personal website. Large IT projects have a history of difficulty in defining the scope of the project and, consequently, have been plagued by budget blowouts and schedule overruns. An example of an IT project of limited complexity would be the designing and hosting of a website for a small business.

Product Development Projects	Product development projects include the design and development of consumer products. In this highly competitive market, project teams need to fast-track the design and development so that they can beat their competitors to market. Examples of product development projects would include the designing, developing and testing of; a prototype car, a washing machine, and the iPod nano. ***'I think this iPod will be a winner!?'***
Advertising and Marketing Project	Most companies launch their new products with an advertising and marketing campaign. These campaigns have the characteristics of an integrated project that needs to work in conjunction with the new product's release schedule and deadlines.
Bank Project	Banks and financial service companies are not usually thought of as project orientated companies, but designing and implementing a new banking product has all the characteristics of a project and, typically, banks have many financial products. Implementing a new mortgage product for a bank would involve co-ordinating the staff training, the IT systems, and the product brochures between hundreds of branches, in conjunction with an advertising and marketing launch date.
Up-Grade Project	Up-grade projects are required to keep products competitive in a world of constantly changing technology. Although repairing and maintaining equipment might be considered ongoing work and business-as-usual, an upgrade or half-life refit has all the characteristics of a project. Examples of an upgrade project would include; implementing the latest computer hardware and software, or installing new generators for a power station.
Event Management	Events, exhibitions, conventions and shows typically have all the characteristics of a project as they are temporary (happen on a certain day) and unique (different from the previous events). They have the distinct lifecycle profile of a long preparation phase, followed by a very short implementation phase. This introduces many special challenges for the project manager and project team members. See *Event Management* chapter.

Sports Events	Sporting events can range from the local school's sports day to the Football World Cup. International sporting events are characterised by large facilities, large crowds, and international media communication. There is an interesting distinction between a one-off sporting event, such as the Football World Cup held every four years, and Manchester United's weekly football matches. Manchester United's football matches attract large numbers of supporters but, as they are regular events, would tend to be managed as business-as-usual rather than as separate projects.
Music Concerts	Music concerts range from a small gig at the local pub to a massive, supergroup rock concert. When the BBC's *Top Gear* team offered to move the WHO's gear to their next gig, they were told they would need 1000 transit vans to move the gear in two days. (Although Pete Townsend does not break his guitars on stage anymore, he still requires 30 guitars.) This type of project is characterised by transporting large amounts of gear to different venues within a tight schedule, and having everything working perfectly when the stage lights up.
Fashion Show	Fashion runway shows are popular events for fashion designers to showcase their latest collections to the fashion buyers and the media. These fashion shows are characterized by a fast turnaround of garment changes by the fashion models against a background of glamour and fashion hype.
Disaster Recovery Project	Disaster recovery projects are the ultimate risk management contingency plan. Although the frequency of disasters might be low, the impacts can be very high. For example, companies that experience a major fire or flood could lose access to their premises for some time. If they do not have backups off site and contingency plans in place they will probably never recover because, by the time they have re-established their premises and supply chain, their loyal clients will have moved on. The development of a disaster recovery plan would be considered a project of limited complexity.

Table 1.1: Types of Projects – shows the different types of projects

Within the context of this book a project may be defined as implementing a change, event, solution, or an opportunity for a new venture that uses a range of special project management techniques. These techniques are used to plan and control the scope of work in order to deliver a product to satisfy the client's and stakeholders' needs and expectations.

4. What is Project Management?

Project management is a special management technique designed to manage projects. The PMBOK and APM BoK both have similar definitions of project management:

The PMBOK 4ed defines **Project Management** as, *the application of knowledge, skills, tools, and techniques to project activities in order to meet project requirements.*

The APM BoK 5ed defines **Project Management** as, *the process by which projects are defined, planned, monitored, controlled and delivered such that agreed benefits are realised.*

The PMBOK and APM BoK definitions introduce a number of new terms which, in the context of project management, mean the following:

Project Requirements: Project requirements refer to the whole purpose of the project – this could be to solve a problem, address a need, or take advantage of an opportunity. These requirements are documented in the statement of requirements (see *Project Charter* in *Project Integration Management* chapter). Determining the stakeholders' requirements is discussed in the *Feasibility Study* chapter as the stakeholders' needs and expectations.

Knowledge: In the context of project management, the application of knowledge is referred to as acquiring an understanding of the body of knowledge's nine knowledge areas (integration, scope, time, costs, quality, procurement, resources, communication, and risk). These topics are developed as separate chapters.

Knowledge refers to the expertise and skills acquired by project managers through their experience and education, and results in a theoretical or practical understanding of project management. Having knowledge in project management implies that the project managers know the relevant facts and information, and have an awareness or familiarity gained by experience working on projects. The term *knowledge* is also used to mean the confident use of project management techniques with the ability to use it to achieve specific objectives.

Skills: A skill (also referred to as a talent) may be defined as an ability or aptitude to perform something well (with the minimum amount of effort and minimum time). A project management expert would have the necessary domain-specific skills to understand how to apply project management techniques. Perfecting a skill requires a certain amount of preparation and practice as 'practice makes perfect'. For example, to be skilled at playing the piano or to play an excellent round of golf requires plenty of practice.

Tools and Techniques: The term 'tools' and the term 'techniques' are often used interchangeably, but a distinction can be made where tools refer to project management templates and checklists that do nothing on their own, and techniques describe how to use the tools or procedures to accomplish a specific activity or task (see the analogy box).

Analogy of Project Management

Consider a cooking analogy to help define the difference between tools, techniques, skills, processes and outputs.

Tools: In your kitchen your cooking tools are the pots, pans, utensils, oven, and microwave etc. All the equipment that is needed to make the meal, but the tools do nothing on their own.

Skills and Techniques: The skills and techniques define how you use the tools. This could involve chopping, carving, dicing and frying, etc.

Process: The process is the transformation the food goes through when it is cooked. For example, when making a cake the raw ingredients are mixed and put in the oven, then during the cooking process the mixture is transformed and the output is an edible cake.

Now relate this analogy to project management.

Tools: The tools could be a WBS template, or planning software that do nothing on their own.

Skills and Techniques: The skills and techniques define how you use the tools to prepare the project management plan.

Process: The process is the collecting of raw data and transforming it into meaningful reports. For example, capturing project progress and transforming it into an earned value report.

'It's all in the tools and techniques!'

5. Types of Management

Although the primary focus of this book is to explain how to use project management techniques, it is helpful to be aware of the other types of management, and appreciate why project management techniques would be used in preference to these other forms of management to manage a project's scope of work.

3 Ps: Besides being a popular marketing term, the field of project management has its own '3 Ps' namely; project management, programme management and portfolio management. The term project management was originally developed as a generic name to include all forms of management associated with managing projects. But, as the field of project management developed so programme management and portfolio management are rightly being recognised as different forms of management.

5.1 Programme Management

The PMBOK 4ed defines a **Program** as, *a group of related projects managed in a coordinated way to obtain benefits and control not available from managing them individually.*

The APM BoK 5ed defines **Programme Management** as, *a group of related projects, which may include business-as-usual activities, that together achieve a beneficial change of a strategic nature for an organization.*

An example of programme management is the development of a new car that includes a number of smaller projects. Although the individual projects can be managed separately with different project managers managing different scopes of work, budgets and schedules, the projects collectively produce the new car. In this situation programme management is used to co-ordinate the resources over all these related projects.

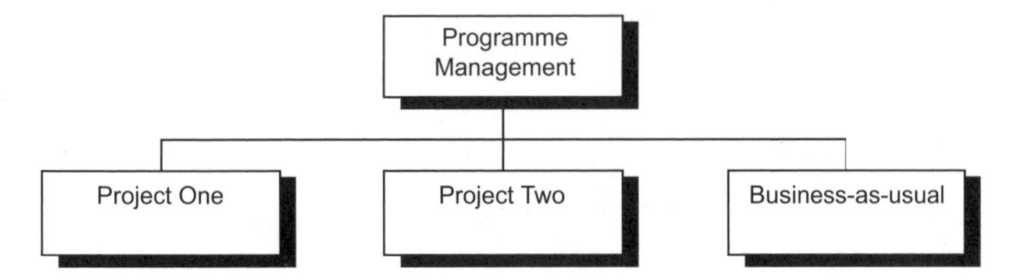

Figure 1.4: Programme Management

5.2 Portfolio Management

The PMBOK 4ed defines a **Portfolio** as, *a collection of projects or programs and other work that are grouped together to facilitate effective management of that work to meet strategic business objectives. The projects and programs of the portfolio may not necessarily be interdependent or directly related.*

The APM BoK 5ed defines **Portfolio Management** as, *the grouping of projects, programmes and other activities carried out under the sponsorship of an organization.*

An example of portfolio management is a project office where the project team are managing a range of unrelated projects, programmes and business-as-usual. For example, a city council's project department might be managing a number of individual house building projects, one large stadium programme and many business-as-usual maintenance and repair tasks (fixing a street light, filling a pothole, or repairing a park bench).

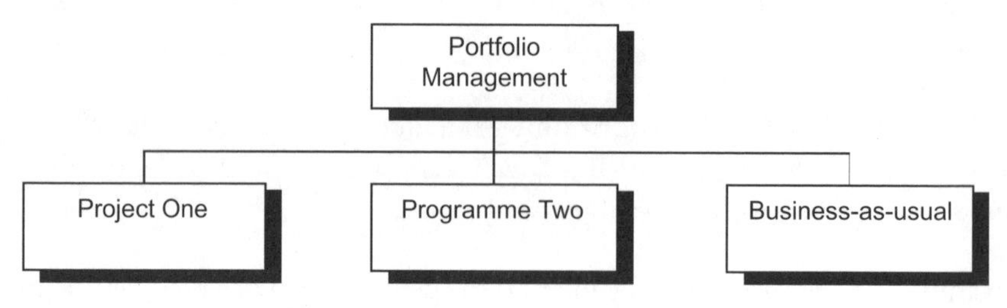

Figure 1.5: Portfolio Management

5.3 General Management

Although the focus of this book is project management, the successful project manager and project team members must also be competent in a wide range of general management skills which includes the following:

- Staffing
- Customer service
- Computer systems / electronic filing
- Legal contracts
- Personnel and human resources
- Sales and marketing
- Accounts and salaries.

Obviously the project manager would not be expected to be an expert in all these fields but, for a project to be successful, all these areas might need to be addressed at one time or another. As the project manager is the single point of responsibility it is his responsibility to delegate the work to team members or outsource as required.

5.4 Technical Management

Technical management skills include the technical skills and product knowledge required to design and manufacture the product or project. Every profession has its unique range of technical skills and competencies which are required to be able to perform the work.

Technical management skills are required to achieve the functionality of the project and, therefore, technical management has a key input into configuration management and scope management which includes the project feasibility study, build-method and approval of scope changes.

On smaller projects, the project manager might be expected to be the technical expert as well as the manager of the project. In fact, an inexperienced person is unlikely to be appointed as project manager unless he is a technical expert in the field.

6. Project Management Software

Today, powerful but inexpensive project management software is readily available for the personal computer. This has essentially moved project management computing away from the data processing department to the project office or project manager's desk. This represents a major shift in the management and communication of project information.

Whilst project planning software will certainly help project managers plan and control their projects, its application will only be effective if the planning and control techniques are clearly understood by the project managers and the project team members. The purpose of this text is, therefore, to develop these planning and control techniques through manual examples and exercises.

7. Project Management Environment

The project environment directly influences the project and how it should be managed. Projects are not carried out in a vacuum; they are influenced by a wide range of stakeholders and issues. Consider the following topics that could influence a project:

- The needs and expectations of a range of stakeholders (and interested parties)
- The client's and project sponsor's requirements
- The company's organization structure
- Market requirements
- Competitors' products and pricing strategy
- New technology
- Rules and regulations (health and safety)
- The economic cycle (bull markets and recessions).

For project managers to be effective they must have a thorough understanding of the project's environment. The project environment consists of the numerous stakeholders and players that have an impact on, or are impacted by the project. All must be managed as any one person could derail the project (see *Feasibility Study* chapter).

8. Role of the Project Manager

Project roles and responsibilities relate to a person's role (position and duties) within the project organization structure. The two key roles considered in this text are the project manager and the project team members. (See *Human Resource Management* chapter for project teams.)

Project Manager: The project manager is the owner of the project management plan and, therefore, responsible for delivering the project on time, within budget, and to the agreed quality. The role of the project manager should be outlined in the project charter together with details of how the project should be managed (see *Project Integration Management* chapter). The following lists some desirable project manager attributes:

Team Creation	Ability to select and develop a project team from a standing start.
Leadership	Ability to lead, motivate and manage all the project participants.
Problem-Solving	Ability to anticipate problems, solve problems and make decisions.
Stakeholders	Ability to integrate the project stakeholders and address their needs and expectations.
Flexibility	Ability to respond flexibly to changing situations and to make the most of opportunities.
Expediting	Ability to plan, expedite and get things done.
Negotiation	Ability to influence and negotiate with the project participants.
Environment	Ability to understand the environment within which the project is being managed.
Control	Ability to monitor and apply control to keep the project on track.
Contract	Ability to administer the contract.
Scope Control	Ability to manage the scope of work and scope changes which includes build-method and configuration management.
Closeout	Ability to learn lessons from the project phase reviews and project closeout reports.
Client	Ability to keep the client happy.

Table 1.2: Project Manager's Role

Experience has shown that the selection of the project manager is a key appointment which can influence the success or failure of the project. As the single point of responsibility, it is the project manager who integrates and co-ordinates all the contributions from the stakeholders, and guides them to successfully complete the project.

9. Benefits of Using Project Management Techniques

For project managers and project team members who are new to project management it is helpful to appreciate the benefits of using project management techniques as opposed to other forms of management.

As the project manager is responsible for achieving the project's objectives and deliverables, he is also responsible for setting up the project's planning and control system to achieve these objectives and deliverables. To do this effectively, the project manager requires accurate and timely information. One of the ways to supply this information is through a project management planning and control system which outlines the scope of work and measures performance against the original plan.

Although the planning and control system will incur additional management costs, it should be appreciated that lack of information could be even more expensive if it leads to poor management decisions, costly mistakes, rework and overruns. Listed below are some of the main benefits associated with using a fully integrated project management planning and control system:

Responsibility	Within a project management system the project manager is made the single point of responsibility. This means only one person is responsible for achieving the project's objectives. This essentially removes underlaps and overlaps.
Client and Project Sponsor	The project management approach is client and project sponsor focused. The project charter gives the project manager the responsibility for achieving the project's objectives, and the authority to use company resources and to represent the company at client meetings. During meetings with the client, the planning and control system will provide information about every aspect of the project. Clients prefer to deal with one person – the project manager – who is accountable, responsible and manages the complete project. Clients do not like being passed around like a football!
Stakeholders	The project management approach is to identify the stakeholders and determine their needs and expectations. This will then form the basis of the project's objectives.
Objectives	Projects have clear goals and objectives which are documented in the statement of requirements and the business case. This helps to focus the project team on the project's scope of work.
Project Scope and the Work Breakdown Structure (WBS)	The work breakdown structure (WBS) subdivides the scope of work into manageable work packages and checklists which are easier to estimate, plan, monitor, control and assign to a responsible person.
Estimating	All plans are underpinned by an estimate. The project management approach is to produce estimates quickly and accurately.
Critical Path Method (CPM)	The critical path method (CPM) calculates the activities' start dates, finish dates and float. The activities with zero float form the critical path which determines the duration of the project. This information enables the project team members to focus on the critical activities that could delay the project.

Fast Track	Changing the logic and crashing activities enables the project manager to get the project to market before the competition.
Scheduled Gantt Chart	The Gantt chart is an excellent document for communicating the what, when and who (scope, schedule and responsibility).
Project Integration	Project integration co-ordinates and integrates the contributions of all the project participants. It limits the underlap and the overlap of work, thus preventing a doubling up of effort and a source of conflict.
Response Time	Timely response on project performance is essential for effective project control. The project planning and control system can adjust the monitoring and frequency of the feedback to address the needs of the project, while the corporate functional systems might be less flexible. Consider the accounts department, for example; it generally uses a monthly reporting cycle where feedback on invoices could be four to six weeks behind timenow.
Trends	Projects are best controlled by monitoring the progress trends of time, cost and performance. This information may not be available to the project manager if the trend parameters are derived from a number of different functional sources. The project manager needs to work through a common data base.
Monitoring and Control Data Capture	If the project progress reporting is based on information supplied by the functional departments, the project manager cannot control the accuracy of this information. The problem here is that it may only become obvious towards the end of the project that the reporting was inaccurate, by which time it might be too late to bring the project back on course in order to meet the project's objectives (see *Project Execution, Monitoring and Control* chapter).
Procedures	The planning and control system enables the project manager to develop procedures and work instructions which are tailored to the specific needs of the project.
Quality Management	A quality management system enables the project manager to set up a quality system to manage the project. The quality control plan can be developed to vary the level of inspection and number of hold points to suit the project. Quality audits enable the project manager to inspect management systems within the company and within the sub-contractor's company.
Closeout Report	The performance of the current project will form the estimating data base for future projects. If this data is not collected by the planning and control system it might be lost forever, and the same mistakes could be repeated in future company projects.

Table: 1.3: Benefits of Project Management

There are many benefits from using a project management approach to managing projects. However, if there is not a culture of managing projects within the company senior management should consider a softly softly approach, as resistance to change could derail future projects.

Exercises:

You have been appointed by the CEO of an international telecommunications company to make a short presentation to the board of directors about the benefit of using a project management approach to manage the company's next large capital project. Your short presentation (written and/or verbal) should consider the following:

1. Explain what project management is, and why it is different to other forms of management.
2. Explain how project management can be applied to your company's projects.
4. Outline the role of the project manager.
5. Suggest a small pilot project on which you can develop and prove your project management systems.

Further Reading:

Burke, R., *Project Management Techniques*, will develop the following topics which relate to this knowledge area:

- Management-by-projects.

Burke, R., **Barron**, S., *Project Management Leadership*, will develop the following topics which relate to this knowledge area:

- Project Leadership
- Team Building.

2

History of Project Management

Learning Outcomes

After reading this chapter, you should be able to:

- Understand the history of project management
- Perform the PERT calculation
- Understand the time, cost, quality triangle trade-off

Modern day project management is associated with Henry Gantt's development of the Gantt chart (early 1900s), and special project management techniques developed during the military and aerospace projects of the 1950s and 1960s in America and Britain. It is these special distinctive project management tools and techniques which are included in the body of knowledge, are used by the planning software, and developed in this book.

Traditionally the management of projects was considered more of an art than a science. With the growing number of project management institutions, associations and academic establishments, project management has become more of a science and an academic discipline as accepted practices are captured and formalised in the global body of knowledge and certificate programmes.

Today, rapidly changing technology, fierce competitive markets and a powerful environmental lobby have encouraged companies to change their management systems - in this sink or swim, adopt or die environment, project management is offering a real solution.

1. Project Management Techniques

Nearly all of the special project management techniques used today were developed during the 1950s and 1960s by the US defence-aerospace industry (DoD and NASA). This includes program evaluation and review technique (PERT), earned value (EV), configuration management, value engineering and work breakdown structures (WBS). The construction industry also made its contribution to the development of the critical path method (CPM) using network diagrams and resource smoothing - the motivation was scheduling urgency. During this period, large capital projects were effectively shielded from the environment, society, and ecology issues. The Apollo space programme and the construction of nuclear power stations typified projects of this period. Some of the key achievements during this time are listed chronologically below:

1950s	Development of PERT and CPM.
1950s	Development of the concept of a single point of responsibility for multi-disciplined projects where one person is made responsible for completing the project. Coupled with this approach came the project team, secondment and resource sharing through a matrix organization structure.
1960	NASA experiments with matrix organization structures.
1963	Earned value adopted by the USAF.
1963	Project lifecycle adopted by the USAF.
1963	The US Navy introduces PERT to plan and control hundreds of sub-contractors on the Polaris submarine project.
1964	Configuration management adopted by NASA to review and document proposed changes.
1965	DoD and NASA move from cost-plus contracts towards incentive type contracts such as firm fixed price or cost plus incentive fee.
1965	The mid 1960s saw a dramatic rise in the number of projects in the construction industry that used modern project management techniques.
1965	The TSR-2 (swing-wing bomber) highlighted the problems of concurrency, i.e. starting the development and production before the design was stable. Increasing the scope of work led to cost overruns and delays - eventually the project was cancelled.
1966	A report in 1966 stated that not enough time was spent on front-end definition and preparation (of the project lifecycle); there were wide variations in standards of cost and schedule control, and inadequate control over design changes.
1967	Founding of the International Project Management Association (IPMA) [formerly called the INTERNET].
1969	Project Management Institute (PMI) formed, certification and the PMBOK (1987, 1996, 2000, 2004) were to follow.

2. Program Evaluation and Review Technique (PERT)

In the late 1950s the US Navy set up a development team under Admiral Red Raborn with the Lockheed Aircraft Corporation, and a management consultant, Booz-Allen & Hamilton, to design PERT as an integrated planning and control system to manage the hundreds of sub-contractors involved in the design, construction and testing of their Polaris Submarine missile system.

The PERT technique was developed to apply a statistical treatment to the possible range of activity time duration. A three time probabilistic model was developed, using pessimistic (p), optimistic (o), and most likely (m) time durations (see figure 2.1). The three time durations were imposed on a normal distribution to calculate the activity's expected time.

In practice one would usually estimate around the most likely time. The optimistic time would be slightly shorter, if everything went better than planned. While the pessimistic time would be extended if everything went worse than planned (late delivery, or a machine breakdown).

The success of the Polaris Submarine project helped to establish the PERT technique in the 1960s as a planning and control tool within many large corporations. At the time the PERT technique was believed to be the main reason the submarine project was so successful, meanwhile CPM was not receiving anywhere near as much recognition even though it also offered a resource allocation and levelling facility.

There were, however, a number of basic problems which reduced PERT's effectiveness and these eventually led to its fall from popularity. Besides the computing limitations, statistical analysis was not generally understood by project managers - they must have been pleased to see the end of standard deviations and confidence limits.

Other features of PERT, however, are seeing a renaissance as the benefits of milestone planning and control is becoming more widely used. By defining the project as a series of milestones, the planner can simplify the planning and control process at his own level and make the sub-contractors responsible for achieving their milestones. Even with the powerful planning software available today there is still a need to empower an increasingly educated workforce.

The early differences between CPM and PERT have largely disappeared and it is now common to use the two terms interchangeably as a generic name to include the whole planning and control process.

Example 1: Using the PERT equation (normal distribution)

Expected time = (o + 4m + p) / 6

o = 6 days

m = 8 days

p = 13 days

T (expected) $\frac{6 + (4 \times 8) + 13}{6} = 8.5$ days

Exercise 2:

o = 12 days

m = 15 days

p = 24 days

What is T (expected)?

See Appendix 1 for solution.

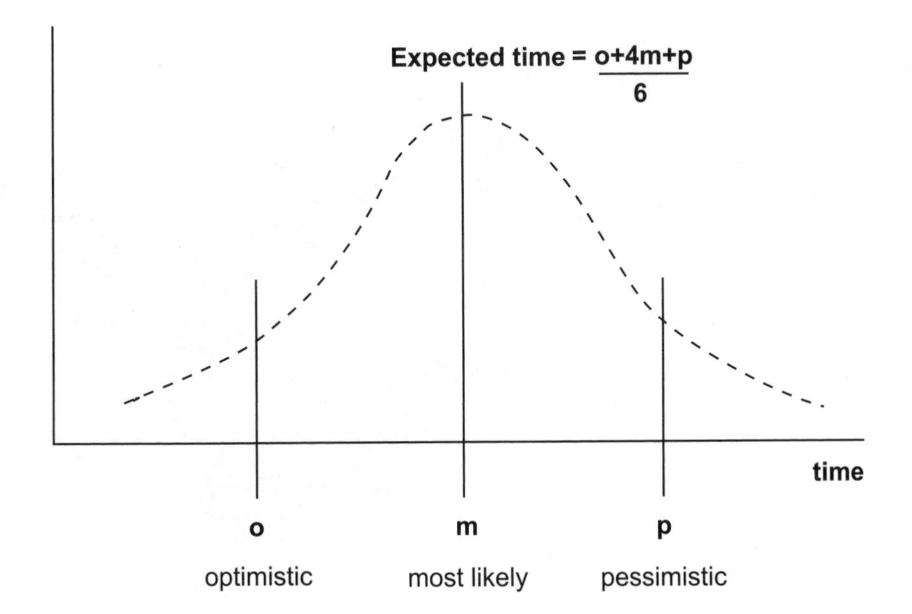

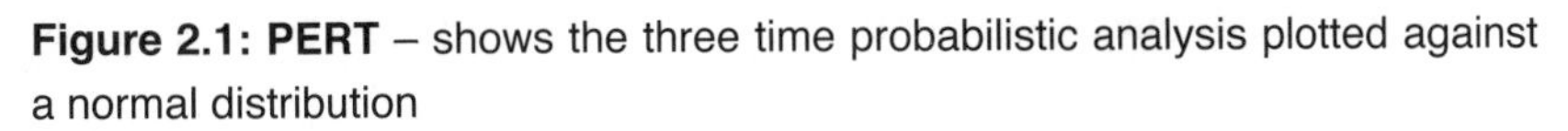

Figure 2.1: PERT – shows the three time probabilistic analysis plotted against a normal distribution

3. Project Organization Structures

Up to the mid 1950s projects tended to be run by companies using the traditional functional hierarchical organization structure, where the project work would be passed from department to department.

In the 1950s, Bechtel was one of the first companies to use a project management organization structure to manage their oil pipeline project in Canada, where responsibility was assigned to an individual operating in a remote location with an autonomous team (see figure 2.2). This is a good example of an organization structure with the project manager as the **single point of responsibility** with autonomous authority over a pool of resources. The norm during this time, (and still the norm for many companies), would be for the head of department or the functional manager to be responsible for the project as it passed through his department. The project management approach is to assign responsibility to one person who would work on the project full-time through the project lifecycle from initiation to completion. In due course, this person came to be called the **project manager**.

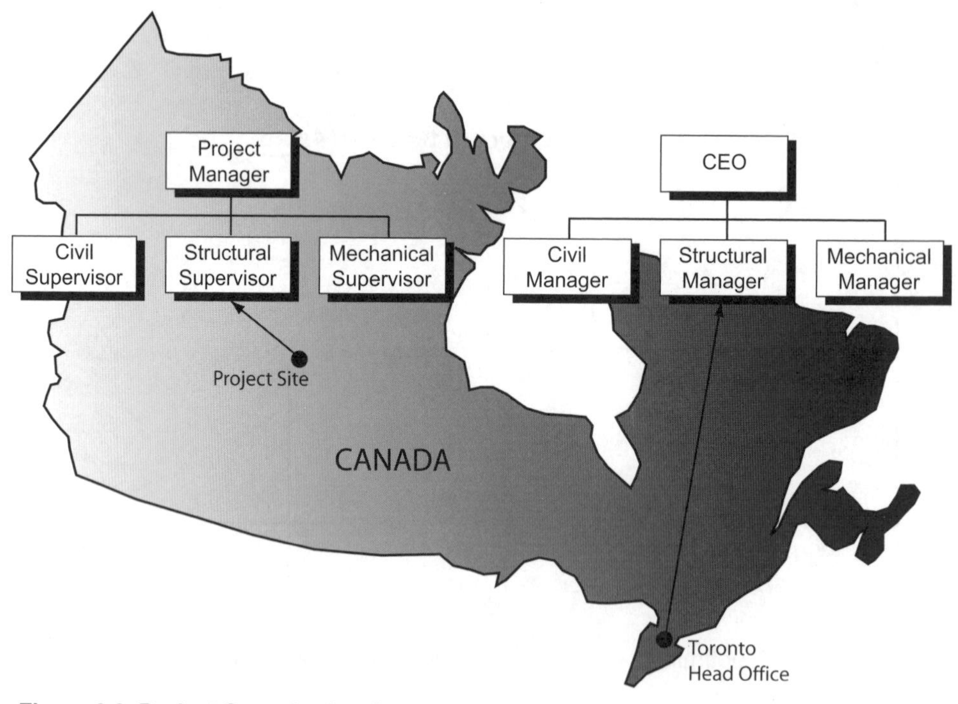

Figure 2.2: Project Organization Structure - shows the project operating in a remote location

4. Project Management Triangle

Project management tools and techniques proliferated in the 1960s, were refined in the 1970s, and were integrated in the 1980s as accepted practices. The integration of time, cost and quality was initially presented as a triangle of balanced requirements; where a change in one parameter could impact on the others (see figure 2.3). In the 1980s there was a significant increase in the influence of external stakeholders, the green issue and the Campaign for Nuclear Disarmament (CND) - this put increasing pressure on project designers to find acceptable solutions for all the stakeholders.

The time, cost quality triangle was later joined by scope and the organization breakdown structure (OBS) to indicate that the scope of work was performed through an organization structure. There was also an increasing awareness of external issues, so the project environment was included (see figure 2.4).

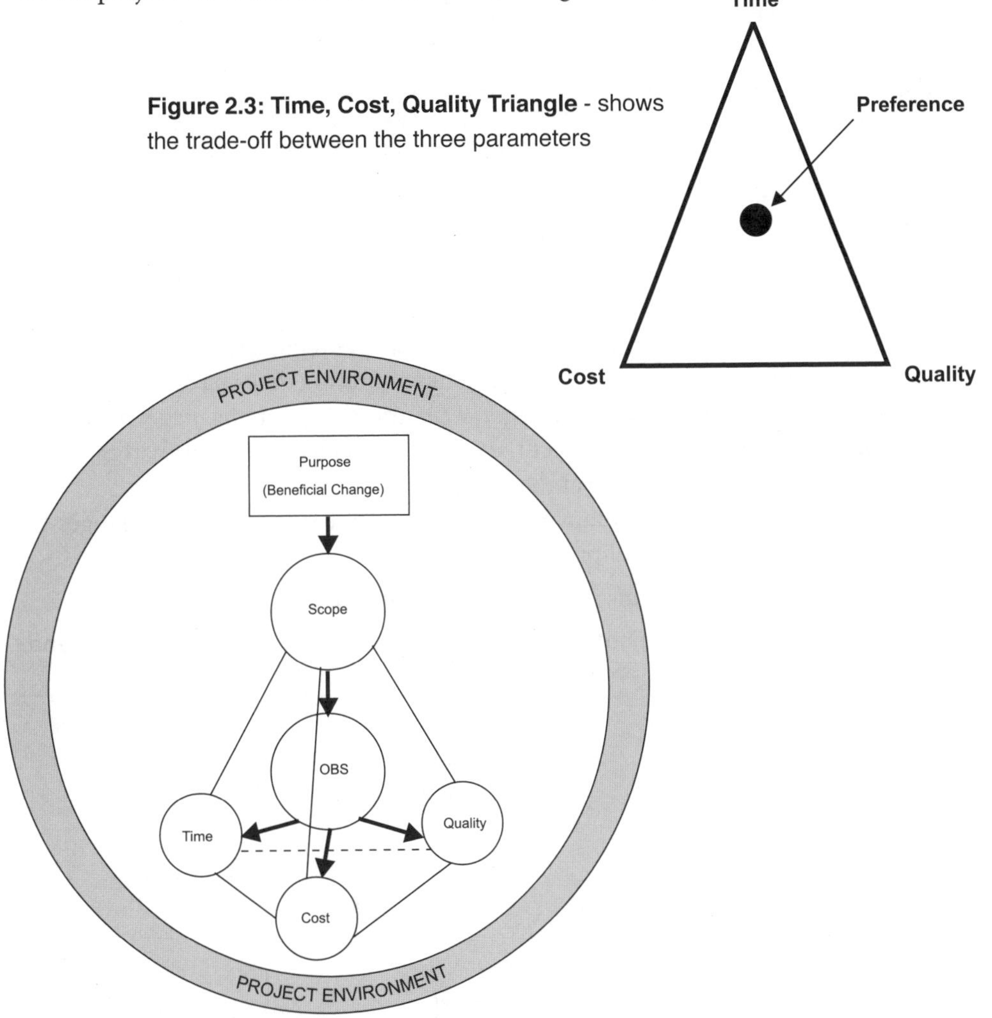

Figure 2.3: Time, Cost, Quality Triangle - shows the trade-off between the three parameters

Figure 2.4: Project Environment Model – shows that the project manager is encouraged to look at the project's bigger picture and consider all the stakeholders' needs and expectations

5. History of Project Management Computing

The development of schedule Gantt charts, network diagrams and other distinctive project management tools were originally developed for manual calculation. These tools were gradually computerised during the 1960s and 1970s using mini and mainframe computers, but it was the introduction of the personal computer (PC) in the 1980s that ushered in a dramatic explosion and proliferation of project management software. Some of the key dates to note are:

1977	Launch of Apple 11 - the first PC.
1979	Launch of VisiCalc, the first spreadsheet - Lotus and Excel were to follow.
1981	Launch of the IBM PC - this established the market standard.
1983	Launch of Harvard Project Manager - the first planning software package.
1990s	Launch of Windows, networks - Internet and email.
2000s	Internet broadband, mobile communication for voice and data; development of web site facilities; B2B (business-to-business) procurement; real-time progress reporting.

The introduction of the PC in the late seventies (Apple 11) and the IBM PC (1981) in the early eighties, with accompanying business software, encouraged the growth of project planning software and the use of project management techniques.

The history of PC based project management computing dates back to 1983 with the launch of the **Harvard Project Manager,** a planning software package. Although this was an isolated event, it does reflect the general development of a broad range of management software taking place at the time.

The development of project management software created two main changes to the manual planning and control process;

1. Change from activity-on-arrow to activity-on-node.
2. Change from departmental planning and control to project planning and control using a common data base.

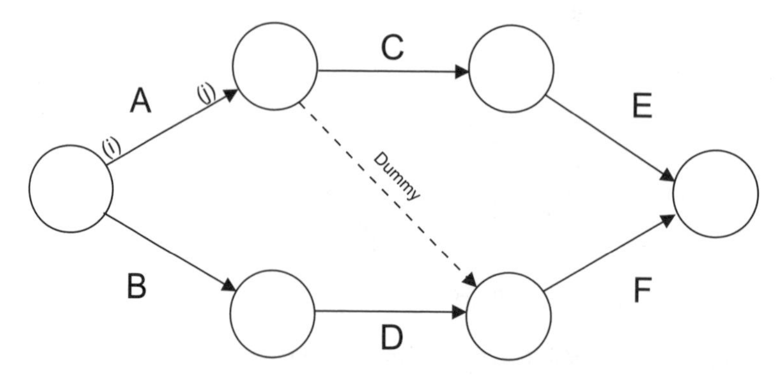

Figure 2.5: Activity-On-Arrow – shows the task's description written on the arrows

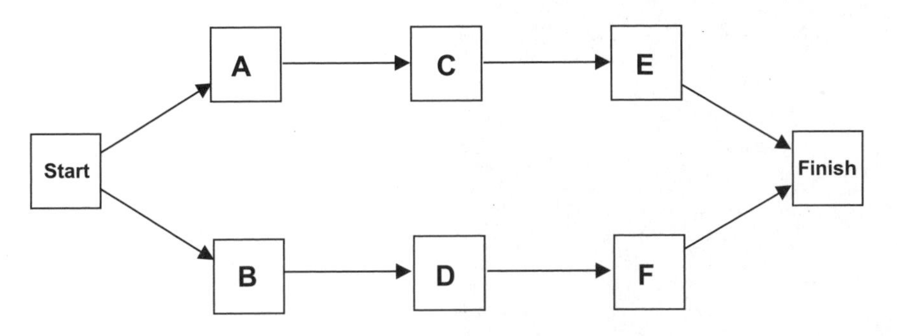

Figure 2.6: Activity-On-Node – shows the task's description written in the boxes

Network diagrams were originally developed as activity-on-arrow. Activity-on-arrow was initially preferred by engineers in the 1960s because it was easier to write the description along the arrow but, with the transition from manual calculations to computer software calculations, the preference changed to activity-on-node. The benefits of activity-on-node (AON):

- AON offers a number of logical relationships between the activities, such as start-to-start, finish-to-finish and lag.
- AON also offers a professional presentation, which is generally known today as a network diagram.

There are many other management tools that present information in a box; WBS, OBS and flow charts.

Common Data Base: The change to a common data base was the most significant change, because this moved the planning and control information into the project office. The project team now had full control over all the project related information.

Exercises:

1. Discuss the history of project management and identify where and why project management techniques were developed.
2. Calculate the PERT exercise from section 2 (solution Appendix 1).
3. Discuss how the time, cost, quality triangle trade-off applies to your projects.

3

Project Management Standards

Learning Outcomes

After reading this chapter, you should be able to:

- Understand the purpose and scope of a body of knowledge
- Define the PMBOK's nine knowledge areas
- Understand the project management certification process

Project managers have traditionally learnt about project management as an extra skill needed to carry out their job. Most project managers would begin their careers in a technical field and as they progressed they would become more involved in the management of their projects. This is when they would develop a need for project management education.

The worldwide trend towards project management has been accompanied by formal project management education and training. There are now many academic and certification programmes available from universities and colleges around the world.

Historically, as the discipline of project management grew and became established, so a number of institutions and associations were formed to represent the project management practitioners with respect to education, professional accreditation, ethics and a body of knowledge.

Internationally, project management associations and institutions have formed chapters to encourage and support the development of project management as a profession. These chapters organise regular meetings and newsletters to keep their members informed about project management issues.

The project management standards are associated with the following:

- Body of knowledge
- Certification of project managers (PMP)
- Unit standards
- Ethics and governance
- Global forum.

Body of Knowledge: Over the past fifty years a considerable body of knowledge has been built up around project management tools, skills, techniques and processes. This data base of information has been developed into what the Project Management Institute (PMI) call the project management body of knowledge (PMBOK).

The PMBOK defines the **Body of Knowledge** as, *an inclusive term that describes the sum of knowledge within the profession .... and rests with the practitioners and academics that apply and advance it.*

There are a number of institutions, associations and government bodies around the world which have produced a body of knowledge, unit standards and competency standards – they all have a presence on the Internet:

- Project Management Institute (PMI) [PMBOK]
- Association for Project Management (APM) [BoK]
- Australian Institute of Project Management (AIPM) [Competency Standards]
- International Project Management Association (IPMA)
- Association for Construction Project Managers (ACPM)
- Cost Engineering Association of South Africa (CEASA)
- Project Management South Africa (PMSA).

There are a number of recognised standards published by APM, PMI, IPMA, Global Performance Standards for Project Management Personnel, American National Standard Institute, International Standards Organization, British Standards and the South African National Standards.

The purpose of the body of knowledge is to identify and describe best practices that are applicable to most projects most of the time, for which there is widespread consensus about their value and usefulness. They are also intended to provide a common lexicon and terminology within the profession of project management – locally and internationally. As a developing international profession there is still a need to converge on a common set of terms.

The PMBOK 4ed describes project management under the following nine knowledge areas:

Knowledge Area	Description
1. Project Scope Management	Project scope management includes the processes required to ensure that the project includes all the work required, and only the work required, to complete the project successfully. It is primarily concerned with defining and controlling what is or what is not included in the project, to meet the sponsors' and stakeholders' goals and objectives. It consists of scope definition, scope verification, and change management.
2. Project Time Management	Project time management includes the processes required to ensure timely performance of the project. It consists of activity definition, activity sequencing, duration estimating, establishing the calendar, schedule development and time control.
3. Project Cost Management	Project cost management includes the processes required to ensure that the project is completed within the approved budget. It consists of cost estimating, cost budgeting, cashflow and cost control.
4. Project Quality Management	Project quality management includes the processes required to ensure that the project will satisfy the needs for which it was undertaken. It consists of determining the required condition, quality planning, quality assurance and quality control.
5. Human Resource Management	Human resource management includes the processes required to make the most effective use of the people involved with the project. It consists of organization planning, staff acquisition and team development.
6. Project Communications Management	Project communications management includes the processes required to ensure proper collection and dissemination of project information. It consists of communication planning, information distribution, project meetings, progress reporting and administrative closure.
7. Project Risk Management	Project risk management includes the processes concerned with identifying, analysing, and responding to project risk. It consists of risk identification, risk quantification and impact, response development and risk control.
8. Project Procurement Management	Project procurement management includes the processes required to acquire goods and services from outside the performing project team or organization. It consists of procurement planning, solicitation planning, solicitation, source selection, contract administration and contract closeout.
9. Project Integration	Project integration management integrates the four main project management processes of initiation, planning, execution and closing - where inputs from several knowledge areas are brought together.

Table 3.1: Project Management Knowledge Areas

The body of knowledge can be subdivided into four core elements which determine the **deliverable** objectives of the project:

- Scope
- Time
- Cost
- Quality.

The other knowledge areas provide the means of achieving the deliverable objectives, namely:

- Human resources
- Communication
- Risk
- Procurement and contract
- Integration.

APM BoK 5ed: The APM BoK takes a broad approach, subdividing project management into 50 + knowledge areas. This incorporates not only inward focused project management topics (such as planning and control techniques), but also broad topics in which the project is being managed (such as social and ecological environment), as well as specific areas (such as technology, economics and finance, organization, procurement, people, and general management).

As an overall scoping guide, the topics are described in the APM BoK at an outline level leaving the details to the texts, listed in their booklist, to explain the working of the knowledge areas (see www.apm.org.uk).

The APM defines its **Body of Knowledge** as,

- *A practical document, defining the broad range of knowledge that the discipline of project management encompasses.*
- *The basis for its various professional development programmes.*
- *Representing topics in which practitioners and experts consider professionals in project management should be knowledgeable and competent.*

Certification of Project Managers (PMP): The certification process offers a means for experienced project managers to gain a formal qualification in project management. There is a trend away from the knowledge based examinations which assess a person's knowledge, towards competence based examinations which assess a person's ability to perform. The PMI's certification is called the Project Management Professional (PMP). There is an increasing recognition of certification and for some projects it is being made a mandatory pre-qualification.

Competence is a mixture of explicit knowledge derived from formal education, tacit knowledge, and skills derived from experience. For young professionals, explicit knowledge is more important, but other competencies will become increasingly important as they progress in their careers. The PMI's (PMP) is a certificate programme, which measures explicit knowledge directly through a multi-choice test, and tacit knowledge and skill indirectly by assessing the candidate's experience. It is, therefore, aimed at an early to mid-career professional.

The IPMA and AIPM (Australian), on the other hand, have developed a multi-stage programme. At the first stage explicit knowledge is measured directly through a multi-question test. This is aimed at professional managers starting their careers. At the second stage tacit knowledge and skill are measured directly - this is early to mid stage certification, equivalent to PMP.

At the third stage, the programmes measure the performance of senior project managers directly; and IPMA has a fourth stage to measure the performance of project directors.

In Europe the integration of the EU is encouraging a growing number of cross-border projects, which not only require collaboration, but also a need to agree on common practices, common legal systems and, not least, a common business language.

Unit Standards: When SAQA (South African Qualifications Authority), the Services SETA, the Project Management Standard Generating Body (SGB) and the PMSA (Project Management South Africa) worked together to register the Further Education and Training Certificate: Project Management (NQF 4, SAQA ID 50080) and the set of unit standards leading towards this qualification, they utilized locally and internationally recognized best practice and standards in project management. This qualification provides an entry point to further learning for NQF level 5 and above qualifications as well as an international qualification in project or general management.

Code of Ethics: An ethical project management style is one where the project manager is honest, sincere and is able to motivate the team, contractors, suppliers and stakeholders for the best and fairest solution.

Ethics in procurement relates to insuring best value to the public in monetary terms. It ensures fairness and, most importantly, deals with conflicts of interest which could influence outcomes and ensure accountability.

Global Project Management Forum: Project management has been an international profession for many years and has a global forum to discuss project management issues. The issues they discuss include:

- What industries or types of projects are the main users of modern project management.
- What industries or areas of application have the greatest need for more or better project management.
- What industries or organizations offer the greatest opportunities for growth in professional project management.

Exercise:

1. Identify the principal bodies of knowledge worldwide and discuss how these relate to your national standards.
2. Discuss how the PMBOK's nine knowledge areas relate to your projects.
3. Discuss how you can achieve your national certification in project management or PMP.

Further Reading;

PMI <www.pmi.org>

APM <www.apm.org.uk>

IPMA <www.ipma.ch>

Australian AIPM <www.aipm.com.au>

South African Qualifications Authority <www.saqa.org.za>

Global PM Forum <www.pmforum.org>

4

Project Integration Management

Learning Outcomes:

After reading this chapter, you should be able to:

- Write a project charter
- Write a project closeout report

This chapter will explain how to develop the project charter and the project closeout report which are two key unifying documents within the *Project Integration Management* knowledge area. The project manager and the project team members use the project charter to initiate the project, and use the project closeout report to document lessons learnt.

Historically each component of a multi-disciplined project was managed separately by different departments. The work was often performed sequentially – from one department to the next - unity of purpose was a lonely voice. But with a better understanding of project dynamics, project integration management has developed as an important knowledge area which **combines and unifies** all aspects of the project.

Figure 4.1 shows how the three main project management system subdivisions (processes, plans and phases) work together to combine and unify the scope of the project. These 3Ps will be introduced in the next three chapters as:

- *Project Management Process* - introduces the key project management processes of initiation, planning, execution and closing of a project.
- *Project Management Plan* - introduces all the project plans that combine to form the baseline plan which is used to plan and control the project's performance.
- *Project Lifecycle* - introduces the project phases, project methodologies and project systems which enable project managers to implement corporate strategy.

Project Integration Management Breakdown Structure

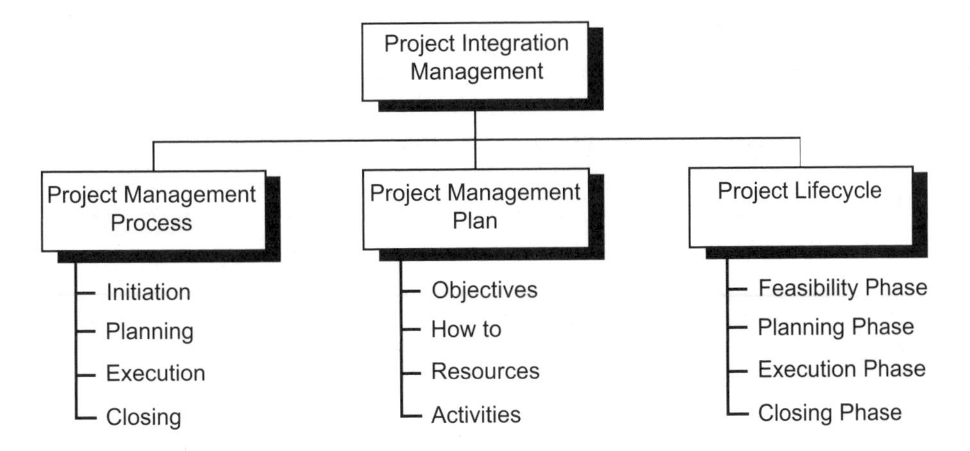

Figure 4.1: Project Integration Management – shows how project integration management can be subdivided into project management processes, project management plans, and the project lifecycle

Body of Knowledge Mapping

PMBOK 4ed: The *Project Integration Management* knowledge area includes:

PMBOK 4ed	Mapping
4.1 Develop a project charter.	This chapter will explain how to write a project charter.
4.2 Develop a project management plan.	See *Project Management Plan* chapter.
4.3 Direct and manage project execution.	See *Project Management Process* chapter.
4.4 Monitor and control project work.	See *Project Management Process* chapter.
4.5 Perform integrated change control.	See *Scope Management* chapter.
4.6 Close project or phase.	This chapter will explain how to write the project closeout report.

APM BoK 5ed: The *Integration* knowledge area includes the following definition:

The APM BoK defines **Integration** as, *the process of bringing people, activities and other things together to perform effectively.*

Unit Standard 50080 (Level 4): There is no unit standard for integration management.

1. Project Charter

The project charter (also referred to as terms of reference, or project mission), officially initiates the project or project phase. The project charter can be a simple document outlining, in a few words, what is required, or it can be a much larger document defining precisely what is required and how it should be carried out. This section will present a project charter template and give a brief description of all the headings.

Ownership: Although the project sponsor owns the project charter, the project manager and project team members often develop it. This involvement helps to ensure the project manager and project team members have a constructive input into the project charter's development, and also ensure they are assigned sufficient authority and power to use company resources to get the job done.

Register Project: The project charter officially initiates the project by formally adding the project to the company's register of projects. The project is given an identity with a project name, a number and a purpose. For example, if the project is to build an offshore supply vessel then the project's name could be the ship's name; the project's number might be the estimate number, or the next number from the register of projects; and the purpose would be no more than a brief statement of the type of project, which could be to build an offshore supply vessel to supply the oil rigs in the North Sea.

Responsibility: The responsibility section identifies the people who are responsible for the management of the project. There are a number of possibilities here:

- The client and/or project sponsor are responsible for initiating the project, paying for the project and appointing the project manager. They are also responsible for acquiring the benefits from the project to implement corporate strategy.
- The project manager is responsible for achieving the project deliverables (time, cost, quality).
- The nominated project experts are responsible for the project design, the build-method and achieving the required operational configuration.

The project charter should assign authority to the responsible parties so that they can issue instructions and use company resources to carry out the work.

Project Charter Template: The project charter template identifies the main headings that form the structure of the project charter.

Project Charter	
Project Name	Gives the project a meaningful name
Project Number	Gives the project an identity
Project Purpose	Explains why the project is being initiated
Client	Identifies the client
Project Sponsor	Identifies who will authorize and pay for the project
Project Manager	Identifies who will manage the project
Background to the Project	Outlines the build up to the project
Statement of Requirements	Outlines the problems, needs and opportunities the project will address
Business Case	Sets out the commercial justification for the project
Project Scope	Outlines the project's deliverables (goods or services)
Feasibility Study	Confirms the project is feasible and will make the best use of company resources
Stakeholders' Analysis	Identifies the stakeholders and analyses their needs and expectations
Project Constraints	Outlines external, internal, and project (build-method and configuration) constraints
Project Assumptions	Logs the assumptions the decisions are based on
Budget	Outlines the project's cost structure
Schedule	Presents all the activities' start and finish dates
Execution Strategy	Explains the 'buy or make' decision (resource and procurement requirements)
Quality	Outlines the required quality and level of inspection
Project Methodology	Outlines the project management systems and the project management office (PMO)
Project OBS	Outlines the organization breakdown structure (OBS) (roles, duties, responsibilities and authorities)
Communication	Outlines the lines of communication and the required level of documentation control
Project Risks	Outlines how project risks will be identified and responses developed.

Figure 4.2: Project Charter Template – shows a typical project charter format

Background: This section introduces a background to the project by briefly outlining the situation – what are the problems, needs or opportunities that have led up to the project being initiated? This should be supported by a statement of requirements and a business case. Where:

- A **statement of requirements** is a high level outline of what is required; for example, a power generation company's requirement could be expressed as being capable of generating X megawatts of power. This requirement would be based on the stakeholders' analysis (needs and expectations), market feedback and economic forecasting.
- A **business case** would then present one or more ways to satisfy the requirements; for example, to address the power generation need (as above) the options might be to build a coal, oil, gas, nuclear, solar, wind or tidal power station. The structure of the business case would also provide financial, technical and commercial justification for each option.

Project Scope: The scope of work section outlines the goods or services the project is to deliver. For internal company projects the scope of work is based on the statement of requirements (problems, needs, opportunities), and the business case (commercial justification). For external projects the scope of work is based on the client's bid document, or contract. The project charter should also outline a high level PBS/WBS to highlight the main deliverables.

Feasibility Study: The issuing of the project charter initiates the project in general, but initiates the feasibility study in particular. This section outlines the scope of the feasibility study and how it should be carried out and documented, who should be responsible (presumably the project manager), when it should be completed (report back date) and its budget.

Project Stakeholders: This section identifies the key project stakeholders and outlines their needs and expectations.

Build-Method: This section outlines how the project will be assembled or implemented. For example, on a house extension project the build-method considers the position of the scaffolding and ladders, together with the storage of the materials. On an IT project the build-method considers how to remove old equipment and install new equipment with the minimum impact on the company's operation.

Configuration Management: This section nominates a group of experts to confirm that all the components of the project should operate as intended, or as the client requires. It is particularly important to include the input from these experts when approving scope changes.

Constraints: This section outlines the key constraints and assumptions which define the boundary of the project.

Budget: This section allocates the project's budgets - the feasibility study budget is assigned, but the overall project budget might only be estimated in outline.

Schedule: This section outlines the project's schedule – the feasibility study's completion date is given, but the rest of the project's schedule might only be estimated in outline.

Execution Strategy: This section outlines the project's execution strategy section – the 'buy or make' decision.

Project Quality: This section outlines the project's quality requirements and how quality assurance should help achieve the required condition, and also how quality control should confirm the required condition has been achieved.

Project Methodology: This section outlines how the project sponsor wants the project to be managed. This would be defined as a project methodology or project system. This section might also outline the size and location of a project office together with the structure of the planning and control system.

Project OBS: This section outlines the organization breakdown structure (OBS) for the project together with the roles, responsibilities and authority of the project participants. This is the section to discuss if a matrix OBS will be used and how the project manager can use company resources.

Communication: This section outlines the lines of communication to satisfy the stakeholders' communication needs, how information will be gathered, processed, disseminated (reported) and stored. This section should also indicate the level of documentation control.

Execution: This section outlines the execution process – how instructions are to be issued, how purchase orders are to be expedited, and how progress is to be monitored and controlled. This section should also outline how scope changes are to be managed and co-ordinated between the nominated experts for their input and approval.

Project Risks: This section outlines the perceived project risks and the process for identifying further project risks, and how to quantify and respond to the risks. This section should also comment on how to make the most of positive opportunities.

Project Charter Example:

This example will set out a project charter for a wind farm:

Project Charter	
Project Name	Highlands Wind Farm
Project Number	1000
Project Purpose	Generate 'x' Gigawatts
Document Number	D100
Prepared By	John Dodd
Client	UK Energy
Project Sponsor	Mark Hackney
Project Manager	Oliver Logan
Background to the Project	New power generation needs to have a low carbon footprint
Statement of Requirements	Outlines future power requirements and power generation
Business Case	Justifies the feasibility of the Highland Wind Farm project
Project Scope	Designed to comply with the building standards
Constraints	Budget $100m
Assumptions	Retail price of electricity will keep in line with inflation
Project Methodology	Fully integrated project management system
Project OBS	Builder to manage all the subcontractors and suppliers
Communication	Weekly site meetings, documentation control
Execution	Written job card instructions
Project Risks	Inclement weather, environmental protests

2. Project Reviews and Project Closeout Report

A project phase review or project closeout report is a project audit of the project's actual work performance and/or the project's condition against the declared (or planned) performance and/or condition. Project phase reviews are officially carried out at the end of each phase, whereas a project closeout report is officially carried out at the end of the project. Nevertheless, the project manager and project team members should informally learn from experience as the project progresses.

The APM BoK 5ed defines **Project Reviews** as, *to check the likely or actual achievement of the objectives specified in the project management plan and the benefits detailed in the business case.*

The APM BoK 5ed defines **Closure** as, *the formal end point of a project, either because it has been completed or because it has been terminated early.*

Responsibility: Although audits should be carried out by an independent third party, certainly outside the project team but preferably outside the project organization, much of the groundwork can be carried out by project team members working in the project office.

Lessons Learnt: It is important to not only learn from the mistakes and successes of previous projects but, to also learn progressively during the present project. The project review and closeout report can be subdivided into three sections:

- Historical data from previous projects to assist with conceptual development, the feasibility study and estimating on future projects.
- Progress reports to help predict future trends and problem areas on the current project.
- A phase review and closeout report which evaluates the performance of the current project against the project objectives and makes recommendations for future projects.

The project phase reviews, closeout reports and lessons learned are documented and communicated to relevant parties.

Historical Data: The search for historical data from previous projects will clearly show the benefit of effective filing and storage of closeout reports. Learning from previous experiences is the most basic form of development, and it is basically free. It is essential to look closely to see what went right and what went wrong, together with any recommendations for future projects, because the same mistakes have an uncanny habit of happening again, particularly if the cause has not been addressed.

3. Questionnaire

Phase reviews and closeout reports can both benefit by using a structured questionnaire. It is advisable to compile the report before the project participants disperse, otherwise it might be difficult to locate them, never mind getting feedback from them. The net should be cast reasonably wide to include a broad range of feedback. The list of participants should certainly include the following:

- Client
- Project manager and project team
- Functional managers and other corporate participants
- Suppliers and subcontractors
- Stakeholders.

The questionnaire would typically be structured to include the following:

Position	Identify your position in the project organization structure and comment on the interface and cooperation with other disciplines.
Delegation	Comment on the delegation of responsibility and authority to you and from you.
Scope of Work	Briefly outline your assigned scope of work.
Planning	Comment on the planning schedules, budgets, quality and manpower performance that relate to your work. Where possible quantify performance with statistical data.
Performance	Give a candid assessment of your performance, analysing what went right and what went wrong. Comment on any non conformance reports (NCR) with reasons for any deviations and the level of re-work. If any audits were conducted comment on their findings.
Build-Method	Comment on the project build-method and make recommendations for future projects.
Communication	Comment on the lines of communication, the issuing of instructions and information, the holding of meetings, reporting, filing and storage.
Technical Changes	Evaluate design and technical changes, as-built drawings and operator manuals.
Scope Changes	Comment on any scope changes and concessions. Evaluate how smoothly the configuration system worked.
New technology	Discuss the use of new technology, computerisation and automation.
Problems	Discuss any unexpected problems and how they impacted on the project, and also how they were resolved.
Procurement	Comment on the performance of procurement suppliers and sub-contractors.

Manpower	Comment on manpower performance, the training and any industrial relation problems.
Budget	Evaluate the accuracy of the budget and list any recommended changes to the company's tariffs.
Schedule	Evaluate the accuracy of the project management plan and list any recommended changes to the company's estimating database.
Contract	Evaluate the contract document.
Recommendations	Make general recommendations for future projects.

This type of questionnaire provides the project manager with an excellent tool for accurate and meaningful feedback. People generally do not like answering questionnaires; to receive feedback it might be best through a debriefing meeting or telephone conversation using the questionnaire as the agenda.

Recommendations for Future Projects: Learning from achievements and mistakes, and improving productivity is a basic economic requirement for sustained commercial competitiveness. The closeout report should highlight any recommendations simply and clearly, because many years from now this might be the only section that is read.

Storage: At the end of the project the files, photographs, all correspondence and particularly the closeout report should not be dumped and lost in the archives, but be readily available for inspection, because this is the data base for future estimating and is an auditable item.

Exercises:

1. Write a project charter to initiate a project you are familiar with.
2. Conduct a phase review at the end of the design phase for a project you are familiar with.
3. Discuss how you would conduct a closeout report for a project you are familiar with.

Further Reading:

Burke, R., *Project Management Techniques*, will develop the following topics which relate to this knowledge area:

- Project lifecycle
- Project stakeholders.

5

Project Management Process

Learning Outcomes After reading this chapter, you should be able to:
Understand Fayol's management process
Understand the Eastonian process
Understand the structure of the project management process
Apply the project management processes

A management process is a linear sequence of steps which are carried out to achieve defined objectives. The *Project Management Processes* knowledge area states that the application of knowledge requires the effective management of appropriate processes. This chapter will discuss these project management processes.

The *Project Integration Management* chapter introduced the 3Ps (processes, plans and phases). The 3Ps will be introduced in this chapter and the next two chapters as:

- *Project Management Process* - introduces the key project management processes of initiating, planning, execution and closing of a project.
- *Project Management Plan* - introduces all the project plans that combine to form the baseline plan which is used to plan and control the project's performance.
- *Project Lifecycle* - introduces the project phases, project methodologies and project systems which enable project managers to implement corporate strategy.

Learners new to project management might initially be confused about the differences between the three topics. This is partly because they have similar names, but also because there is a certain amount of overlap between the three topics. These three chapters have been set out in this sequence because the project management process underpins both the project management plan and the project lifecycle. It is, therefore, essential to gain a robust understanding of the project management process first.

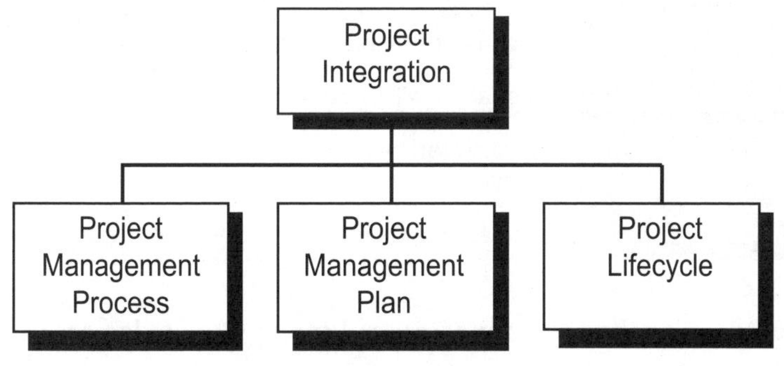

Figure 5.1: Project integration

Summary of Management Processes

There are a number of management processes and a number of project management processes which all have a similar format. Consider the following summary:

Fayol's Management Process	Eastonian Process	APM BoK 5ed's Project Management Process	PMBOK 4ed's Project Management Process
Planning	Input	Starting and Initiating	Initiating
Organizing	Process	Defining and Planning	Planning
Commanding	Output	Monitoring and Controlling	Executing
Directing		Learning and Closing	Monitoring and Controlling
Controlling			Closing

Table 5.1: Management Processes – shows a number of management and project management processes

As a starting point these processes can be seen as generic processes that apply to most projects most of the time, and apply to all levels of the project (see *Process Levels* section later in this chapter).

This chapter will start by introducing Fayol's management process from a historical perspective, and then introduce the Eastonian Process which is the simplest but most widely used process.

The chapter will then explain how to use the PMBOK and APM BoK's recommended project management processes to manage projects of limited complexity.

Body of Knowledge Mapping

PMBOK 4ed: The *Project Management Process* knowledge area includes the following:

PMBOK 4ed	Mapping
3.3: Initiating Process	This chapter will explain the initiating process.
3.4: Planning Process	This chapter will explain the planning process.
3.5: Execution Process	This chapter will explain the execution process.
3.6: Monitoring and Controlling Process	This chapter will explain the monitoring and controlling process.
3.7: Closing process	This chapter will explain the closing process.

APM BoK 5ed: The *Project Management Process* knowledge area includes the following:

APM BoK 5ed	Mapping
1.1: Starting and Initiating	This chapter will explain how to start and initiate a project.
1.1: Defining and Planning	This chapter will explain how to define and plan a project.
1.1: Monitoring and Controlling	This chapter will explain how to monitor and control a project.
1.1: Learning and Closing	This chapter will explain how to close a project and learn from the experience.

Unit Standard 50080 (Level 4): The *Project Management Process* knowledge area includes the following:

Unit 120372: *Explain fundamentals of project management*

Learning Outcomes	Mapping
120372/2.2: The major project management processes are described and explained according to recognised best practice.	This chapter will explain how to use the project management process.
120372/5.1: Key processes and activities that take place to manage a project are described from beginning to end.	This chapter will explain how to use the key project management processes; initiate, plan, execute and close.

Unit 120372 range statement says, *the processes and activities may include, but are not limited to, start up, initiating, planning, controlling, monitoring, execution, implementing, closing, evaluating.*

1. Fayol's Management Process

The first recorded management process can be traced to a French industrialist, Henri Fayol, who in 1916 presented his management process that consisted of the following:

- Planning
- Organizing
- Commanding
- Directing
- Controlling.

Where each process was described as:

Planning	Planning was described as forecasting the work to be done.
Organizing	Organizing was described as staffing.
Commanding	Commanding was described as motivating the staff.
Directing	Directing was described as co-ordinating the staff and the work.
Controlling	Controlling was described as monitoring and receiving feedback on the progress in order to make necessary adjustments.

Fayol's principles are fundamentally universal and apply to most types of management, whether it is ongoing work or project work. The PMBOK and APM BoK use the nucleus of Fayol's management process, to which they add the initiating process and the closing process.

Fayol was very forward thinking for his time; his management process suggested that it is important to have 'unity of command' with each worker reporting to only one supervisor, so that the workers were not receiving commands from more than one person. In modern project management speak this would be referred to as having a 'single point of responsibility' which is one of the corner stones of project management.

Fayol's work has stood the test of time and has been shown to be relevant and appropriate to contemporary project management.

'Fayol on the road to success.'

2. Eastonian Process

The Eastonian process (input-process-output), is named after David Easton (not a European country!). Although the Eastonian process was originally applied to political systems, its structure is widely used within the PMBOK either as **input-output** or **input-process-output**. The Eastonian process is the simplest of the four examples (see figure 5.1, the summary of management processes), and is used extensively by the body of knowledge to explain many of the process groups within each knowledge area. For example, the CPM analysis can be shown as follows:

Input	Process	Output
1. List of activities 2. Logical relationships between activities 3. Activity durations 4. Work calendar 5. Start and finish dates	1. CPM analysis	1. Network diagram 2. Activity table 3. Critical path 4. Gantt chart

Table 5.2: CPM Process – shows the CPM process using the Eastonian model

The table headings refer to the following:

Input: Input lists all the documents or information required to do the CPM analysis.

Process: The process is the CPM calculation to develop the network diagram and perform a forward pass and backward pass.

Output: The output includes the network diagram, the activity with all the start and finish dates, the critical path and the Gantt chart.

3. Project Management Process

The PMBOK states that project management is accomplished through **Processes** which it defines as a set of interrelated actions and activities that are performed to achieve a pre-specified set of objectives, products, results or services.

There is general agreement between the PMBOK and APM BoK that the project management process should be subdivided into five key processes which are linked by the results they produce - the outcome from one process is usually the input to the next process. In some presentations monitoring and controlling are shown as separate processes, while in other models they are part of the execution process. In practice both models go through the same actions.

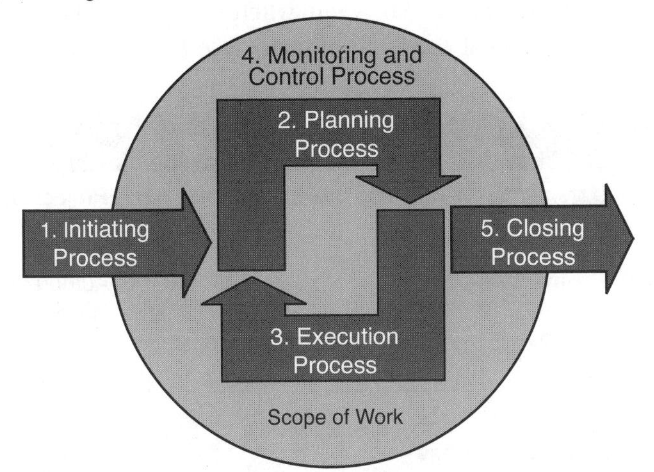

Figure 5.2: Project Management Process – shows the relationship between the five project management processes

The processes are:

Initiating Process	The initiating process starts the work or project – this would usually include the project charter and feasibility study.
Planning Process	The planning process selects and develops the best course of action to attain the stated objectives (deliverables) that the project is to achieve. This would usually include the project management plan.
Execution Process	The execution process integrates, instructs and co-ordinates people and resources to implement and carry out the project management plan and make-it-happen.
Controlling Process	The controlling process ensures the project objectives are met by regularly monitoring and measuring progress. This process will identify any variances from the project management plan so that corrective action can be taken as and when necessary. Included in this process is the scope change controlling process.
Closing Process	The closing process formally accepts the project and brings it to an orderly end. This involves commissioning the product and confirming the objectives have been achieved to the required standard (as per the project management plan). The project is then handed over to the client for operation. This section would also include a closeout report to document the lessons learnt.

4. Initiating Process

The initiating process includes the processes required to define and authorize the start of a new phase or a new project. The initial scope of the project is developed along with the associated costs and benefits of pursuing the project. Some of the key documents of the initiating process include the project charter, feasibility study and the stakeholders' analysis.

Project Charter: The project charter officially initiates the project by formally adding the project to the company's register of projects. The project is given an identity with a project name, a number and a purpose. The project charter (owned by the project sponsor) sets out the why, what, who, when, where, how to and how much the project aims to achieve to implement corporate strategy (see *Project Integration Management* chapter).

Input	Process	Output
1. Statement of requirements 2. Business case 3. Contract 4. Enterprise environmental factors 5. Organizational process assets.	1. Develop project charter.	1. Project charter 2. Authorize feasibility study 3. Authorize stakeholder analysis.

Table 5.3: Project Charter Process

Feasibility Study: The feasibility study (owned by the project manager) confirms the project is feasible and will solve the identified problem or make the most of the identified opportunity. The feasibility study should also offer alternative solutions, confirm the project will make the best use of the company's resources and maintain the company's competitive advantage (see *Feasibility Study* chapter).

Stakeholder Analysis: The stakeholder analysis identifies all the project stakeholders and determines their needs and expectations. Table 5.4 shows the stakeholder analysis process.

Input	Process	Output
1. Project charter 2. Procurement document package 3. Enterprise environment factors 4. Organizational process assets.	1. Develop stakeholder analysis.	1. Stakeholder register 2. Stakeholders' needs and expectations 3. Stakeholder management strategy.

Table 5.4: Stakeholders' Analysis Process

5. Planning Process

The planning process includes the processes required to define the scope of the project, to develop a detailed design of the project, and to develop a detailed project management plan outlining how to achieve the goals and objectives of the project. Table 5.5 shows the process to produce the project management plan.

Input	Process	Output
1. Scope plan 2. Time plan 3. Cost plan 4. Quality plan 5. Procurement plan 6. Resources plan 7. Human resource plan 8. Communication plan 9. Risk plan.	1. Develop project management plan.	1. Project management plan.

Table 5.5: Project Management Plan Process

The table headings refer to the following:

Input: The input consists of all the individual plans which roll up to form the project management plan.

Process: Developing the project management plan is an iterative process of trade-offs, compromises and alternatives. As all the knowledge areas are interlinked, so a development in one area might influence or change another area. The project management plan might need to be revisited a number of times before an agreed plan is produced.

Output: The output is a fully integrated project management plan or baseline plan.

6. Executing, Monitoring and Controlling Process

The execution process includes the processes to execute, monitor and control the work defined in the project management plan. As explained in the previous section, the project management plan is developed to achieve the project's objectives and expectations.

The first step is to co-ordinate the project team members, the internal resources, external contractors, suppliers, and equipment. This involves authorizing the work by issuing instructions and awarding contracts.

Input	Process	Output
1. Project charter 2. Project management plan 3. Scope change approvals.	1. Issue instructions 2. Expedite procurement 3. In process quality inspection 4. Monitor progress 5. Replan to incorporate changes.	1. Project progress report.

Table 5.6: Execution Process

Monitor and Control: Once the work has started the following processes are used to monitor and control the project:

- The instructions and orders are expedited (particularly procurement) to confirm the orders have been received, are being acted upon and are on track to be delivered on time.
- In-process quality inspections are used to confirm the project is being made to the required quality.
- Monitoring provides the project team members with insight into the dynamic nature of the project; it measures the amount of actual work to quantify the project's progress.
- Performance reports provide information on the project's progress with respect to; scope, time, cost, quality, procurement, resources and risk. Monitoring includes status reporting, progress measurements, and forecasting to confirm the project is on track to achieve its objectives.
- Scope change control includes the process to; log change requests, assess impact by the nominated experts, and to update the build-method and the configuration management systems.

7. Closing Process

The closing process includes the processes required to confirm the project has been made to the pre-defined objectives as outlined in the project management plan.

Input	Process	Output
1. Project management plan 2. Apply for acceptance 3. Verification and acceptance criteria.	1. Commissioning.	1. Certificate of completion 2. Owner and operator manuals.

Table 5.7: Closing Process

The table headings refer to the following:

Input: The inputs to the closing process are all the documents which define the project's objectives and acceptance criteria.

Process: The closing process confirms the project has reached the required condition. This might be performed by commissioning and testing the project.

Output: The output confirms the project has been made to the required condition and hands over the project to the client for operation.

When the objectives of the project have been achieved, the project manager will close down the project. This will involve some financial closure tasks, as well as the archiving of the project materials. A lessons-learned document will be developed to benefit future projects.

Closeout Report: A closeout report is developed at the end of the project to document lessons learnt for future projects.

Input	Process	Output
1. Project management plan 2. Commission and acceptance certificates.	1. Produce closeout report.	1. Project closeout report 2. Lessons learnt.

Table 5.8: Closeout Report Process

When the closeout report is finished it is circulated as agreed with the stakeholders and then archived together with all the relevant documents. Ease of retrieval will be particularly helpful for future projects.

Project Management Process Worked Example 1: House Extension Project

Consider the project management processes which can be used to manage a house extension project. Note that execution and controlling have been shown as two different processes.

1. Initiating Process	The initiating process is officially started by the project charter. The project charter authorises the feasibility study to investigate the house extension proposal and confirms it is feasible and the best option to address the needs for an expanding family.
2. Planning Process	The planning process designs the house extension and develops a project management plan to achieve the objectives.
3. Execution Process	The execution process instructs and co-ordinates the builder and sub-contractors to build the house extension as per the design and project management plan.
4. Monitor and Control Process	The monitor and control process ensures the house extension objectives are met by monitoring and measuring progress regularly to identify any variances from the project management plan so that corrective action can be taken as necessary. The controlling process also expedites the progress (particularly procurement), and manages any scope changes.
5. Closing Process	The closing process formally accepts the house extension and brings the project to an orderly end. This involves commissioning all the new house systems; electrical, plumbing and heating, before handing over to the homeowner.

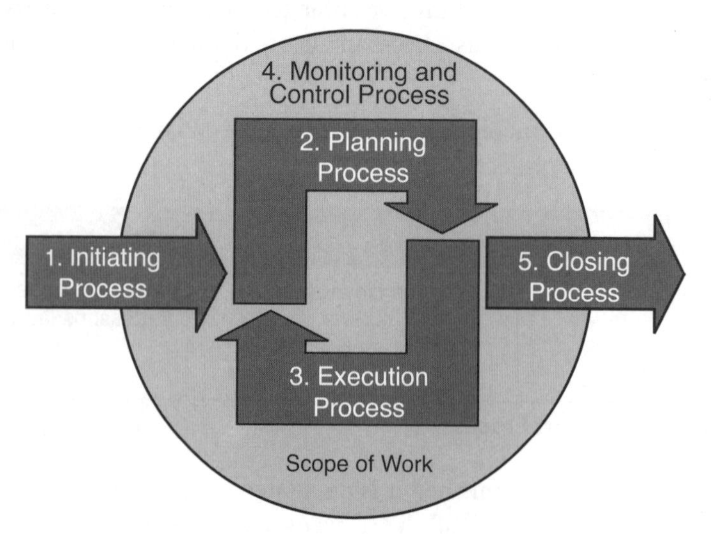

Figure 5.3: Project Management Process – shows the relationship between the five project management processes

8. Process Levels

The format of the project management process can be applied at different levels within the project. Consider the three levels below:

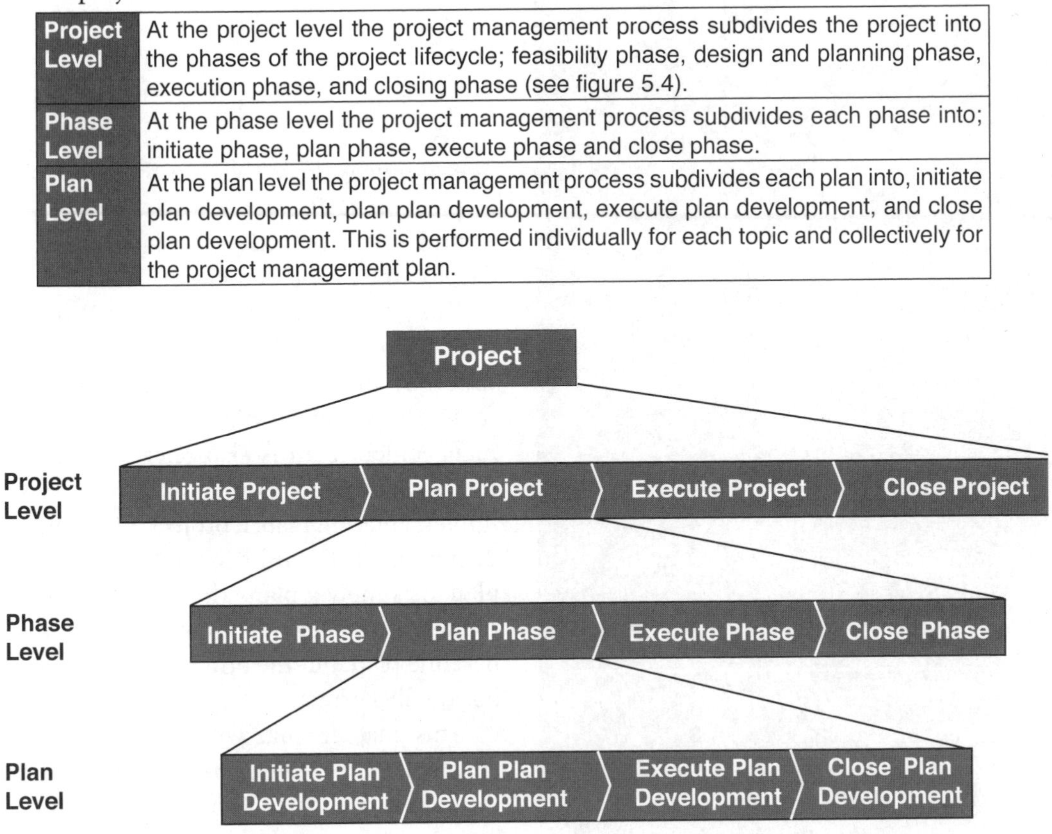

Project Level	At the project level the project management process subdivides the project into the phases of the project lifecycle; feasibility phase, design and planning phase, execution phase, and closing phase (see figure 5.4).
Phase Level	At the phase level the project management process subdivides each phase into; initiate phase, plan phase, execute phase and close phase.
Plan Level	At the plan level the project management process subdivides each plan into, initiate plan development, plan plan development, execute plan development, and close plan development. This is performed individually for each topic and collectively for the project management plan.

Figure 5.4 Process Levels

At the beginning of the chapter it was suggested that learners new to project management might initially be confused by the differences between the three topics; processes, plans and phases. After completing this chapter the learner should now have a clear understanding of the project management process. And, as this section has shown, the project manager can now see how the project management process is interrelated with the project management plan and the project lifecycle.

Exercises:

1. Show how you can use the Eastonian process to manage any section or topic of a project you are familiar with.
2. Develop a project management process for a project you are familiar with.
3. Develop the project process levels for a project you are familiar with.

Further Reading:

Burke, R., *Project Management Techniques*

6

Project Management Plan

Learning Outcomes
After reading this chapter, you should be able to:
Understand what is a plan
Understand how to develop the project management plan
Develop the project management plan as a: • Checklist • Flowchart • Iterative spiral

The *Planning Process* within the *Project Management Process* knowledge area includes the techniques required to develop the *Project Management Plan.*

The project management plan (PMP) is a summary document which brings together all the individual plans to form the baseline plan. The project management plan is also referred to as the project execution plan, a project implementation plan, a project initiation document (PID) in PRINCE2, a project plan, or simply a plan. This text will use the terms project management plan, or baseline plan, but the other terms would be equally correct.

This chapter links with the previous chapter (*Project Management Process*) and the following chapter (*Project Lifecycle)* to introduce the second of the 3Ps (processes, plans and phases). Where:

- *Project Management Processes* - introduces the key project management processes of initiation, planning, execution and closing of a project.
- *Project Management Plan* - introduces all the project plans that combine to form the baseline plan which is used to plan and control the project's performance.
- *Project Lifecycle* - introduces the project phases, project methodologies and project systems which enable project managers to implement corporate strategy.

The figure below (figure 6.1) shows the relative position of the planning process with respect to the other processes. The figure also shows that the purpose of the planning process is to develop the project management plan.

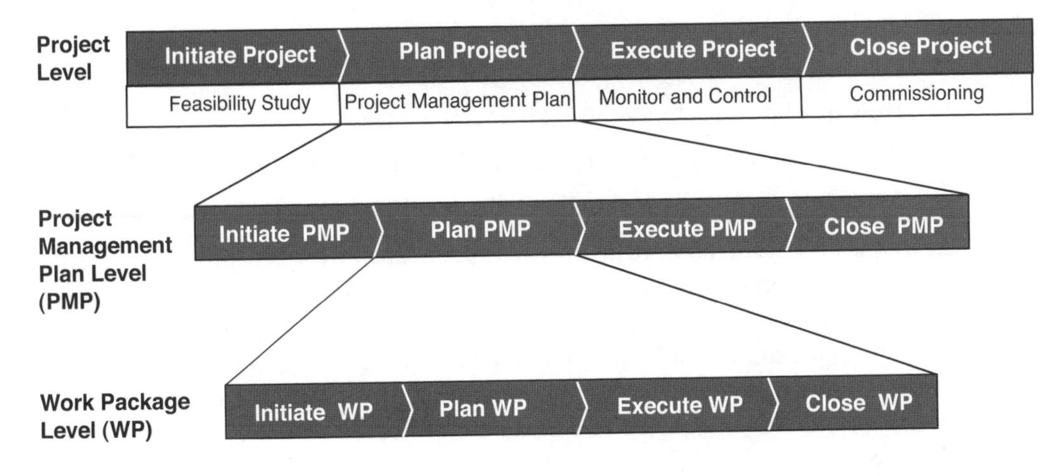

Figure 6.1: Planning Process – shows the relative position of the project management plan

Body of Knowledge Mapping

PMBOK 4ed: The *Project Management Plan* knowledge area includes the following definition:

> The PMBOK 4ed defines the **Project Management Plan** as, *the process of documenting the actions necessary to define, prepare, integrate, and co-ordinate all the supporting plans. The project management plan then becomes the main source of information for the project management process (initiation, planning, execution, and closing).*

APM BoK 5ed: The *Project Management Plan* knowledge area includes the following definition:

> The APM BoK 5ed defines the **Project Management Plan** (PMP) as, *a plan which brings together all the plans for a project. The purpose of the project management plan is to document the outcomes of the planning process and to provide the reference document for managing the project. The project management plan is owned by the project manager.*

Unit Standard 50080 (Level 4): There are no units for this knowledge area.

1. What is a Plan?

The APM BoK 5ed defines a **Plan** as, *an intended future course of action.*

A plan in its simplest form usually contains the following:

- **Objectives** – what are the goals and objectives?
- **How to** – how to achieve the goals and objectives.
- **Resources** – what people, materials and equipment are required/available.
- **Activities** – sequence of activities with details of time, cost and quality.

A plan is typically a set of intended actions or steps, through which the project manager expects to achieve a predefined goal or objective. A plan could be as simple as a checklist of activities or as complicated as the critical path method (CPM). Planning is the (psychological) process of thinking about the activities required to achieve predefined objectives – the output of which is a plan. A plan can be created for individual topics (time, cost, or quality), or a plan could be for the project as a whole, in which case it is called the project management plan, or baseline plan.

The planning process prompts the project manager to ask the questions; why, what, when, where, who, how to, and how much, trying to leave as little as possible to guesswork by those responsible for the project execution. In turn, this increases awareness amongst the team members and stakeholders; it helps to solve problems, and strives for decisions based on consensus, logging assumptions and constraints. Consider the following:

Why	The 'why' refers to the purpose of the project which includes a definition of the problem, need or opportunity. This is developed in the statement of requirements, and the business case.
What	The 'what' describes the project's objectives, a description of the scope of work, the deliverables and their acceptance criteria and verification. The 'what' needs to take into consideration the project's constraints (internal and external), assumptions and dependencies.
When	The 'when' defines the timescales, including the schedule (CPM and Gantt Chart), milestones and deadlines. Tight timescales might have an impact on 'how much'.
Where	The 'where' defines the geographical location(s) of where the work is planned to take place. This could be the head office, the design office, the factory, on site, or outsourced overseas. The location has an impact on communication, resources, equipment, logistics, schedules and costs.
Who	The 'who' includes a description of the organization structures and team structures that are used to manage the scope of work. The 'who' includes a description of the key project roles, duties, responsibilities and authorities, and a plan to balance resource availability with the workload.
How to	The 'how to' defines the project methodology or project system for managing the project. The 'how to' defines the build-method for assembling the project, and the configuration management to confirm the project will perform as required.
How much	The 'how much' defines the estimating process and the project budget.

Table 6.1: The '*Serving Men*'

"I had six honest serving men. (They taught me all I knew); their names are ***What*** *and* ***Why*** *and* ***When*** *and* ***How*** *and* ***Where*** *and* ***Who.****"* - Rudyard Kipling (from *The Elephant's Child)*

The objective of a project management plan is to define the approach to be used by the project manager to deliver the intended scope of the project. This will include policies and plans for managing scope changes, project communication, governance, configuration, health and safety, and the environment.

Part of the planning process is to identify the communication needs of the stakeholders - what information they require, when it should be communicated and how it should be communicated.

The planning process communicates planning information to the stakeholders, encouraging them to participate in the process and obliging them to 'sign-on' and pledge their support. When the plans are drawn up by those who are implementing them the stakeholders should be more obliged and committed to complete the work as planned. Conversely if people are not involved in the planning process this might lead to plans being misinterpreted, or even being ignored. The behavioural side of project management is an important component of the planning process.

This chapter will outline how the various project plans discussed in this book relate to each other, and how the development of the project management plan involves making trade-offs and compromises between competing objectives and alternatives in order to meet the stakeholders' requirements (needs and expectations). It is important to appreciate how a change in one plan will impact on other plans. For example, a resource overload (shown on the resource histogram) might impact on a critical activity (shown on the Gantt chart) even though the material and equipment are available.

2. Project Planning Steps

Although the planning steps are outlined here as a sequence of discrete operations, in practice, other factors could influence the sequence. This means there will almost certainly be a number of iterations, compromises and trade-offs before achieving an optimum plan that all the stakeholders can agree on.

There are a number of formats used to present the project management plan:

- A checklist (this chapter)
- An iterative spiral (see *Execution, Monitoring and Control* chapter)
- A flow chart (this chapter).

Consider the following presentations:

Baseline Plan Checklist	Document Number	Date Raised
Project Trigger		
Statement of Requirements		
Business Case		
Project Charter		
Feasibility Study		
Stakeholders' Analysis		
Build-Method		
Configuration Management		
Scope Management		
Work Breakdown Structure (WBS)		
Estimating Time		
Critical Path Method (CPM)		
Gantt Chart		
Execution Strategy		
Procurement Schedule		
Contract		
Resource Plan		
Project Budget (Estimating Costs)		
Cashflow Plan		
Risk Management Plan		
Quality Control Plan		
Communication Plan		
Organization Breakdown Structure (OBS)		
Responsibility Assignment Matrix (RAM)		
Project Team		
Baseline Plan		

Table 6.2: Baseline Plan Checklist – shows, in a suggested sequence, all the project management topics to be considered

3. Project Planning Documents

This section will outline a number of plans and documents which combine to form the project management plan or the baseline plan.

Project Trigger: New creative ideas and opportunities evolve and develop into projects in different ways. The trigger to start a project might be an automatic response to a problem or equipment failure, or it could be the identification of a commercial opportunity, technology driven upgrade, or an entrepreneur spotting a gap in the market for a new product.

Statement of Requirements: A statement of requirements is a high level outline of what is required; for example, a power generation requirement could be expressed as x amount of megawatts. This requirement would be based on the stakeholders' analysis (needs and expectations), market feedback and economic forecasting.

Business Case: A business case would then present one or more options to satisfy the requirements. For example, to address the power generation need above, the options might be to build a coal, oil, gas, nuclear, solar, wind or tidal power station. The structure of the business case would provide financial, technical and market justification for each option.

Project Charter: The project charter is the document which officially formalises the existence of a project and gives the project an identity (name and number) so that budgets and responsibilities can be assigned. The project charter should outline the purpose of the project, the beneficial changes and the key objectives, together with the means of achieving them (see *Project Integration Management* chapter).

Feasibility Study: The project charter initiates the feasibility study to conduct an in depth study on the feasibility of performing the project. It offers a structured approach for identifying the **stakeholders** and assessing their needs and expectations. It reviews closeout reports, together with investigating a range of options and alternatives to support the new venture's business viability, and confirm it is making the best use of the company's resources (see *Feasibility Study* chapter).

Build-Method: The build-method outlines how the project will be assembled or implemented. For example, on a house extension project it considers the position of the site office, the position of the scaffolding and ladders, and the storage of the materials. On an IT project it considers how to remove old equipment and install new equipment with the minimum impact on the company's operation.

Configuration Management: The configuration management process nominates a group of experts to confirm that all the components of the project will operate as intended. This is particularly important when approving scope changes.

Estimating Time: All the individual plans are underpinned by an estimate or forecast of what might happen. The accuracy of the planning is, therefore, directly dependent on the accuracy of the estimate. The time estimate is an input to all plans which have a scheduling component. Estimating an activity's duration is a trade-off between the amount of work, and the resources available, together with productivity.

Scope Management: Scope management defines what work the project includes and, just as importantly, what work is not included in order to meet the stated objectives. On an engineering project, for example, the scope of work would be developed into a list of drawings, bill of materials (BOM) and specifications.

Work Breakdown Structure (WBS): The WBS is one of the key scope management tools used to subdivide the scope of work into manageable work packages which are easier to estimate, plan, assign and control (see *WBS* chapter).

Critical Path Method (CPM): The CPM uses a network diagram to present the work packages and activities in a logical sequence of work which is developed from the build-method and other constraints (internal and external). Activity durations and work calendars are estimated while the availability of procurement, resources and funds are initially assumed. The CPM time analysis (forward pass and backward pass) calculates the activities' early start, early finish, late start, late finish, float and the critical path (see *CPM* chapter). This information is often presented in an activity table and a schedule Gantt chart.

Schedule Gantt Chart: The schedule Gantt chart is one of the best documents for communicating schedule information. It enables the project participants to see what is happening, at a glance, making it is easy to walk through the sequencing of the project's work. The planning structure can be further simplified by focusing on hammocks, milestones and time horizons (see *Gantt Chart* chapter).

Execution Strategy: The project's execution strategy considers the ***buy or make*** decision. If components of the project are to be purchased this becomes a procurement issue. If components of the project are to be made in-house, this becomes a resource issue. The execution strategy considers the availability of in-house resources and expertise, and the benefits of outsourcing work.

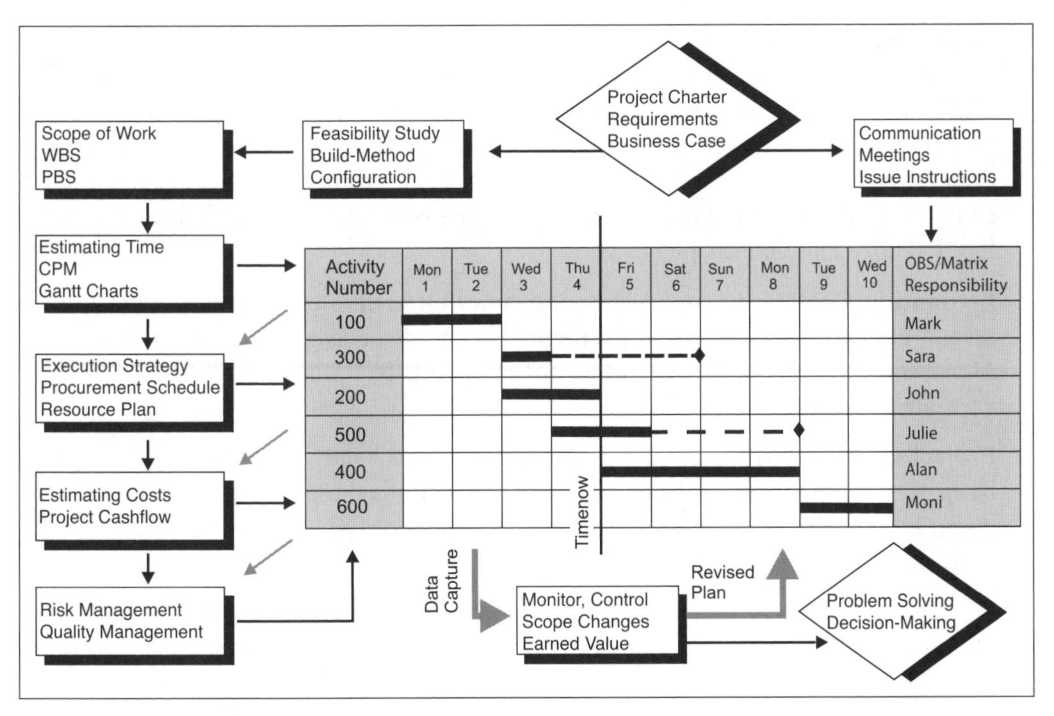

Figure 6.2: Project Management Plan Flowchart – shows the relative position of the project's parameters

Procurement Schedule: The procurement schedule integrates the procurement list with the project schedule. The procurement function is to supply all the bought-in items at the best price and the required quality to meet the project schedule. Long lead items need to be identified early on to assess their impact on the project schedule. If it looks like any items will be delivered late, the project manager needs to trade-off the specification with the delivery transport method and revise the scheduled Gantt chart (see *Procurement Schedule* chapter).

Contract: The contract document outlines the terms and conditions of the agreement. This could be for the main contact or sub-contracts. There are a number of types of contracts to consider; fixed price contract, reimbursable contract, turnkey contract, or a partnership.

Resource Histogram: The resource function is to supply a skilled workforce, machines and equipment to complete the work as outlined in the project schedule (Gantt chart). The manpower process forecasts and compares the resource loading with resource availability. Resource overloads, or resource underloads, need to accommodate both project and company requirements. The resource smoothing needs to consider other company projects and outside contractors, before revising the scheduled Gantt chart (see *Resource Planning* chapter).

Estimating Costs: The cost estimating function estimates all project costs and assigns budgets. These are usually estimated at a work package level. The *Project Cost Management* chapter outlines a number of estimating techniques which are quick and accurate to carry out.

Project Cashflow: The project accounting process not only establishes and assigns budgets to all the work packages but, also, determines the project's cashflow. There might be cashflow constraints restricting the supply of funds which will require the scheduled Gantt chart to be revised (see *Project Cashflow* chapter).

Earned Value: The project costs or man hours can be integrated with the project's schedule to produce the planned value (PV) which forms the baseline plan for the earned value calculation.

Risk Management Plan: The risk management function includes the process of identifying, analysing, and responding to project risk and opportunities (see *Project Risk Management* chapter).

Project Quality Plan: The project quality plan outlines a quality management system (quality assurance and quality control), which is designed to guide and enable the project to meet the required condition. This might include pre-qualifying project personnel and suppliers, developing procedures, quality inspections and quality documentation – all under the umbrella of total quality management (TQM) (see *Quality Management* chapter).

Communication Plan: The communication plan includes the process required to ensure proper collection, storage and dissemination of project information. It consists of communication planning documents, information distribution (lines of communication), a schedule of project meetings, progress reporting and administrative closeout (see *Project Communications Management* chapter).

Organization Breakdown Structure (OBS): The OBS and responsibility assignment matrix (RAM) links the WBS work packages to the company, department, project team or person who is responsible for performing the work.

Baseline Plan (Project Management Plan): The baseline plan is a portfolio of plans and documents which outline how to achieve the project's objectives. The level of detail and accuracy will depend on the project phase and complexity. The baseline plan should be a coherent document to guide the project through the execution and project control cycle.

4. Planning and Control Cycle

The baseline plan completes the planning phase of the planning and control cycle. See the *Execution, Monitoring and Control* chapter for a detailed explanation on how to manage the execution or implementation phase of the project.

Exercises:

1. Develop a why, what, when, where, who, how to, and how much template for a project you are familiar with.
2. List the planning documents you use on your projects. Highlight any logical sequence of development.
3. Develop a baseline plan for your project. Show how you name and number the documents.

Further Reading:

Burke, R., *Project Management Techniques,* will develop the following topics which relate to this knowledge area:

- Decision making
- Ownership and responsibility.

7

Project Lifecycle

Learning Outcomes
After reading this chapter, you should be able to:
Understand how a project can be subdivided into a number of phases
Calculate and plot a line graph of the level of effort
Understand the link and overlap between phases (fast tracking)
Understand how to subdivide phases into input–process–output
Understand the structure of the eight phase product lifecycle

The *Project Lifecycle* knowledge area includes methods of looking at the project environment that is broader than the project itself. The project manager and the project team members need to appreciate this broader context so that they can align the project to the goals of the enterprise.

This chapter will explain how to subdivide the project's scope of work into a number of manageable project phases. This will enable the project manager and project team members to achieve better planning and control.

This chapter links with the previous two chapters (*Project Management Process*) and (*Project Management Plan*) to introduce the last of the 3Ps (processes, plans and phases). Where:

- *Project Management Process* - introduces the key project management processes of initiation, planning, execution and closing of a project.
- *Project Management Plan* - introduces all the project plans that combine to form the baseline plan which is used to plan and control the project's performance.
- *Project Lifecycle* - introduces the project phases, project methodologies and project systems which enable project managers to implement corporate strategy.

Versions of this humorous project lifecycle sketch are often seen on project office notice boards to show the subdivision of the project into a number of phases of increasing mismanagement – hopefully this sketch does not apply to your projects!

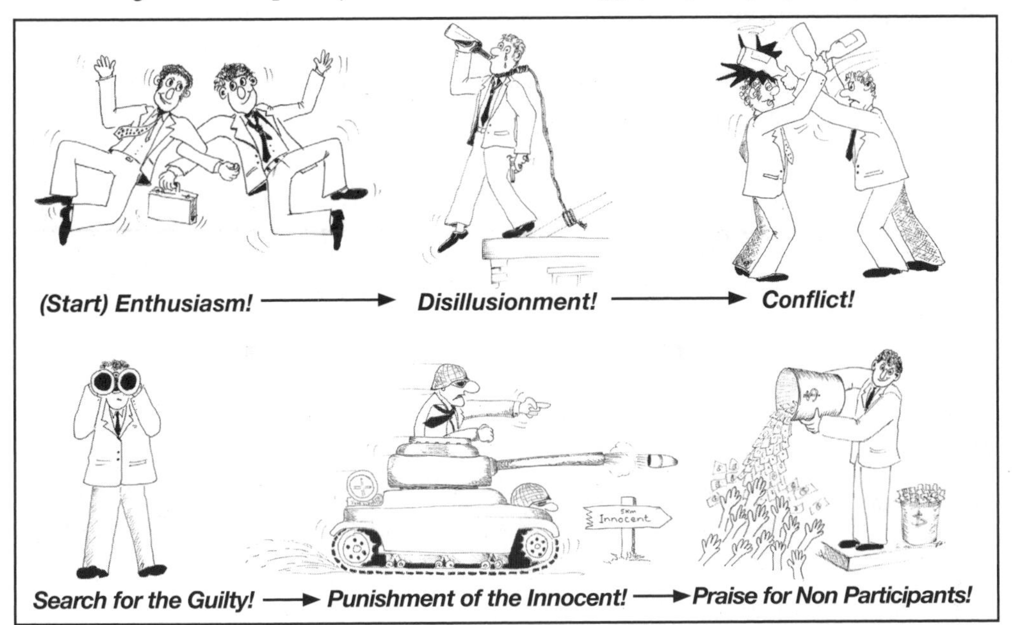

Figure 7.1: Project Lifecycle – shows a lifecycle of increasing mismanagement

Body of Knowledge Mapping

PMBOK 4ed: The *Project Lifecycle* knowledge area includes the following:

PMBOK 4ed	Mapping
2.1.3: Project phases	This chapter will explain how to subdivide a project into a number of phases.
2.1.3: Phase-to-phase relationships	This chapter will explain how to link phases, including fast-tracking.

APM BoK 5ed: The *Project Lifecycle* knowledge area includes the following:

APM BoK 5ed	Mapping
6.1: Distinct phases	This chapter will explain how to subdivide a project into a number of distinct phases.

Unit Standard 50080 (Level 4): The unit standard for the *Project Lifecycle* knowledge area includes the following:

Unit 120372: *Explain fundamentals of project management*

Specific Outcomes	Mapping
120372/SO1.3: A basic project life cycle is explained with examples of possible phases.	This chapter will explain how to subdivide a project into a number of phases.

1. Project Lifecycles

There is general agreement that most projects pass through a number of distinct phases. The purpose of the project lifecycle is to define these phases so that each phase can be defined and managed in a structured manner. All the BoKs have their own project lifecycle terminology but they all align with the basic structure of the generic project lifecycle (see figure 7.2 below).

> The PMBOK 4ed states, *because projects are unique and involve a certain degree of risk, companies performing projects will generally sub-divide their projects into several project phases to provide better management control. Collectively these project phases are called the project life-cycle.*

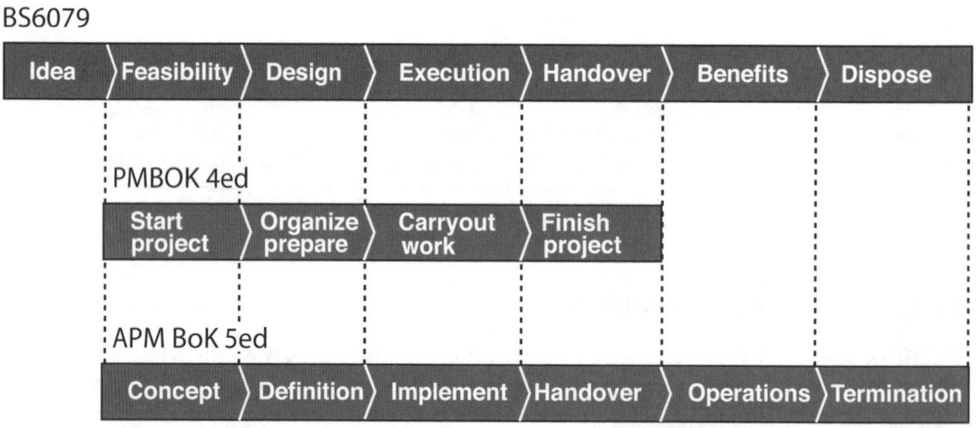

BS6079

Idea | Feasibility | Design | Execution | Handover | Benefits | Dispose

PMBOK 4ed

Start project | Organize prepare | Carryout work | Finish project

APM BoK 5ed

Concept | Definition | Implement | Handover | Operations | Termination

Figure 7.2: Project Lifecycles Alignment – shows how the different project lifecycles, BS6079, PMBOK 4ed and APM BoK 5ed, have a similar structure

All projects start with an idea which triggers a 'quick look' feasibility check. If the idea looks promising the project sponsor authorizes a more 'detailed look' which delivers detailed designs and specifications, and a detailed plan for the project. On approval the work is executed and then handed over for commissioning. On acceptance the project sponsor gains the benefits from the project. At the end of the lifecycle the project is decommissioned and disposed of.

2. Project Lifecycle (4 Phases)

The four phase project lifecycle is typically presented as below (figure 7.3) together with its associated level of effort.

Accumulative Effort

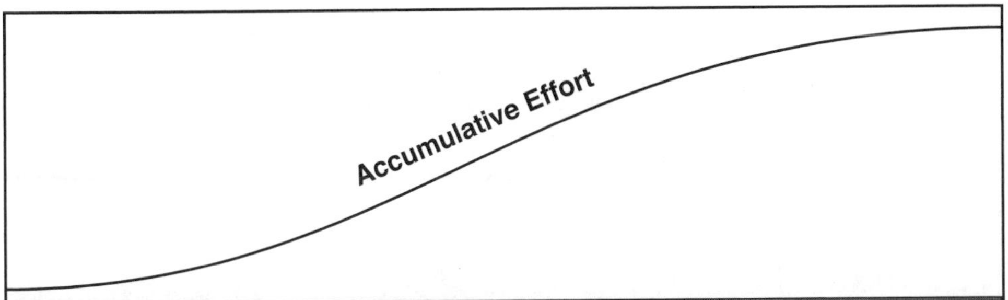

1. Feasibility Phase	2. Design and Development Phase	3. Execution Phase	4. Commissioning and Handover Phase
Identify the need for a project. Initiate the project and carry out a feasibility study.	Develop the project's concept and produce detailed designs and specifications, and a detailed project management plan outlining how to make the project.	Make the project, facility or product (goods or services) as per the design and project management plan.	Confirm the project has been made to the design and plan. Confirm the project works within the intended configuration. Handover the project to the client for operation.

Figure 7.3: Project Lifecycle – shows the widely accepted four phase project lifecycle with its associated level of effort

The APM BoK 5ed defines a **Project Lifecycle** as, *consisting of a number of distinct phases and allows the project to be considered as a sequence of phases which provides the structure and approach for progressively delivering the required outputs.*

The APM BoK's definition emphasises that the project phases are sequential and each phase can be considered as mini projects with their own objectives. As the phases are completed they will progressively develop and deliver the project.

It is logical that the output from one phase becomes the input for the next phase. For example, the approval of the output from the feasibility phase would become the input for the design phase, and when the design is complete this would become the input for the execution phase.

Phase Names: The phases are typically named after the deliverable from each phase. For example, the feasibility phase delivers the feasibility study, and the design phase delivers the project's design. This helps the project team members associate the phase name with the phases' deliverables.

3. Project Lifecycle (Generic)

This section will develop the four phase model a step further and explain in more detail what happens during each of the four phases. This is presented generically so it should apply to most projects most of the time.

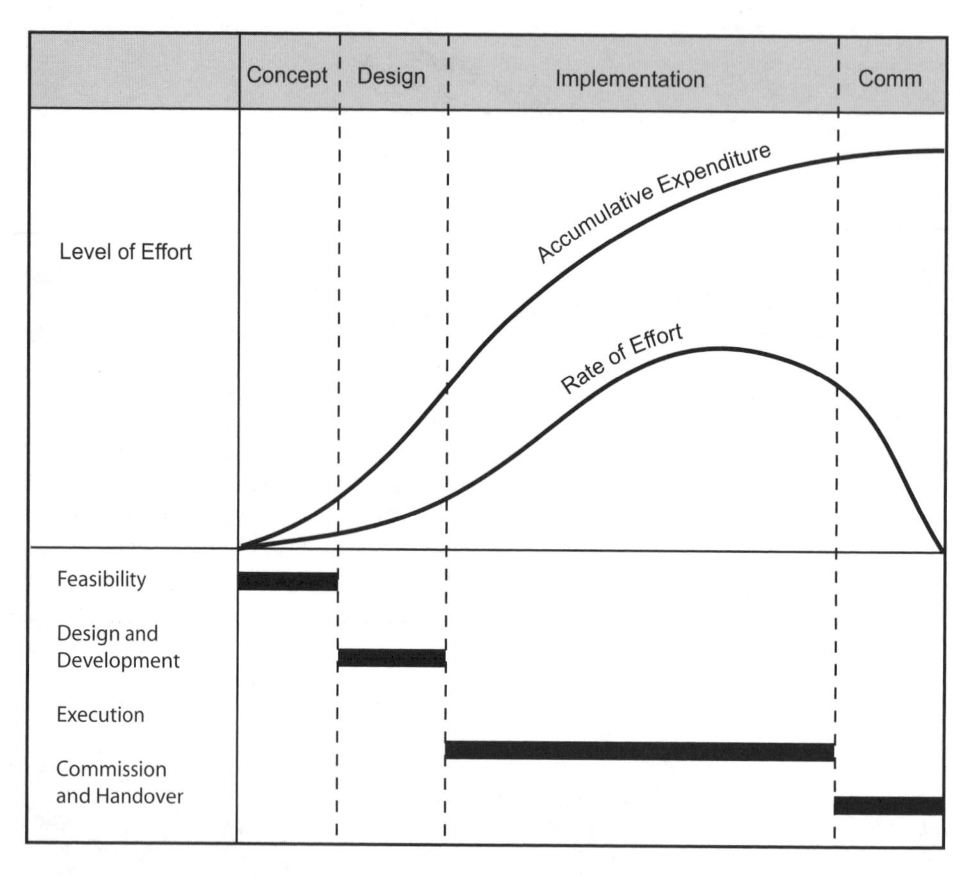

Figure 7.4: Project Lifecycle – shows the relationship between the timing of the four phases presented as a Gantt chart and its associated level of effort

1. Feasibility Phase

The first phase starts the project by establishing a need or opportunity for the product, facility or service. The start of the project is usually formalised by establishing a **project charter** which gives the project an identity (usually a name and number) so that budgets and responsibilities can be assigned. This is where new ideas and options are considered and tested **(feasibility study** and **build-method)** to ensure the product can be made and is making the best use of the company's funds and resources. The output from this phase is an understanding of the risks and opportunities of pursuing the project.

2. Design and Development Phase

On acceptance of the project proposal, the project moves into the second phase to design and develop the project. A budget is allocated to produce detailed designs and specifications of the project and to develop detailed scope and planning documents. These all roll-up into the **project management plan**. The negotiation for long lead items and contracts would begin in this phase.

3. Execution or Implementation Phase

On acceptance of the project management plan, the third phase allocates a budget to implement the project's project management plan to make the facility or solve the problem. This is usually the biggest phase of the project in terms of the level of effort and expenditure but, in principle, this phase should only implement the project as per the project management plan and detailed design from the design and development phase.

4. Commission and Handover Phase

The fourth and last phase of the project lifecycle confirms the project has been implemented or built to the project management plan. It uses a range of inspection and testing techniques to verify compliance. It commissions the equipment to confirm everything is working before handing over the facility to the client. This phase may also involve the training of the client's operators. Following acceptance by the client the project is terminated and a **closeout report** produced.

4. Level of Effort

The project lifecycle is often presented with its associated level of effort. The level of effort gives an indication of the amount of activity or work being performed during each phase. This could be any parameter that flows through the project, but it is most commonly expressed as man hours or expenditure. These parameters can be presented as a line graph of level of expenditure per day (or unit of time) and/or as the accumulated expenditure (or S curve).

How to Draw the Level of Effort Curve: First draw the project's Gantt chart to show when the work is scheduled (figure 7.5). Then insert the values of the parameter to be tracked, per day (or preferred unit of time). These values are added vertically to give a total per day, and then accumulated to give the running total. Finally, plot the total per day and running total to give the level of effort and the accumulated effort (figure 7.6).

Level of Effort		1	2	3	4	5	6	7	8	9	10	11	12
Feasibility	1	2	2	2									
	2		2	2									
	3			5									
Design	1				10	10							
	2				5	5	10						
	3					3	10						
Execution	1							15	20	15			
	2							15	20	20			
	3									6			
Commission	1										10	2	
	2											5	2
	3												2
Daily Total		2	4	9	15	18	20	30	40	41	10	7	4
Accumulated			6	15	30	48	68	98	138	179	189	196	200

Figure 7.5: Level of Effort Example - shows level of effort example

S Curve: From figure 7.6 the level of effort profile shows a slow build-up of effort during the feasibility study phase and the design phase as the project is being designed and developed. The build up of effort accelerates during the execution phase to a maximum, as the work faces are opened-up and resources assigned. On completion of the work there is a sharp decline in the level of effort as the work is commissioned and the project draws to a close and is handed over to operation.

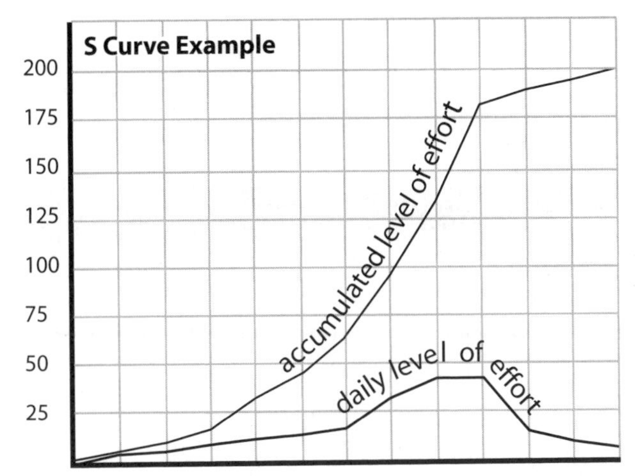

Figure 7.6: Level of Effort S Curve - shows the daily and the accumulative level of effort

Level of Effort Exercise: Calculate and draw the level of expenditure per day and the accumulated expenditure (S curve) for the project below. See Appendix 1 for solution.

Level of Effort		1	2	3	4	5	6	7	8	9	10	11	12
Feasibility	1	1	2	3									
	2		2	2									
	3			2									
Design	1				9	9	9						
	2					3	4						
	3						4						
Execution	1							30	33	24			
	2								20	9			
	3									6			
Commission	1										12	8	3
	2											2	2
	3												1
Daily Total													
Accumulated													

Figure 7.7: Level of Effort Exercise - shows level of effort exercise

5. House Extension Project Lifecycle (4 phases)

A good example of the project lifecycle is to show how a house extension project passes through the four project phases, see figure 7.8 below.

Figure 7.8: Project Lifecycle – shows the project lifecycle for a house extension

3. Execution Phase	
On acceptance of the detailed baseline plan, the contracts are negotiated, materials are procured and the construction is executed. The house is built to the detailed plans developed in the previous phase. Changes may be made to the original baseline plan as problems arise or better information is available (for example, up-spec bathroom fittings).	
4. Commissioning and Handover Phase	
On completion the building is inspected and approved by the client and responsible authorities. The house is now ready to be handed over for occupation. The project is terminated and a closeout report is produced to document lessons learnt.	

6. Phase-to-Phase Relationships (fast-tracking)

The project phases are shown here in sequence, implying that the feasibility phase must be complete before the design and development phase can start. And, further, the design and development phase must be complete before the execution phase can start.

However, in practice, there is usually some degree of overlap between the phases meaning the following phase can start before the preceding phase is totally complete. For example, the construction of a house extension could start before the design is totally compete – if the plans for the foundation, walls and roof are complete and approved, the construction can go ahead even though the design of the lighting and plumbing fittings might not be finalised. This practice is called **fast-tracking**, where the deliverables from the preceding phase are progressively approved so that work can start early on the next phase.

When fast-tracking is practised throughout the project this speeds up the completion and gets the project to market before the competition. This is particularly important for commercial companies operating in a competitive environment where coming second will significantly devalue the project. For example, Apple was the first to market with the iPod to download music and captured the market share.

Fast-Tracking Risk: Fast-tracking the project obviously increases the project's risks as the amount of float between activities is reduced. The level of monitoring and control might need to be increased with an associated cost. Procurement costs might also increase as components are air freighted rather than shipped to speed up the delivery. However, the benefit of fast tracking is the reduction of the project's overall duration, which should reduce all the time related costs, such as, overheads, salaries and plant hire.

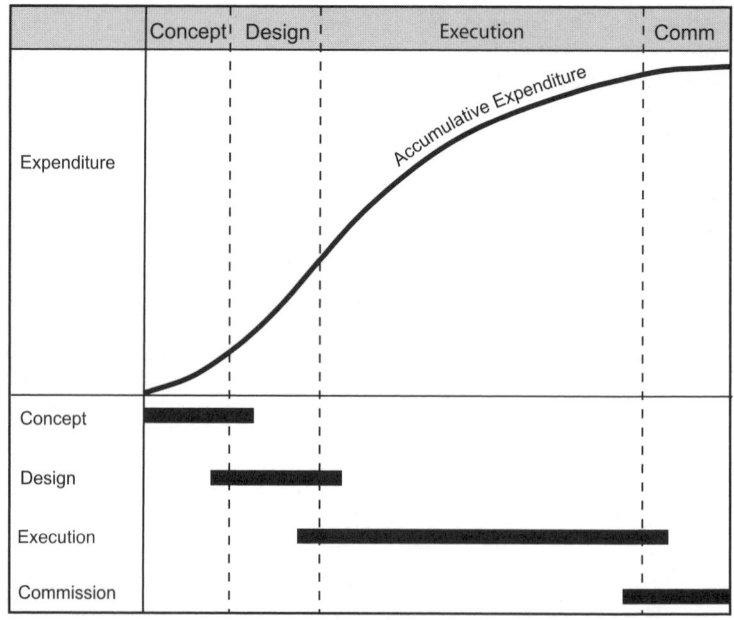

Figure 7.9: Project Lifecycle Overlap – shows the overlap (fast-tracking) between phases on a house extension project

7. Lifecycle Methodology (Input, Process, Output)

So far in this chapter the text has shown how the project can be subdivided into a number of sequential phases. This section will explain how each project phase can be further subdivided by the project management processes within each phase.

Accumulative Effort

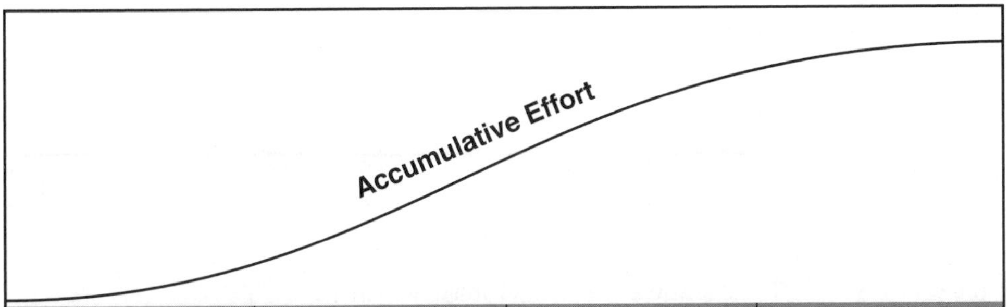

1. Feasibility	2. Design and Planning	3. Execution	4. Commission
A. Initiation	**Initiation**	**Initiation**	**Initiation**
- Project charter (initiation) - Appoint project manager	- Design charter - Appoint project manager	- Execution charter - Appoint project manager	- Commission charter - Appoint project manager
B. Planning	**Planning**	**Planning**	**Planning**
- Plan feasibility study	- Plan the design project - Plan the planning project	- Plan execution phase	- Plan commissioning phase
C. Execution	**Execution**	**Execution**	**Execution**
- Feasibility study - Stakeholders' analysis - Cost benefit analysis	- Design project - Model testing - Prototype testing - Detailed planning (WBS, estimate, risk analysis, CPM, Gantt chart, OBS)	- Award contracts - Expedite - Monitor and Control - Scope change control	- Test, run-up equipment - Certificate of completion
D. Closing	**Closing**	**Closing**	**Closing**
- Project proposal (solution) - Review phase - Go/No Go	- Project Management plan - Review phase - Go/No Go	- Ready for commissioning - Review phase - Go/No Go	- Project handover for operation - Closeout report

Figure 7.10: Project Lifecycle Phase Processes – shows the project management processes within each phase

8. Product Lifecycle (8 Phases)

The typical project lifecycle only considers the project from feasibility to handover. However, if the project is to build a facility, a factory, a computer system or sports stadium then, looking at the project from the client's perspective, the efficient operation of the facility and the return on investment should also be considered. To look at the wider picture we use what is termed as the ***product lifecycle***, which considers the facility from ***the cradle to the grave***.

1. Strategy	2. Feasibility	3. Design	4. Execution
Problem, need, opportunity triggers project	Select best option, confirm the project is feasible	Design project, produce the project management plan	Execute project, monitor and control

Figure 7.11: Product Lifecycle (8 Phases) – shows the eight phases of the product lifecycle

1. Strategy Phase: Projects usually evolve from within the work environment or market the company operates in. The event which triggers the project is typically a problem, a need or an opportunity. Consider the following:

- R&D generates new ideas, innovation and creativity
- A computer system needs upgrading to take advantage of new technology
- Market research identifies market changes
- A response to the competitor's new product
- Facilities need expanding to meet increased demand.

Managing the pre-project environment is important for a company's long term survival in a changing world.

2. Feasibility Phase: Select best option, confirms the project is feasible.

3. Design Phase: Design project, produce the project management plan.

4. Implementation Phase: Execute project, monitor and control.

5. Commissioning Phase: Commission project, handover to operation.

6. Operation Phase: The starting up of the operation phase can be considered as a mini-project. This will include training the operators and ironing out teething problems.

7. Up-Grade, Half-Life Refit and Expansion Phase: At some point all facilities require a major overhaul, refit, up-grade, or expansion to keep them running efficiently, to incorporate new technology and to maintain competitive advantage. New technology, competition, market requirements, rules and regulations are all factors influencing up-grades. These types of projects are characterised by tight time scales and working around the clock to get the facility up and running again and reduce the downtime.

5. Commissioning	6. Operation Start Up	7. Up Grade	8. Disposal
Confirm the project has met the required condition	Start to operate the facility	Retro-fit new technology	Decommission and dispose of project

8. Decommission and Disposal: The final part of the product lifecycle is the decommissioning and disposing of the facility. Depending on the type of product, disposal could mean a simple trip to the scrap heap, or a more involved decommissioning process. For example, as people become more aware of the environment, a product that contains any substances which could have a negative impact on the environment (asbestos, chemicals) dictates that disposal could be an elaborate project in itself.

Exercises:

1. Draw the level of effort for the exercise in section 4 - see Appendix 1 for solution.
2. Discuss how your projects can be subdivided into four phases.
3. Discuss how each phase can be subdivided by the project management processes.
4. Discuss how your projects can be subdivided into an eight phase product lifecycle.

Further Reading:

Burke, R., *Project Management Techniques*, will develop the following topics which relate to this knowledge area:

- Level of influence vs cost of changes
- Phase subdivision
- Entrepreneur lifecycle
- Procurement lifecycle.

8

Feasibility Study

Learning Outcomes
After reading this chapter, you should be able to:
Identify all the project stakeholders
Determine the stakeholders' needs and expectations
Evaluate constraints (project, internal and external)
Consider alternative options
Identify further opportunities

This chapter will explain how to carry out a feasibility study. A feasibility study is a key topic within the *Project Scope Management knowledge* area. Scope management uses the feasibility study to identify the project's stakeholders so that their needs and expectations can be determined as a means of quantifying the project's objectives and scope of work.

Ideas, needs and problems crystallize into projects in different ways. The process of project formulation varies in each company and on different types of projects. Whichever way the new venture or project develops there should, at some point, be a feasibility study to not only ensure the project is feasible, but also ensure the venture is making the best use of the company's resources.

Project Lifecycle

The project lifecycle (figure 8.1) indicates the relative position of the feasibility phase (and therefore the feasibility study) with respect to the strategy phase and the design phase. The project charter which is owned by the project sponsor initiates the feasibility study.

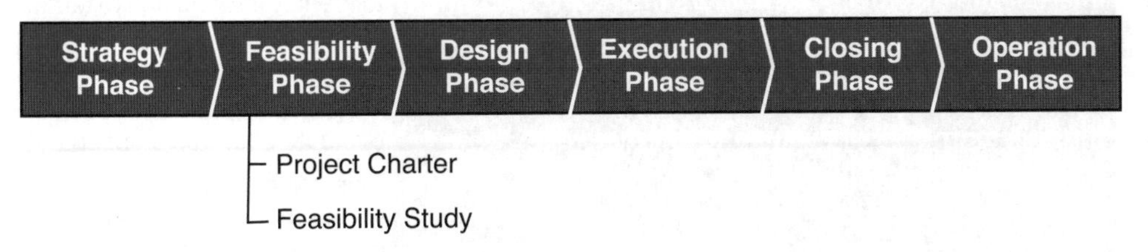

Figure 8.1: Project Lifecycle – shows the relative position of the feasibility study with respect to the strategy phase and the design phase

Project Management Plan Flowchart

The project management plan flowchart (figure 8.2) shows the relative position of the feasibility study with respect to the other techniques. It is logical that the project charter initiates the feasibility study. It is also logical that the detailed scope of work follows the feasibility study.

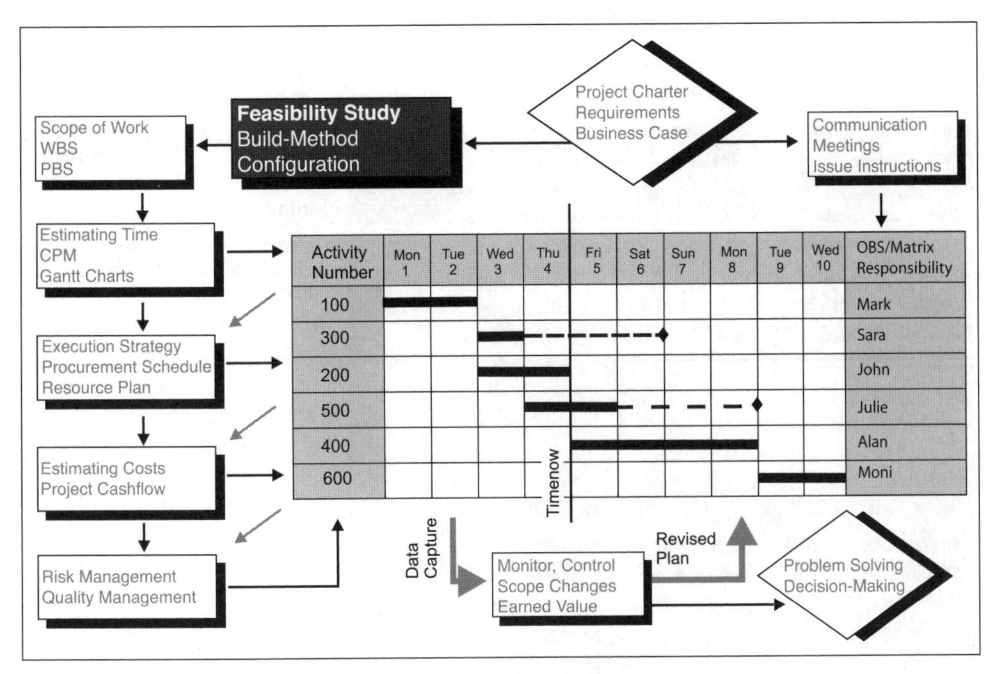

Figure 8.2: Project Management Plan Flowchart - shows the relative position of the feasibility study with respect to the other techniques

Body of Knowledge Mapping

PMBOK 4ed: The *Feasibility Study* is part of the *Project Scope Management* knowledge area that includes the following:

PMBOK 4ed	Mapping
2.3: The project management team must identify both internal and external stakeholders in order to determine the requirements and expectations of all parties involved.	This chapter will explain how to identify internal and external stakeholders.
3.3: The feasibility of a new undertaking may be established through a process of evaluating alternatives.	This chapter will explain how to evaluate alternative options.
5.1: Collecting requirements is the process of defining and documenting the project and product features and functions needed to fulfil stakeholders' needs and expectations.	This chapter will explain how to determine the stakeholders' needs and expectations.

APM BoK 5ed: The *Feasibility Study's* knowledge area includes the following definition:

The APM BoK 5ed defines the **Feasibility Study** as, *an analysis to determine if a course of action is possible within the terms of reference of the project. Work carried out on a proposed project or alternatives to provide a basis for deciding whether or not to proceed.*

Unit Standard 50080 (level 4): The unit standard for the *Feasibility Study* knowledge area includes the following:

Unit 120372: *Explain fundamentals of project management*

Unit 120373: *Contribute to project initiation, scope definition and scope change control*

Specific Outcomes	Mapping
120372/SO4.4: Stakeholders are explained with examples of at least six different stakeholders.	This chapter will explain how to identify the stakeholders.

Specific Outcomes	Mapping
120373/SO1: Contribute to the identification and co-ordination of stakeholders, their roles, needs and expectations.	This chapter will explain how to identify the stakeholders' roles, needs and expectations.
120373/SO2: Contribute to the identification, description and analysis of the project needs, expectations, constraints, assumptions, exclusions, inclusions and deliverables.	This chapter will explain how to identify stakeholders' needs and expectations, constraints, assumptions, exclusions, inclusions and deliverables.
120373/SO1.1: Project stakeholders are identified and their roles on achievement of project outcomes are recorded and / or explained with examples.	This chapter will explain how to identify the stakeholders.
120373/SO1.2: Project stakeholders' needs and expectations are identified and documented according to agreed format.	This chapter will discuss the stakeholders' needs and expectations.
120373/SO1.3: Project deliverables are verified against the needs of stakeholders.	See *Scope Management* chapter.

1. Stakeholder Analysis

Projects are not performed in a vacuum – they are performed within a company, within an industry, and within a market. They involve a wide range of people who have a wide range of needs and expectations. These people are called stakeholders, and the purpose of this chapter is to present a structured approach, which the project manager and project team members can use, to identify the stakeholders and determine their needs and expectations.

There is a great temptation by newly appointed project managers and team members to rush into a project and try to complete everything in record time. But if they do, they will soon find out that project success depends on a certain amount of cooperation from a range of stakeholders. That is why it is essential to identify these stakeholders and build a working relationship with them. For example, when using a matrix OBS one of the key group of stakeholders are the functional managers because they own the resources (labour, machinery and equipment). In the real world, project managers need to negotiate with the functional managers for the use of their resources. If they do not negotiate and expect the functional managers to jump to their demands, in preference to their own work commitments, they will soon be disappointed.

The project team members should create an environment where the stakeholders are encouraged to contribute their skills and knowledge so their needs and expectations can be integrated with the other stakeholders' needs and expectations. Consider the following types of stakeholders:

- Those who are actively involved in the new venture or project
- Those whose interests are impacted by the new venture while it is being implemented
- Those whose interests are impacted after the new venture has been implemented
- Those who could have an impact on the new venture, for example, the Green lobby.

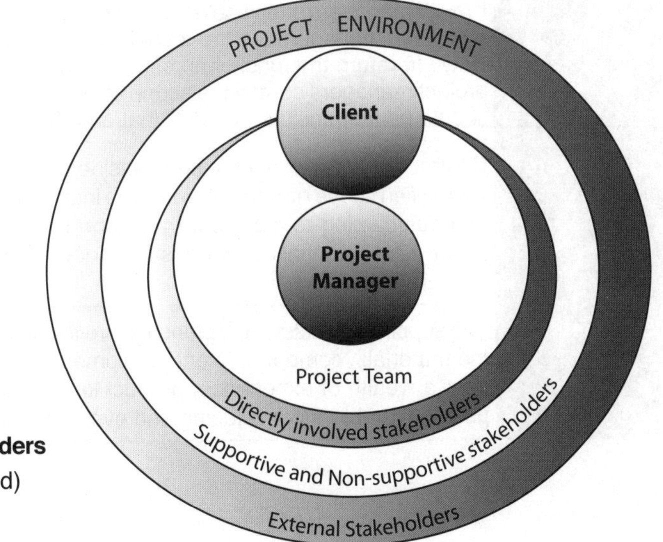

Figure 8.3: Project Stakeholders (developed from PMBOK 3ed)

Consider the following types of stakeholders:

Originator	The originator of the project is the person with entrepreneurial skills who suggests the creative and innovative idea for the project, or spots the opportunity in the first place. The project manager needs to work closely with the originator as the originator is the source of the project's vision and market opportunity. *'I get my best ideas in the shower.'*
Owner	The owner is the person, department or company whose strategic plan creates the need for the project. The project manager needs to work closely with the owner to ensure the project benefits satisfy the owner's needs and expectations.
Sponsor	The project sponsor is the client or internal executive who authorises the expenditure on the project. The sponsor is responsible for acquiring the benefits from the project. The sponsor initiates the project with the project charter which outlines the objectives of the project and how the project is to be managed. The sponsor appoints the project manager and approves the appointment of the project team members. The sponsor can provide the project manager with vital support and nurture high-level contacts. The project manager needs to keep the sponsor informed on all aspects of the project's progress.
Functional Managers	The functional managers, also referred to as line managers, usually 'own' the resources (labour, machines and equipment) that the project manager needs to complete the project. The project manager / functional manager interface is one of the weak links in the matrix type structure, because the project manager is not in a position to order and demand the supply of resources but, rather, has to request and negotiate with the functional managers for the use of their resources. The quality of this relationship will obviously influence the success of the project. Project managers, therefore, need to have a negotiation strategy. Project managers often use a trade-off approach where they acknowledge the functional managers are the technical experts responsible for the 'Who and How'. In return the functional managers are asked to acknowledge that the project managers control the scope of work and the schedule and, therefore, should be responsible for the 'What and When'.
Contractors	Contractors are a workforce from outside the company, employed to carry out a specified scope of work. Projects are increasingly using external contractors and outsourcing to carry out the scope of work. Project managers should identify the potential contractors to smooth the labour peaks in the resource histogram.
Suppliers	The suppliers, vendors and plant hire are the companies and people supplying the materials, components and equipment required by the project, and can offer a wealth of experience, product knowledge, give advance warning of potential problems developing and even offer solutions.

Support Companies	Support companies provide goods and services to enable the facility or product to be manufactured. For example, this could be the suppliers of telecommunication or electricity, the postal service and even the corner shop. Financial support through the banking system should be included under this heading.
Users	The users are the people who will operate the project or facility on behalf of the owner (not to be confused with the customer). For example, the project may be to build an airport terminal, a power station, a manufacturing plant or a transport facility. The project manager should involve the users as early as possible in the design stage so that they can discuss items of the project that impact on them personally, such as, the ergonomics of the operators' seating arrangement and controls. This involvement will help to address any resistance to change and also gain the users' commitment to accept the new facility.  ***'No problem Sir, we have the latest high tech luggage handling system in the world!'***
Customers	The customers are the people who receive and pay for the benefit of consuming the project's product or facility. For example, we are all customers for electricity, telephones, travel and clothing. It is the project manager's responsibility to ensure the project meets the customer's requirements - this can be quantified using market research (see **Burke**, R., *Entrepreneurs Toolkit*). Ultimately the commercial success of a project depends on the **customer buying the product**.

Table 8.1: Project Stakeholders

There are other external stakeholders who may not be directly involved with the project, but can influence the outcome:

- Regulatory authorities - health and safety
- Trade unions
- Special interest groups (environmentalists) who represent the society at large
- Lobby groups
- Government agencies and media outlets
- Individual citizens.

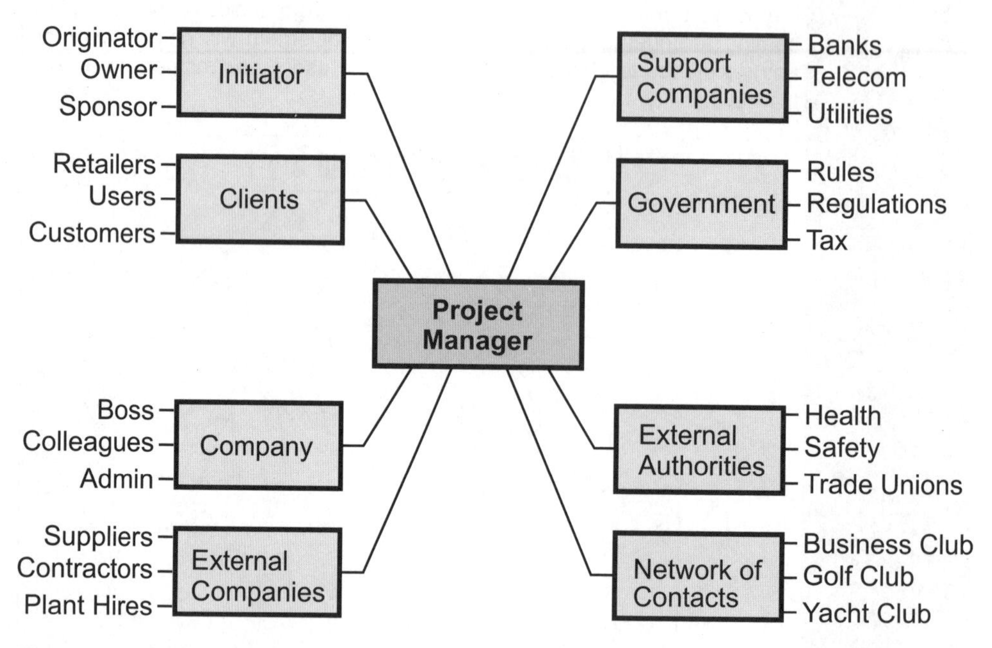

Figure 8.4: Project Stakeholders – shows the project manager's web of stakeholders

2. Define the Client's Needs

The starting point for a new venture or project usually begins with the client or project sponsor wanting to solve a problem, or take advantage of an entrepreneurial opportunity which could be internal or external to the company. The project manager's challenge is to translate the client's needs from something quite vague to something tangible, which will then serve as the basis of the project management plan.

The client's needs are constraints which relate directly to the scope of the project and ask basic questions about the configuration, operation and feasibility of the project delivering the required benefits. These constraints could be structured as follows:

Functions	The project must carry out certain functions at a predefined rate (produce 'x' units per hour).	**Priority**
Environment	The project must operate in a specific environment (summer or winter conditions).	
Life Span	The project must have a working life of so many years (say, guaranteed for 5 years).	
Budget	The project's budget must not exceed $ x.	
Efficiency	The project must be energy efficient. A car would quantify this requirement as 'x' miles per gallon, or 'x' kilometres per litre.	
Design	The design ergonomics must be consistent with the latest accepted practices and trends. *'It's another world spinner!'*	
Risk	The project must achieve reliability requirements. These may be quantified as mean time between failures (MTBF). This enables companies to use a planned maintenance programme.	
Maintenance	Ease of maintenance and repair must be incorporated into the design. For example, it would be expensive to have to dismantle a car to replace its cam belt.	
Redundancy	A predetermined level of system redundancy and inter-changeable parts must be achieved.	

Specifications and Standards	The project must meet certain specifications, national and international standards (BS, ABS, DIN).	
Regulations	The product must meet statutory health and safety regulations.	
Local Content	The product must be manufactured with a predefined value of local content. For example, certain countries require their cars to have 25% local content by value.	
Manpower	The operational requirements must achieve predetermined manpower levels and automation.	
Expansion	The product must be flexible and provide opportunities for future expansion and up-grade. For example, a computer must have expansion slots.	
Schedule	The project must be operational by a predefined date. For example, the building of a holiday resort would need to be completed in time for the tourist summer season.	
Approved Suppliers	Approved and accredited suppliers must manufacture the product. If necessary, the suppliers should be pre-qualified by an audit.	
Quality	All suppliers must have implemented an approved quality management system which can be audited.	
Suppliers	All suppliers must have a good track record, supported by references, and must be financially stable, supported by a bank reference.	

Table 8.2: Stakeholders' Needs Analysis – shows a column for ranking by priority so that the most important needs can be identified

Many of the above items could be mutually exclusive, which means there will have to be a trade-off and a priority list. For example, it is generally not possible for a car to achieve both maximum power and maximum efficiency. These items of conflict need to be discussed and resolved during the early stages of the project, with all decisions recorded to form the basis of the design philosophy. This key document must be structured in such a way as to facilitate an audit trail of the decisions. If the field of the project is highly specialised the client might employ consultants and specialists to assist in defining the scope and specifications.

3. Internal Project Constraints

The internal project constraints relate directly to the scope of the project and ask basic questions about the build-method and the feasibility of making the project, product or facility:

Feasibility	Can the project physically be made? (See *build-method* below.)
Execution Strategy	How will the company make the project? This could be a combination of; internal resources, partnership, contractors, procurement or outsourcing.
Technology Transfer	Does the company have the technology to make the project? If not, can the technology be acquired through a technology transfer; if so, with whom? For example, on multi-disciplined projects, companies might not have all the skills required to make all of the project.
Best Time	Should the project start now with the present technology or is it better to wait until new and probably better technology is available? For example, should the new computer system be installed now or is it better to wait for Microsoft's next operating platform?
Technology Risk	How big is the new technology component? If there are too many new and untested components this might increase the level of risk and uncertainty to an unacceptable level.
Design Freeze	At what point in the development phase should a design freeze be imposed?

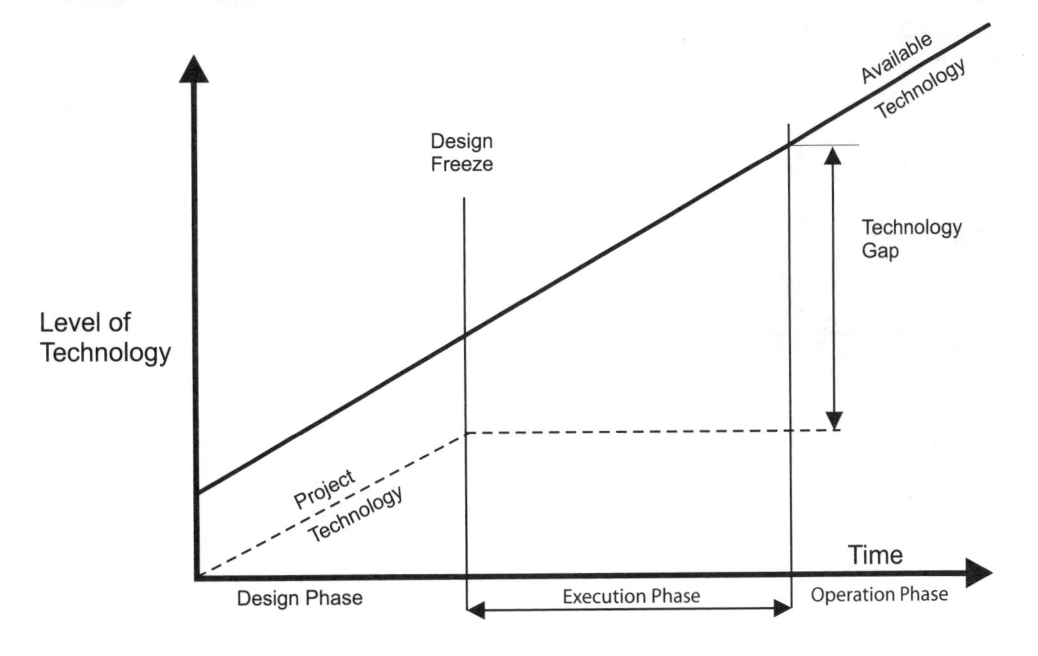

Figure 8.5: Technology Change Against Time - shows that at some point a design freeze will have to be implemented so that the product can be built

Facilities	Are there sufficient space and manufacturing facilities available to make the project?
Resources	Can the resources be trained up to the required level of ability to make the project, or should contractors be employed to meet the forecasted skills requirement?
Multi-Projects	Will there be sufficient resources available for all projects? If the company is running a number of projects concurrently, the multi-project resource analysis will need to consider the overall impact on the supply of internal resources.
Machines	Are special machines, equipment or plant required? If yes, are they available within the company or can they be hired or procured externally?
Schedule	Can the project meet the milestones and completion date with the resources available?
Transport	Are there any special transport requirements? Can the product be transported in one piece, or does it need to be broken down and assembled on site?
Management	Are there any new management systems to be introduced? Will they be compatible with existing systems they interface with?
Budget	Can the project be completed within the assigned budget?
Quality	What is the quality management requirement? For example, is accreditation to ISO 9000 required? Is the present quality system (assurance and control) sufficient?
Specifications	Can the company meet the project's requirements and specification tolerances?
Procedures	Are there company and project procedures in place? If not, is there time to develop them?
Project Office	Is the project office (PO) set up for the project? Has the project team been selected? Has the office space been allocated and the office equipment been installed?
Penalties	Can the company accept the time penalties?
Risk	Is the level of project risk and uncertainty acceptable?
Terms	Can the company accept the terms and conditions outlined in the contract document?

Table 8.3: Internal Project Constraints

Build-Method: The build-method explains how the project will be made. It is an important part of the feasibility study because it asks the hard questions about how the project will be made. The build-method supplements the CPM and the Gantt chart which are the documents normally associated with outlining the logical sequence of work. But the build-method requires a greater level of detail and asks the following type of questions:

What	What machines and equipment will be used and where will they be positioned?
Where	Where will the project work be manufactured? Will it be outsourced?
How	How will the work be carried out (method)? Is there a recommended or prescribed procedure to follow?
Responsibility	What departments or managers will be responsible for the work?
Material Storage	Where will the materials and components be stored? Are there any special handling and storage requirements? For example, welding rods need to be kept in a temperature and humidity controlled environment.
Inspection	How will the product be inspected, and who will inspect the product? Will there be an in-process inspection? Will there be hold point inspections? Will there be operational testing?
Transport	What type of transport will be used to move the materials and finished product (road, rail, air)?

Table 8.4: Build-Method – shows how the project will be made

Many of these build-method items might be discussed formally or informally, in project meetings or conversations. The aim of the build-method is to capture all the relevant points into one coherent document which can be agreed upon by the stakeholders and signed off. This will help to ensure that all the stakeholders, particularly the client and contractors and suppliers, are working to the same build-method.

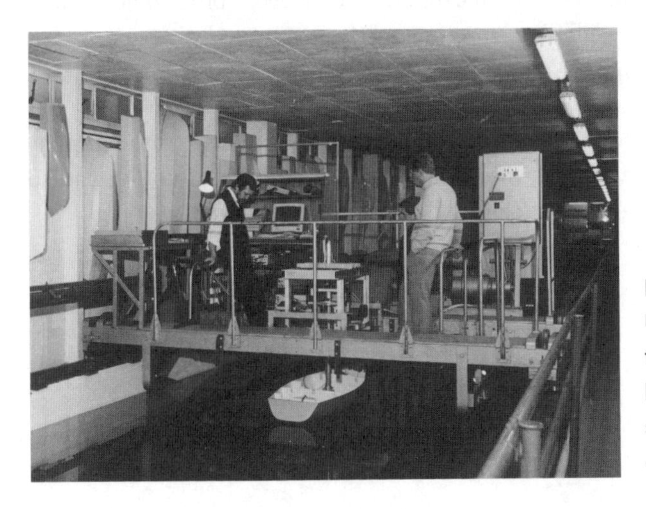

Model Testing at Solent University - shows how model testing is used to confirm the boat can achieve the design specifications and so reduce design risks

4. Internal Corporate Constraints

The company itself may impose further quasi constraints on the project. Corporate policy and strategy usually relates to long term issues which, indirectly and unintentionally, may impose limitations on the project. Consider the following:

Financial Objectives	The financial project selection criteria might be based on corporate requirements expressed as; payback period, breakeven point and return on investment.
Cashflow	The company might want the project to maintain a positive cashflow through all the phases.
Marketing	The company might want to diversify its product range and enter new markets. As a new venture, the project might be to implement the technology transfer for the company to operate in this new market.
Estimating	During a down turn in the economy, the company's main priority might be to keep the workforce together. In this situation the project manager would be encouraged to lower the quotation to secure the contract.The lower the bid the greater the probability of being awarded the next contract. The lowest a company can bid is to cover variable costs with the fixed costs being written-off.
Partner	The company might want to take on a partner who has previous experience in the field of the project (technology transfer) to share and lower the risk.
Industrial Relations	Industrial unrest is often caused by conflict over pay and working conditions. The project manager might have little power to influence these conditions as they are determined at the corporate level.
Exports	The finance director might influence the estimate in an effort to acquire exports to enter new markets overseas or take advantage of export incentives.
Outsourcing	The company might want to take advantage of lower labour costs overseas by encouraging the project manager to outsource work to China, for example. This will have an impact on co-ordination, procurement lead times and quality control.

Table 8.5: Internal Corporate Constraints

Where these company objectives are in conflict with the project objectives, the company objectives usually take preference. This internal corporate constraint could lead to increased project costs which must be included in the project's budget.

5. External Constraints

Stakeholders outside the company and beyond the project's sphere of influence can impose external constraints. Many of these constraints will not be negotiable at the project level. Consider:

Regulations	National and international laws and regulations - right of way, planning permission, licences, permits, covenants.
Procurement	Material and component delivery lead times.
Contractors	Limited number of sub-contractors who can perform the work.
Resources	External resources are unavailable due to other large projects being carried out at the same time. For example, it is usually difficult to find carpenters after there have been floods or storm damage, or find welders while there are offshore projects and sports stadiums being built.
Transportation	Logistic constraints, availability of transportation, routes avoiding low bridges.
Foreign Currency	Availability of foreign currency and currency fluctuations when items are procured overseas.
Market	Market forces influence the supply and demand curve of products and hence their value. For example, on fixed price contracts the budget is at risk from material price increases.
Environment	Environmental issues, government legislation and pressure group activities; for example, Greenpeace and Campaign for Nuclear Disarmament (CND). The nuclear, chemical, mining and transport industries have been particularly affected in the past.
Climate	Climatic conditions, inclement weather, rain, wind, heat and humidity.
Political	Political unrest.
Insurance	Insuring certain risks might impose constraints on the project.

Table 8.6: External Constraints

To help manage the different types of constraints a summary table collating all the constraints should be considered:

WBS	Description	Internal Project Constraints	Internal Corporate Constraints	External Constraints
1001	Technical	Need new computer		
1002	Budget		Need to make 15% profit	
1003	Regulation			Health and Safety

Table 8.7: Summary Table of Constraints – shows a template to capture and present constraints

These headings should not be seen as comprehensive, but as the forerunner of a company checklist that ensures all the necessary questions are asked which, in turn, should reduce the level of risk and uncertainty.

6. Evaluate Options and Alternatives

Having identified the client's needs and the constraints (project, internal and external), the next step is to consider alternative ways of producing the project. Consider the following questions:

Time	Can the project be completed sooner?
Cost	Can the budget be reduced?
Quality	Can the project be made to a lower level of quality which would be acceptable to the client but more cost effective and quicker to produce?
Resources	Can the work be automated to reduce the manpower requirement?
Technology	Has the latest technology been considered?
Design	Is there a simpler design configuration?
Materials	Can cheaper materials be used?
Equipment	Has the use of different equipment and machinery been considered?
Build-Method	Is there a simpler build-method?
Trade-off	Has the trade-off between cost, delivery schedule and technical performance been quantified?
Management	Have alternative management systems been considered?

Table 8.8: Options and Alternatives

To help structure all the options and the alternative suggestions, a summary template can be set up which links in with the WBS.

WBS	Description	Alternatives and Options
2001	Design	Use carbon fibre to reduce weight.
2002	Manufacture	Outsource manufacturing to China to reduce costs.
2003	Transport	Use airfreight to reduce the delivery time.

Table 8.9: Table of Options and Alternatives

The technical definition should aid the direct comparison between options and alternatives. With a machine, for example, the capital costs should be compared with the operating costs. Although this process should be on-going during the project, the design freeze would usually signal the end of this phase. Once the manufacturing phase starts the emphasis would shift to considering manufacturing alternatives.

7. Feasibility Study Summary Template

After developing all the separate areas of the feasibility study, a summary template which gathers together all the main findings should be produced. Summary documents help to give an overall picture of the project and should make it easier to highlight areas of concern, and also the areas of opportunity.

WBS	Idea	Stakeholders' Needs	Constraints	Alternatives
1000	Design a tourism website	To present information to potential and existing customers	1 GB host site	Other providers
2000	Build yacht	To go bluewater cruising	Comply with Category 1 requirements	Different materials; GRP, Steel, Timber
3000	Your project???			

Table 8.10: Feasibility Study Summary Template

Exercises:

1. Discuss the purpose of a feasibility study in a project environment.
2. Discuss how you identify all the stakeholders on your project and assess their needs and expectations.
3. Discuss how you walk through the build-method of your project to confirm its feasibility.
4. Discuss how you check the feasibility of the project working effectively in the operating environment (include the configuration management).

Further Reading:

Burke, R., *Project Management Techniques*, will develop the following topics which relate to this knowledge area:

- Project viability check
- Value management
- Cost-benefit analysis
- Project selection using numerical, non-numerical and discounted cashflow techniques.

Burke, R., *Entrepreneurs Toolkit*

- Spot opportunities
- Market research.

Burke, R., **Barron**, S., *Project Management Leadership*

9

Scope Management

Learning Outcomes After reading this chapter, you should be able to:
Define the project's scope of work
Verify and accept the completed scope of work
Administer the scope change control process

The *Scope Management* knowledge area includes the processes required to ensure that the project includes all the work required, and only the work required, to complete the project deliverables successfully. Managing the project's scope of work (SOW) is primarily concerned with **defining**, **verifying** and **controlling** the scope of work.

Effective scope management is one of the key factors determining project success. Failure to accurately interpret the client's problems, needs or opportunities will produce a misleading definition of the scope of work. And, if this causes rework and additional effort, there could be implications for project time, cost and quality. Therefore, project success will be self-limiting if the project's scope of work is not adequately defined.

Scope management defines what the project will achieve, what it will deliver, what it will produce and where the work packages start and finish. Since most projects seem to be riddled with fuzzy definitions, scope management takes on a greater importance to avoid **scope creep** (unnecessary expansion of the scope of work). To prevent scope creep, effective scope change control helps to avoid adding features and functionality to the product that were not part of the original contract. Scope changes should only be made with an appropriate (approved) increase in time and budget.

Project Management Plan Flowchart

The project management plan flowchart (figure 9.1) shows the relative position of the scope of work to the other topics. The scope of work (defined using the feasibility study and the WBS) should be considered after the corporate strategy has been established, but before the CPM and the Gantt chart (project schedule) are calculated. It is logical to develop the scope of work and the list of activities before they can be scheduled.

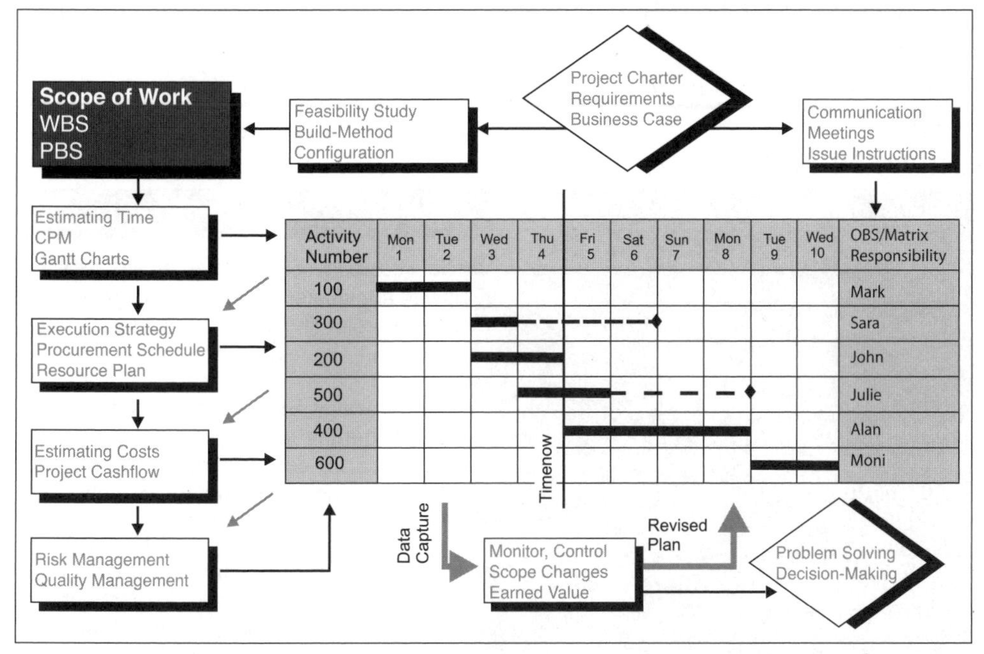

Figure 9.1: Project Management Plan Flowchart – shows the relative position of scope management with respect to the other topics

Body of Knowledge Mapping

PMBOK 4ed: The *Scope Management* knowledge area includes the following:

PMBOK 4ed	Mapping
5.1: Collect requirements	The *Feasibility Study* chapter will explain how to administer the process of collecting the requirements from the project stakeholders. This is usually expressed as their needs and expectations.
5.2: Define scope	This chapter will explain how to administer the process of developing a detailed description of the project.
5.3: Create WBS	The *WBS* chapter will explain how to subdivide the scope of work into smaller and more manageable work packages.
5.4: Verify scope	This chapter will explain how to administer the process of formally accepting the completed project deliverables.
5.5: Control scope	This chapter will explain how to administer the scope change control process.

APM BoK 5ed: The *Scope Management* knowledge area includes the following definition:

The APM BoK 5ed defines **Scope Management** as, *the process by which the deliverables and the work to produce them are identified and defined.*

Unit Standard 50080 (Level 4): The unit standard for the *Scope Management* knowledge area includes the following:

Unit 120373: *Contribute to project initiation, scope definition and scope change control*

Unit 120381: *Implement project administration processes according to requirements*

Specific Outcomes	Mapping
120373/SO1: Contribute to the identification and co-ordination of stakeholders, their roles, needs and expectations.	The *Feasibility Study* chapter will explain how to administer the process of identifying and co-ordinating the stakeholders' roles, needs and expectations.
120373/SO2: Contribute to the identification, description and analysis of the project needs, expectations, constraints, assumptions, exclusions, inclusions and deliverables.	The *Feasibility Study* chapter will explain how to identify stakeholders' needs and expectations, constraints, assumptions, exclusions, inclusions and deliverables.
120373/SO3: Contribute to preparing and producing inputs to be used for further planning activities.	The *WBS* chapter will explain how to subdivide the scope of work into manageable work packages and activities which are the input to the CPM and Gantt charts for time scheduling.
120373/SO4: Contribute to the monitoring of the achievement of the project's scope.	This chapter will explain how to administer the process of verifying the scope to formally accepting the completed project deliverables.

Specific Outcomes	Mapping
120381/SO1: Execute processes and standards to support project change control.	This chapter will explain how to manage the scope change control.
120381/SO2: Update and communicate status of change requests.	This chapter will explain how to inform stakeholders of the changes to the scope of work.

The scope management topics are subdivided into three chapters. This chapter will explain how to define and control the scope of work. The next chapter will explain how to manage a feasibility study, and the following chapter will explain how to develop the work breakdown structure (WBS).

1. Scope Definition

Defining the project's scope of work is the process of developing a detailed description of the project. The scope definition outlines the content of the project, how the project will be approached, and explains how the project will solve the client's problems, needs or opportunities. Scope definition establishes a method to identify all the items of work that are required to be carried out to complete the project and, by implication, the activities which are not included in the project.

The scope definition uses the output from a number of other project documents which are explained in other chapters:

- **Project Charter** (*Project Integration Management* chapter): The project charter, which is owned by the project sponsor, aims to implement corporate strategy by setting out the why, what, who, when, where, how to, and how much.
- **Feasibility Study** (*Feasibility Study* chapter): The feasibility study, which is owned by the project manager, identifies the stakeholders and determines their needs and expectations.
- **WBS** (*WBS* chapter): The WBS, which is owned by the project manager and the project team members, is used to subdivide the scope of work into manageable work packages for better planning and control.

Input	Process	Output
1. Project charter 2. Stakeholders' analysis 3. Statement of requirements 4. Business case	1. Expert judgement 2. Stakeholders workshop	1. Project scope definition statement

Figure 9.2: Scope Definition Process – shows the scope definition using the Eastonian model (input-process-output)

The scope definition is an input into the following:

Stakeholders: The scope definition defines the boundary of the project and confirms common understanding of the project scope amongst the stakeholders. For example, on a house extension project the key stakeholders will be the owner, architect, builder, suppliers and local council.

Input	Process	Output
Scope definition	Determine project boundary	List of stakeholders

Agreement: The scope definition forms the basis of the **agreement** between the client and the contractor by identifying the project's objectives as achieving defined benefits and defined deliverables. For example, the defined benefits would be the implementation of corporate strategy (or owner's strategy) and the defined deliverables would be quantified at the work package level as time, cost and quality.

Input	Process	Output
Scope definition	Determine agreement	Project objectives

Build-Method: The scope definition is an input to the **build-method** constraint which considers how to build the project and, by doing so, confirms it is feasible to build the project.

Input	Process	Output
Scope definition	Build-method analysis	Build-method

Configuration: The scope definition is an input into the **configuration management** constraint which considers how the project will perform in its operational situation and, by doing so, confirms the project will meet the required operational condition.

Input	Process	Output
Scope definition	Configuration analysis	Configuration

Scope Change Control: The scope definition is an input into the **scope change control** management which approves all the scope changes. This process confirms the changes can be built and will not negatively impact the operation of the project. Effective scope change control limits the opportunity for uncontrolled **scope creep**.

Input	Process	Output
Scope definition	Scope change control process	Approved changes

Verification: The scope definition is an input into scope verification (next section), which sets out the approval criteria for the client to formally accept the completed work. This is part of the commissioning process to confirm the project has been executed to the required condition and will be accepted by the client.

Input	Process	Output
Scope definition	Determine verification	Verification statement

Estimate: The scope definition is an input into the project's estimate. By using the WBS to subdivide the scope of work into work packages, this will help to improve the level of accuracy of the estimate.

Input	Process	Output
Scope definition	Estimating process	Project Estimate

Organization Structure: The scope definition has an input into the design of the organization structure. By using the WBS to subdivide the scope of work packages, responsibility can be assigned to each work package.

Input	Process	Output
Scope definition	OBS design	Assign responsibility

The scope definition process develops a written statement which acts as the basis for future decisions and helps to establish a criteria for the completion of an activity, completion of a project phase, or the completion of the project itself and, therefore, project acceptance (this is developed in the verification section).

As the project progresses, the scope statement may need to be revised or refined to reflect better information, approve the changes to the scope of the project, and to respond to external market and environmental conditions.

2. Scope Verification

Scope verification is the process through which the client and the stakeholders formally accept the completed project deliverables. Verifying the completed work is the process of checking the work has been completed to the approved design and specification.

Verification must not be confused with quality control. The subtle difference between verification and quality control, is that **quality control** is concerned with correctness, whereas, verification is concerned with acceptance of the deliverable being **'fit for purpose'**. For example, the Millennium Bridge in London was built correctly to the design and specification, but it initially failed its **'fitness for purpose'** test because its excessive swaying movement made it difficult to walk across the bridge.

Scope verification can be related to the phases of the project lifecycle. The scope of work should be formally approved at the end of each phase, for example:

1. Feasibility Phase	2. Design and Development Phase	3. Execution Phase	4. Commissioning and Handover Phase
Approve the feasibility study.	Approve the detailed design and the project management plan.	Approve all the changes and additional work. Confirm the project is ready for commissioning.	Commission and test the project, and confirm it is ready for handover to the client.

Figure 9.3: Scope Verification

It is essential to establish the verification controls and acceptance criteria before starting to execute the scope of work because, once the project has started, if there is a dispute it will be difficult to agree on what was the original agreement.

3. Scope Change Control

All projects are subjected to scope changes from time to time as the project progresses. These might be prompted by design errors, by building errors, or by changes to improve the project as new opportunities present themselves.

It is the project manager and project team members' responsibility to manage the administration of these changes through the approved scope change control system. This includes logging, monitoring, evaluating and approving the changes (by the designated people) before the changes are incorporated into the project management plan. This will ensure that the project management plan always reflects the current status of the project and limits scope creep.

Configuration Management: Configuration management is the process of identifying and managing scope changes to the deliverables. The configuration management process helps to ensure that the proposed changes are necessary, appropriate and possible, and that the integrity of the project's operational system is maintained.

The configuration management system includes:

Process	A scope change control system which formally outlines a process that defines the steps by which official project documents may be changed.
Authority	A list of designated experts who have the authority to make changes to the scope of work, in both the client's and the contractor's organizations. And, by implication, the other stakeholders who do not have authority to make changes to the scope of work.
Communication	A communication system which disseminates the current and up-to-date description of the project to all the identified stakeholders. This might be the latest revisions of drawings and specifications, and the status of all change requests.
Document Control	A document control process which identifies who should receive what documents, how and when. The document control process uses a transmittal system to confirm delivery and, where appropriate, removes old documents from the project.
Traceability	A traceability of all previous project baseline configurations. This includes a record and, therefore, an audit trail of all the approved changes.
Monitor	A framework to monitor, evaluate and update the scope baseline to accommodate any scope changes. This will ensure that the project baseline always reflects the current status of the project.
Fast-Track	A system to fast-track scope change approvals in an emergency situation.

Scope change control is also concerned with influencing the factors which initially create scope changes to ensure that the changes are beneficial to the project. The scope change approval system should be agreed by all stakeholders at the outset and confirmed at the handover meeting.

The scope change control system should allow for scope changes to be initiated by any of the project stakeholders. Consider the following (figure 9.4):

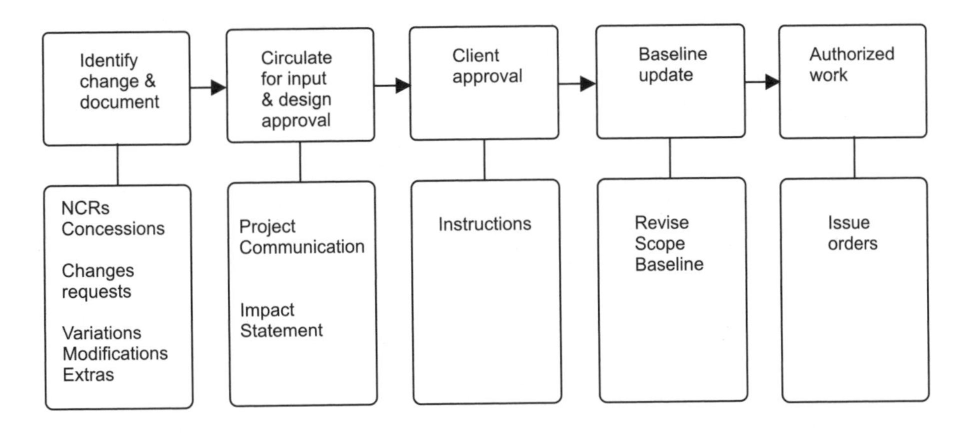

Figure 9.4: Configuration Control Flow Chart

Scope Change Initiation: Figure 9.4 shows that the scope change control system should allow for anyone working on the project, or any stakeholder for that matter, to initiate a change. Consider the following:

Non Conformance Report (NCR)	**Non conformance report** (NCR) is usually initiated by quality control when someone has worked outside the accepted boundaries of a procedure, or the project is outside the required condition. The non-conformance would either be corrected and fall away or motivate a concession.
Concession	A **concession** requests the client to accept an item which has been built and is functional, but is outside the design specification. If approved, this would be shown as a scope change for this project only; it would be shown on the as-built drawings, but not on the baseline drawings which might be used on other projects.
Change Requests	**Change requests,** also referred to as modifications and variations, request the client to approve a change to the scope baseline.
Project Communication	Anyone working on the project should be able to raise a concern or make a suggestion about the project. These comments and suggestions should be captured and logged into the configuration system so that they can be processed.

Project Office (PO): The project office is the ideal venue for the project team members to administer the scope change control process. The project office becomes the central point for communication and the filing of project documentation. The project team needs to keep a register of all the approved changes (signed) and their status. These documents could be required later to support any claims or disputes.

Change Request Form: Figure 9.4 shows the relative position of the change request form with respect to the other documents. The change request forms should be consecutively numbered for ease of registration, planning and control. The forms should describe the scope change, list associated drawings and documents, together with the reason for the change.

CHANGE REQUEST

NUMBER:		DATE RAISED:	
PROJECT:		INITIATED BY:	
CHANGE REQUESTED (related drawings / work packages):			
REASON FOR CHANGE :			
APPROVAL:			
NAME	POSITION	APPROVAL	DATE

Figure 9.5: Change Request - shows the framework of a change request form

Project Communication: Figure 9.4 shows the relative position of the project communication form with respect to the other documents. The project communication document allows any stakeholder working on the project to make a formal statement which is logged in a project register (consecutively numbered); this could be a question, highlighting a problem, or making a suggestion. Once the document has been logged into the scope change control system the configuration management system will ensure that it is acknowledged and actioned (see figure 9.6). This essentially puts the onus on the client and stakeholders to give clear written instructions.

PROJECT COMMUNICATION

NUMBER: DATE RAISED:
PROJECT:
DESCRIPTION (related drawings/work packages): INITIATED BY:

COMMENTS / INSTRUCTIONS:

WE ACKNOWLEDGE YOUR ENQUIRY / INSTRUCTION:

VERBAL FROM: TO:
WRITTEN FROM: TO:
DATE:

PLEASE ADVISE HOW WE ARE TO PROCEED:

1. START IMMEDIATELY AND QUOTE WITHIN 7 DAYS
2. START IMMEDIATELY ON UNIT RATES
3. DO NOT START, QUOTE WITHIN 7 DAYS
4. OTHER

REQUEST FROM: INSTRUCTION FROM:

CONTRACTOR CLIENT
PROJECT MANAGER PROJECT MANAGER

Figure 9.6: Project Communication - shows the framework of a project communication form

Impact Statement: Figure 9.4 shows the relative position of the impact statement with respect to the other documents. The impact statement quantifies the implications of making the proposed changes. An information pack is compiled by the project office to collect input, information, comments and approval from the nominated experts in the configuration system. Consider the following:

Department	Comments
Design Team	The design team need to consider the design impact. What design changes are required? Can the change be included in the design? Will the change affect the design stability and performance (configuration) in the operational environment?
Production	The production department needs to consider the manufacturing impact. Can the change be made? What is the impact on the build-method? What resources and equipment are required, and are they available?
Procurement	The procurement department needs to consider the procurement impact – from the scope change they need to develop a procurement list. Can the item(s) be procured, where, how much and when (delivery cycle)?
Planning	The planning department needs to consider the schedule impact. Will the scope change impact on the schedule? And particularly, will the change delay the project?
Cost	The project accounts department needs to consider the budget impact. Will the scope change impact on the budget? And particularly, will the change increase the overall project budget?
Quality	The quality department needs to consider the quality impact. What level of inspection is required?
Risk	The project manager needs to consider the risk impact. What are the risks associated with this change; are they acceptable?
Governance	The project sponsor needs to consider the corporate governance impact. Will this scope change broach the company's agreed ethics?
Legal	The legal department needs to consider the contractual impact. Will the scope change affect the contract?
Project Manager	The project manager needs to consider the input from all parties and approve or reject the scope change.
Client / Project Sponsor	The project sponsor and client need to consider the input from all parties and approve (and pay for) or reject the scope change.

IMPACT STATEMENT

NUMBER: DATE RAISED:
PROJECT:
DESCRIPTION (related drawings/work packages): INITIATED BY:

REFERENCE PROJECT COMMUNICATION:

IMPACT ON PROJECT:		IF YES QUANTIFY
TECHNICAL:	YES/NO	
PROCUREMENT:	YES/NO	
PRODUCTION:	YES/NO	
SCHEDULE:	YES/NO	
COST:	YES/NO	
QUALITY:	YES/NO	
CONTRACT:	YES/NO	
RISK:	YES/NO	

PLEASE ADVISE IF WE ARE TO PROCEED: YES/NO

REQUEST FROM:	INSTRUCTION FROM:
CONTRACTOR PROJECT MANAGER	CLIENT PROJECT MANAGER

Figure 9.7: Impact Statement - shows the framework of an impact statement form

The impact statement feedback within the scope change control process might require a few iterations before the nominated experts converge on an acceptable scope change. The scope change control process is a good example of the type of project administration work the project team members would be expected to perform. The project manager's and client's approval formally authorises the scope change into the production system – this completes the configuration management control cycle.

Scope Change Control Example: Consider this example of a simple scope change that went wrong.

After working ten years in the Middle East, Bill decides to build a yacht. While the yacht is being built Bill meets his knowledgeable friend Kirk in the local yacht club. Kirk advises Bill to go for the biggest fuel tankage the yacht can accommodate.

Bill drives back to the shipyard and asks Bob, the yacht builder, if he can increase the size of the tankage. Bob says they have space in the starboard locker to fit an extra large tank. This is duly done.

When the yacht is launched and all the equipment and provisions are fitted Jim, the naval architect who designed the yacht but did not know about the new fuel tank, notices the yacht is heeling over five degrees to starboard!!!! On inspection Jim finds that the additional fuel tank is causing the yacht to heel over.

Moral of the story: If the shipyard had a configuration management system the project manager would have ensured that this change would have been approved by all the nominated parties.

Bill, the client, would have approved and paid for the change - *'Yes I will pay for the change'*.

Bob the builder would have approved the build-method – *'Yes we can make the change'*.

Jim the naval architect would have checked the impact on the hydrodynamics which, in this case, he would **not** have accepted in the present arrangement - *'No this change will negatively impact on the yacht's weight distribution'*.

Exercises:

1. Discuss what techniques you use to define the scope of work for your projects, indicating what is included and what is excluded.
2. Discuss how you verify the work has been completed as required.
3. Discuss how you manage the change of scope on your projects to ensure the project will still operate as planned.
4. Discuss how you control scope creep.
5. Develop a scope management checklist (see Appendix 1 solution).

Further Reading:

Burke, R., *Project Management Techniques*, will develop the following topic which relates to this knowledge area:

- Scope control through the phases of the project lifecycle.

10

Work Breakdown Structure (WBS)

Learning Outcomes After reading this chapter, you should be able to:
Produce a scope of work checklist
Develop a WBS
Define the structure and content of a work package
Develop a product breakdown structure (PBS)

This chapter will explain how to develop a *Work Breakdown Structure* (WBS). The WBS is a key technique within the *Project Scope Management* knowledge area. Scope management uses the WBS to subdivide the scope of work of complex projects into a number of smaller manageable units of work called work packages.

The project management philosophy is to subdivide the project into a number of small units which are easier to estimate, easier to plan, easier to assign the work to a responsible person or department for completion, and easier to control. This in turn gives the project manager and the project team members the tools to effectively plan and control the project.

Generally, the project manager's role is to facilitate the creation of the WBS, while the team members are the ones who are best equipped technically to decompose the project deliverables into smaller, and more manageable components.

Project Management Plan Flowchart

The project management plan flowchart (figure 10.1) shows the relative position of the WBS technique with respect to the project charter and the CPM. It is logical that the WBS follows the project charter because it is not possible to subdivide a project that has yet to be initiated or defined. And it is logical that the WBS precedes the CPM because the activity list for the CPM calculation is developed from the WBS's work packages.

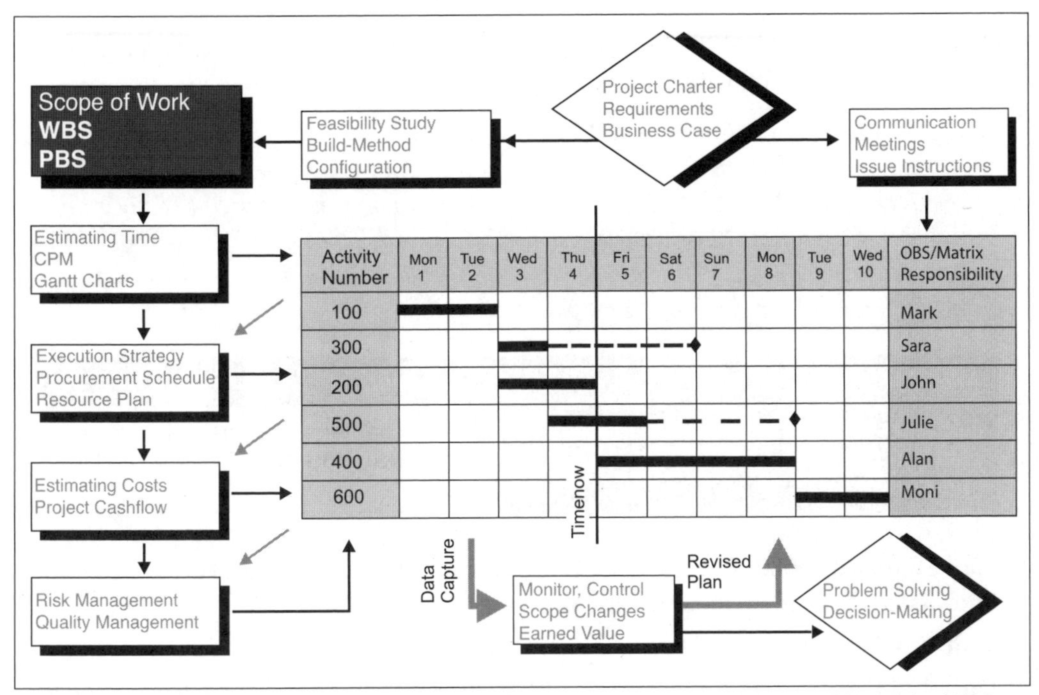

Figure 10.1: Project Management Plan Flowchart – shows the relative position of the WBS technique with respect to the other techniques

Body of Knowledge Mapping

PMBOK 4ed: The WBS technique is part of the *Scope Management* knowledge area which includes the following:

PMBOK 4ed	Mapping
5.3: The WBS identifies the deliverables, the work packages and budgets used to measure project performance.	This chapter will explain how to identify deliverables (PBS), work packages (WBS) and budgets (CBS).

APM BoK 5ed: The *WBS* technique is part of the *Scope Management* knowledge area which includes the following:

The APM BoK 5ed defines the **WBS** as, *a way in which a project may be divided by level into discrete groups for programming, cost planning and control purposes. The WBS is a tool for defining the hierarchical breakdown of work required to deliver the products of a project. Major categories are broken down into smaller components. These are sub-divided until the lowest required level of detail is established. The lowest units of the WBS are generally work packages.*

APM BoK 5ed	Mapping
3.1: The scope in the project management plan is refined using the PBS, WBS, OBS and CBS.	This chapter will explain how to use the PBS, WBS, OBS (or RAM) and the CBS.

Unit Standard 50080 (Level 4): The *WBS* technique is part of the *Scope Management* knowledge area and includes the following:

Unit 120372: *Explain fundamentals of project management*

Unit 120373: *Contribute to project initiation, scope definition and scope change control*

Specific Outcomes	Mapping
120372/SO3.3: The purpose of decomposing a project into manageable components or parts is explained with practical examples.	This chapter will explain how to decompose a project into a number of manageable work packages.
120372/SO3.4: The concepts of breakdown structures for product, work and cost are explained in simple terms.	This chapter will explain the concept of breakdown structures.
120373/SO2.3: Work packages are developed and further elaborated to present an overall view of the project scope.	This chapter will explain how the WBS presents an overall view of the project's scope of work.
120373/SO2.4: A work breakdown structure is developed and documented, within agreed time frames.	This chapter will explain how to develop and document a breakdown structure.

1. Checklists

Checklists – love them or hate them – do provide an effective management tool to confirm there is a complete list of all the items of work, materials and resources. '*Why try to remember everything in your head when checklists never forget*.' Checklists are an excellent planning and control tool – even NASA uses them!

'Hey Buck, I'm just checking we have everything.'

'Okay Neal.'

The PMBOK 4ed defines a **Checklist** as, *a structured tool, usually component-specific, used to verify that a set of required steps has been performed.*

Checklists are a form of expert system which captures the experts' knowledge and experience and presents it in a structured format for others to use. For example, checklists are created for:

- Fire fighting safety drills
- First aid procedures
- Recipes/menus

Shopping List: A shopping list is a good example of a checklist that is commonly used. Consider the following shopping list which is subdivided into food types.

Veg	Fruit	Cereal	Jams/Spreads
Potatoes	Apples	Porridge	Marmalade
Onions	Oranges	Weetabix	Jam
Broccoli	Kiwi fruit	All-Bran	Honey
Carrots	Bananas	Rice Krispies	Marmite
Tomatoes	Avocado	Cornflakes	
Cucumber	Ginger	Bran	
Mushrooms	Lemon		
Leeks	Lime		
Pepper			

Figure 10.2: Shopping Checklist – shows a portion of a shopping checklist subdivided by food type

2. WBS Methods of Subdivision

Producing a shopping checklist, even for a large family, is a relatively simple and straightforward process. Producing a structured WBS for a project, even a project of limited complexity, is a more challenging task which requires a more structured approach.

Managers from companies that do not traditionally run projects might find the WBS terminology strange initially, and wonder why the breakdown is not called a checklist as this would be something they would be more familiar with and understand. In some ways they would be right, but they need to appreciate that the WBS is more than just a simple checklist (as the other sections of this chapter will explain) and, consequently, it has become accepted project management terminology.

Project managers new to project management might initially feel overwhelmed by the sheer size of the project as they try to grasp all the details of the scope of work. There is a tendency to want to identify all the items of work at a high level of detail. The psychologists say we can only comprehend a limited number of items simultaneously (about ten). Therefore, a project of several thousand tasks would go way over the project managers' comprehension and ability.

This is where the WBS comes to the rescue and encourages the project manager to divide the project into a number of chunks which can be understood simultaneously. If each chunk of work is then progressively subdivided into manageable work packages, this will enable the project manager to expand the breakdown in a controlled manner. Figure 10.3 shows how the WBS interlinks the projects with the work packages.

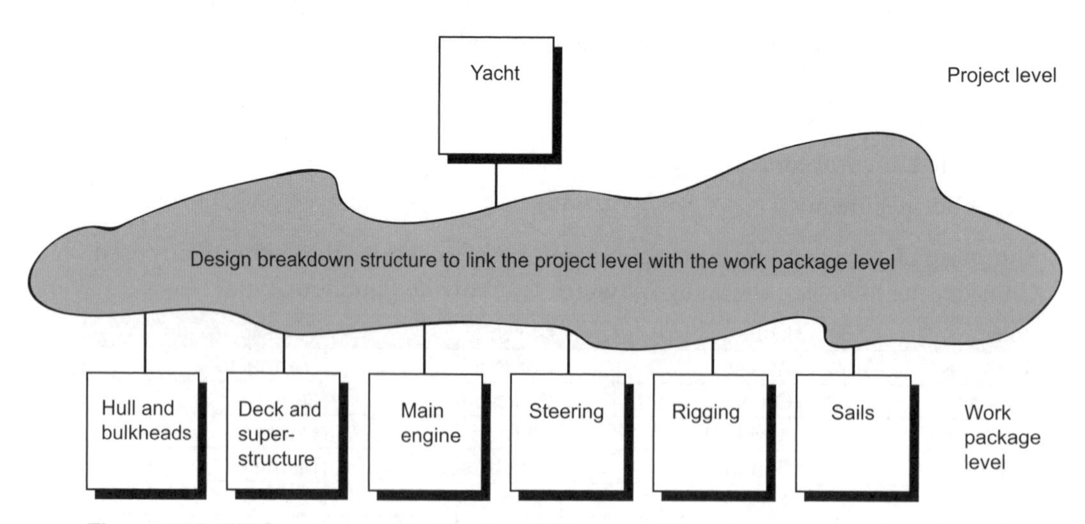

Figure 10.3: WBS – shows a need for the WBS to provide a structure to get from the project level to the work package level

The WBS not only helps to fully define the scope of work but it also forms the key part of the project management plan. It must be stressed – the **WBS is the backbone of the project.** Without an effective backbone the project has no structure to plan, interlink and control all the parameters of time, cost, quality, procurement, resources, risk and communication.

The WBS is an excellent tool for establishing the full scope of work as a list of independent work packages. It is an essential tool for ensuring that the project estimate or quotation includes 100% of the complete scope of work (identifying and clarifying the exclusions), because the cost of making anything that is not included in the scope of work on a fixed price contract will be taken straight off the profit margin.

There are many methods of using the WBS to subdivide the scope of work – limited only by imagination. Designing the WBS requires a delicate balance to address the needs of the various departments and the needs of the project. There is not necessarily a right or wrong breakdown structure; what might be an excellent fit for one department could be an awkward burden for another. And further, as the project progresses, other breakdown subdivisions might become more appropriate.

The only way to determine which is the best breakdown structure for the project is to experiment with a few different methods of subdivision. With practice and experience the project team members should be able to develop a number of workable templates.

This chapter will develop a number of graphical breakdown structures which are known as:

- Product Breakdown Structure (PBS) to identify deliverables
- Work Breakdown Structure (WBS) to quantify the work to produce the deliverables
- Organization Breakdown Structure (OBS) to identify the people responsible for producing the deliverables
- Cost Breakdown Structure (CBS) to identify budgets to pay for the deliverables.

The breakdown structure should strive to group similar activities together to improve:

- Productivity and the efficient use of company resources
- Cost estimating and management control
- Identification of all the work packages with a unique numbering system
- Integration of the WBS and the OBS to assign responsibility.

When all the work packages are complete, it should be confirmed and agreed with the client and stakeholders that all the deliverables are included and there is no overlap between work packages.

3. Product Breakdown Structure (PBS)

Historically all the methods of breaking down the scope of work were called work breakdown structures, but as the technique developed so a number of breakdown structures have been assigned their own names – this section will explain how to use the product breakdown structure (PBS).

There are two ways of explaining the PBS. It can be explained as either the physical subdivision of the project into its component parts; these are referred to as its deliverables, or as the total list of deliverables that are required to meet the project's objectives.

The PBS of a car might be subdivided into the physical assemblies, sub-assemblies, and components needed to manufacture the product. Consider the subdivision of a car (see figure 10.4):

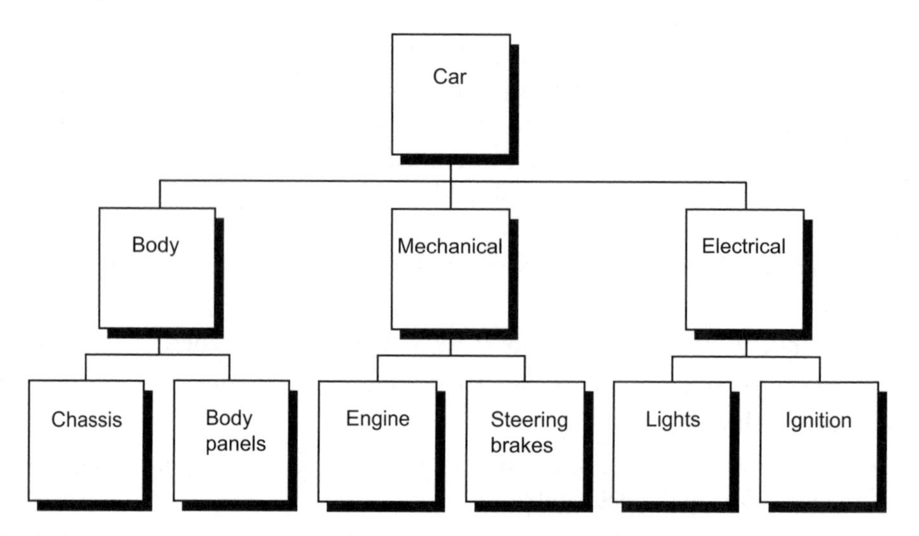

Figure 10.4: Product Breakdown Structure – shows the subdivision of a car (product) into a number of deliverables

'Just keep turning Fred.'

Many project managers prefer to start with a PBS to present a complete list of all the project deliverables, and then develop the WBS as the work required to produce the product or deliverables. This is supported by the APM BoK 5ed.

The APM BoK 5ed defines the **relationship between the PBS and the WBS** as, *the scope in the project management plan (PMP) is refined using a product breakdown structure (PBS) and work breakdown structure (WBS):*

- The PBS defines all the products (deliverables) that the project will produce. The lowest level of a PBS is a product (deliverable).
- The WBS defines the work required to produce the deliverables. The lowest level of detail normally shown in a WBS is a work package.

Developing the PBS first would be a logical starting point to start subdividing a project into its tangible deliverables. This will help to:

- Make sure all the stakeholders understand what the product will be like when it is completed, and help to gain their agreement.
- Review the components to be built with the team members and identify the dependencies between components.

PBS Example: Consider a fashion project to design and make a range of garments. Figure 10.5 shows how the PBS can be used to produce a list of deliverables and the WBS identifies the work required to produce the deliverables.

PBS and WBS: It is helpful to note that the PBS deliverables are usually a noun, while the WBS work packages are usually a verb.

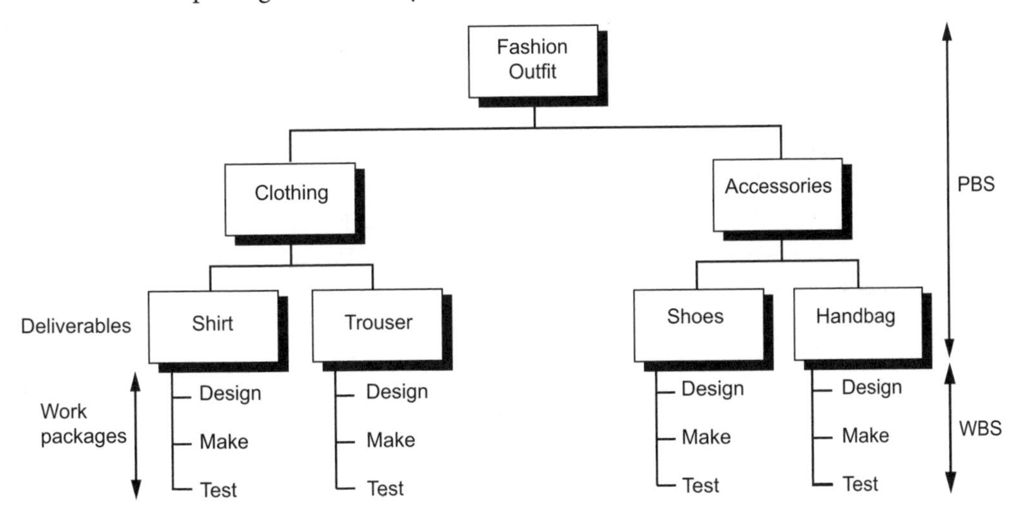

Figure 10.5: PBS/WBS for a Fashion Project – shows the project deliverables and the work to make the deliverables

4. WBS

The title of this chapter, *Work Breakdown Structure*, is used as a generic name to include all the different types of breakdown structures, but this section will focus on the WBS as a technique to break down the work of the project.

Using the WBS method of subdivision enables the project manager to experiment with a mix of breakdowns. Figure 10.6 uses location, system and discipline as three methods of subdivision.

Experienced planners prefer this approach because the location breakdown helps to physically subdivide the projects into different parts which are easy to visualise. The systems breakdown is similar to the PBS and relates to the tangible deliverables, and the discipline split identifies the trades and work required to produce the components of the systems.

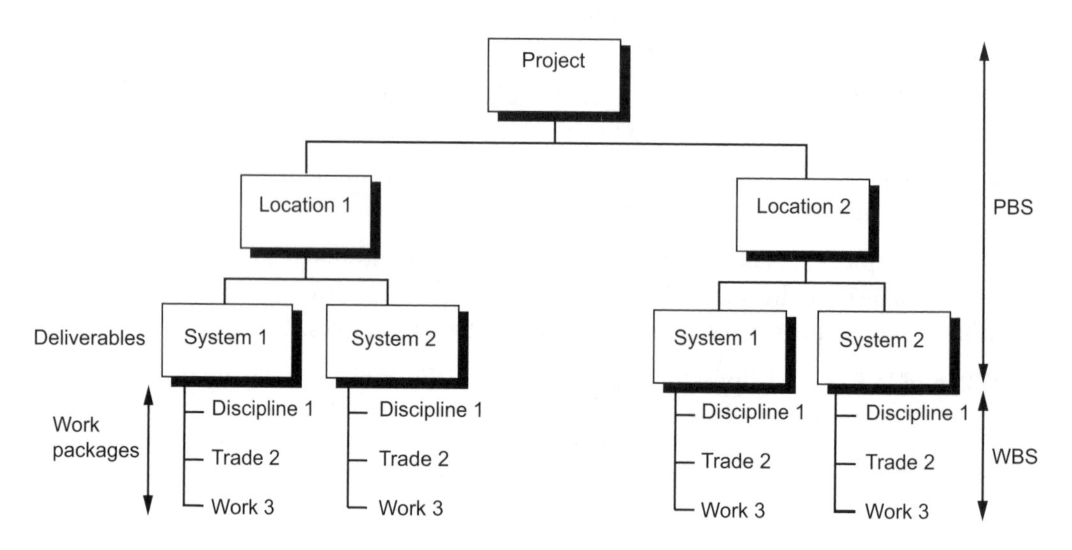

Figure 10.6: WBS – shows the WBS using a mix of breakdowns

The lowest level of the WBS is a **work package** which is the work required to make the deliverables. The work package contains a wealth of information which is described in the next section.

5. Work Packages

A project can be subdivided into a number of work packages or tangible deliverables. The structure and content of the work packages and deliverables can be defined as follows:

Ownership	Who is responsible for the work package? It is a key project management philosophy to assign ownership to all aspects of the project.
Deliverables	The PBS breaks down the scope of work into tangible deliverables, such as, the individual components of a car.
Scope of Work	The scope of work describes what has to be made or performed and, by implication, what is not included.
Independent	The work packages should be independent of each other so that they can be managed separately.
Specifications	The specifications define what the work packages need to achieve and outline what is the required standard.
Quality Requirements	The quality requirements outline the required condition, the level of inspection and the qualifications of the workforce.
Estimate (man hours)	The estimate of man hours is a measure of the work content per work package.
Duration	The duration is the trade-off between work content and resources available.
Budgets	A budget is the amount of money assigned to complete the work package. Some companies might limit the value of a work package to, say, $1,000 or $10,000.
Procurement	The procurement lists all the bought-in items per work package.
Resources	The resources identify all the machines and people required to complete the work.
Equipment / materials	The equipment list identifies all the equipment and plant hire that are required per work package.
Similar size	The work packages should be of a manageable size, typically between 20 to 80 man hours.
Number	The work packages should be uniquely identified with a number.

With better work package definition comes a higher level of accuracy and control.

6. Cost Breakdown Structure (CBS) [Top Down vs Bottom Up Estimating]

If a company wins contracts by competitive bidding it is important to have an efficient system for generating quotations quickly and accurately. The CBS offers a top down estimate at the project level that can be subdivided to offer a bottom up costing at the deliverable and work package level. The **accuracy** of the estimate will increase progressively as the work package level of detail increases.

For example, a house extension project could be quoted on a surface area rate, but managed at the work package level. If the house extension is 200 m^2 @ \$1000 per sq. metre, then the quote will be \$200,000.

Problems can arise when a contract is based on an estimate at the project level but the budgets are assigned at the work package level – will the two amounts meet in the middle?

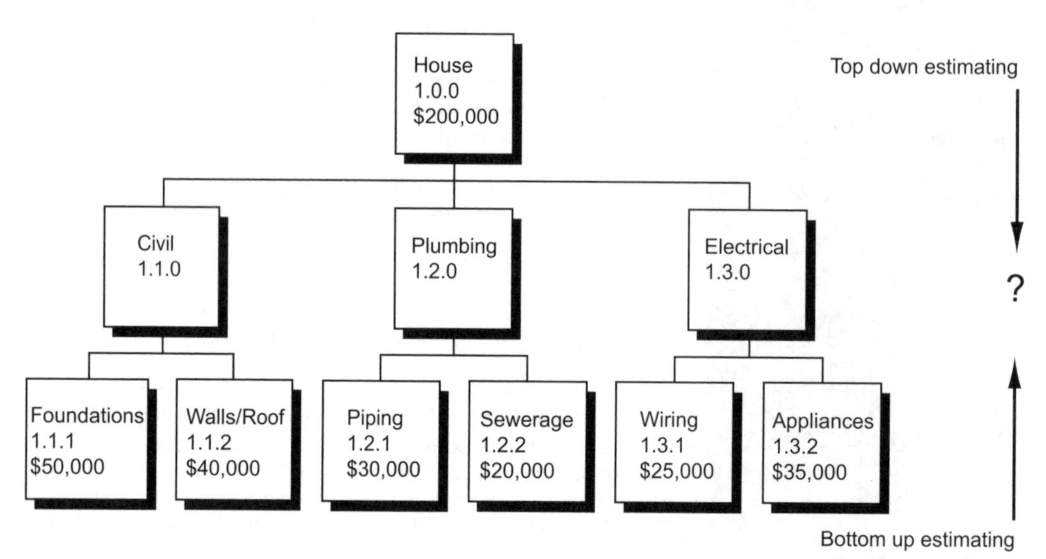

Figure 10.7: Top Down Estimating vs Bottom Up Estimating - shows the problem when a company quotes on a top down estimate, but expenses are incurred at the work package level

7. Organization Breakdown Structure (OBS)

Assigning responsibility for performing project work is one of the key project management functions. This can be achieved through the interface between the work (WBS) and the organization breakdown structure (OBS). The WBS / OBS interface clearly links the work packages to the department or person responsible for carrying out the work.

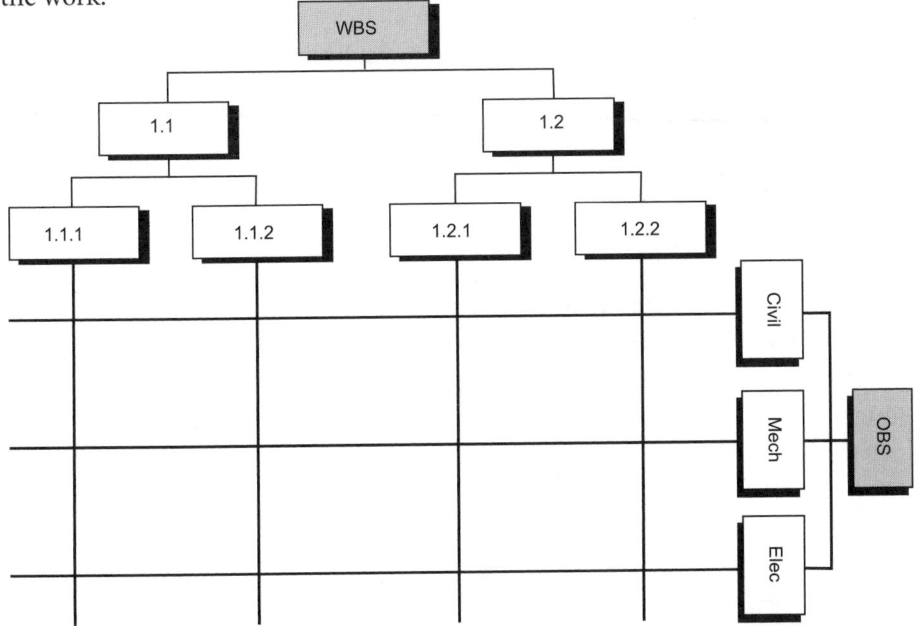

Figure 10.8: WBS / OBS Links – shows a link between work packages and responsible department or person

Responsibility Assignment Matrix (RAM): The WBS/OBS interface (above) is a rather cumbersome presentation, particularly as the level of detail increases. For this reason it is best suited to only high-level links. The RAM offers a more practical tool for linking the scope of work to the responsible parties. (See **Burke,** R., **Barron,** S., *Project Management Leadership,* for more information on the RAM and RACI diagrams).

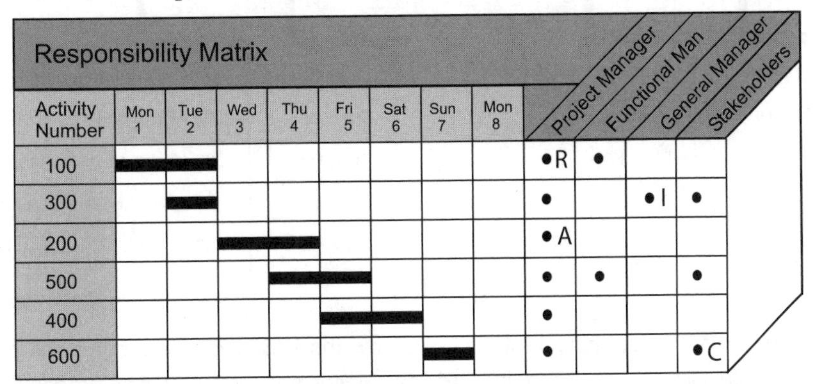

Figure 10.9: Responsible Assignment Matrix (RAM) – shows the link between the work and the person responsible for carrying out the work

8. Activity List

The WBS can be used to subdivide the scope of work into a list of activities. In the discussion so far the scope of work has been subdivided into work packages for the purposes of identifying deliverables and work, and assigning responsibility and a budget. This section will show that the WBS can also be used to subdivide the work packages into a list of activities which are required for the CPM calculation and Gantt chart presentation.

Although the work packages may not be logically linked, the CPM and Gantt chart activities have to be logically linked to perform the time calculations. This adds another dimension and adds complexity to the WBS subdivision.

The house extension project, figure 10.10, shows how the decorating work package can be subdivided into a number of activities.

Activities: The work packages can be further subdivided into a list of activities. The activities are a list of jobs which are required to be done to complete the work package. Where the project manager would develop the WBS to the work package level, the team member or supervisor responsible for the work package would develop the list of activities. This effectively pushes responsibility down the organization structure and empowers the supervisors.

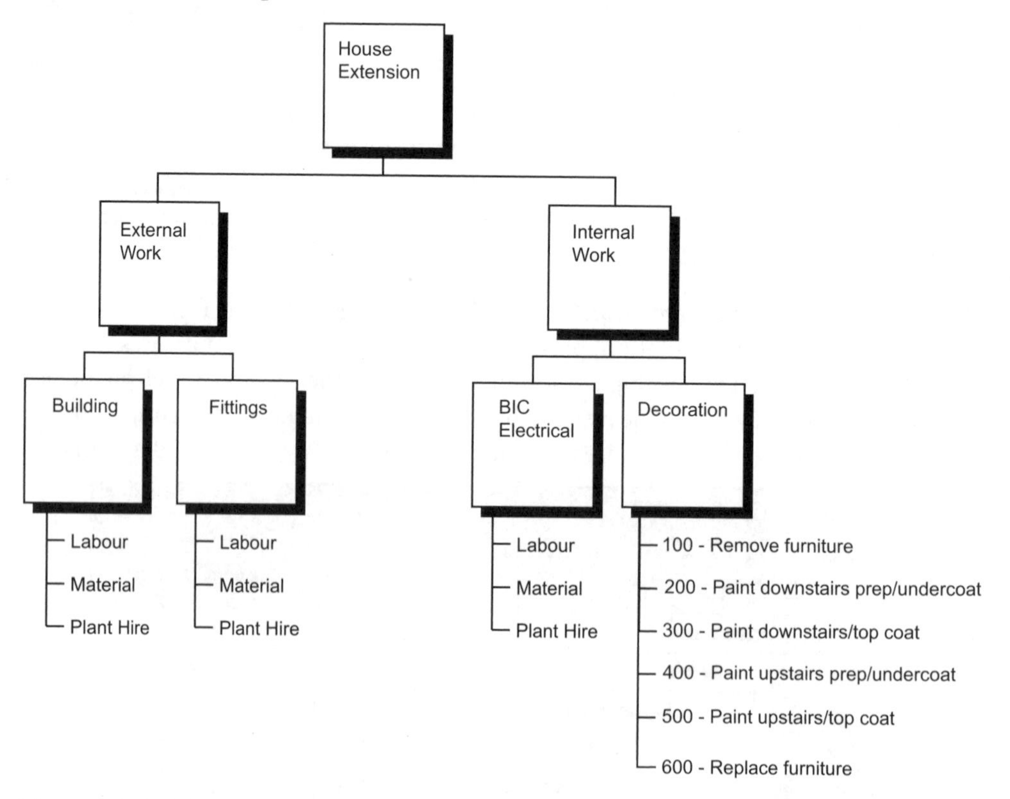

Figure 10.10: WBS / Activity List – shows how the WBS work packages can be further subdivided into a list of activities for the CPM calculation and the Gantt chart presentation. This list of activities is transferred to table 12.1 for the CPM calculation.

9. Numbering System

One of the beneficial features of the WBS is its ability to uniquely identify all the elements of work in a numerical and logical manner by using a number or code. With a unique number, all the work packages can be linked to the project's schedule, purchase orders, resources, accounts, together with the corporate accounts and the client's accounts.

The numbering system can be alphabetical, numerical, or alphanumeric (letters and numbers). In most of the examples here the numbering systems will be numerical.

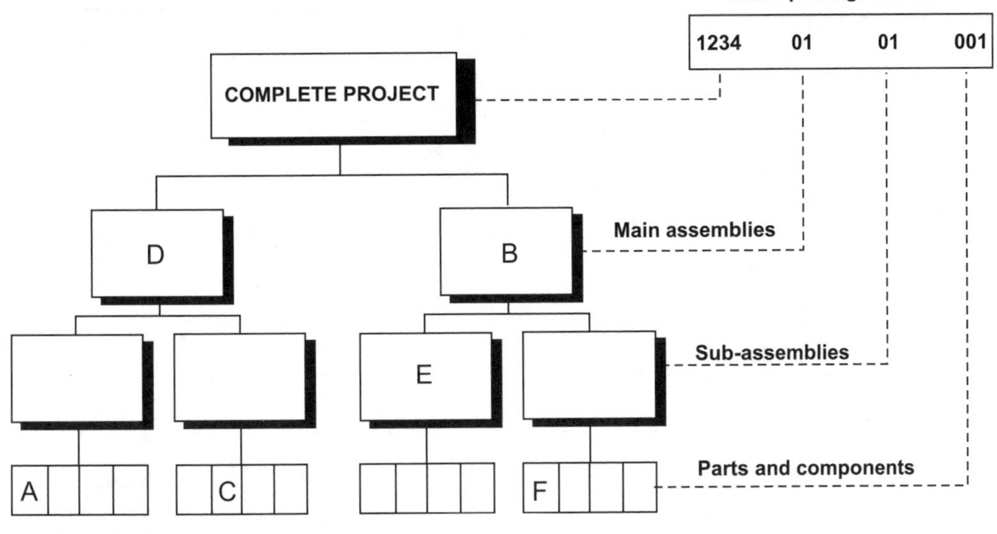

Consider the following example:

Figure 10.11: WBS Numbering System – shows how each work package can be uniquely identified

Numbering Exercise:

Referring to figure 10.11, if A, B, and C are the following work package numbers, what are D, E and F?

A = 1234 01 01 001

B = 1234 02 00 000

C = 1234 01 02 002

Exercise: (Solutions see Appendix 1)

D =

E =

F =

10. PBS, WBS, OBS, CBS Interfaces

In the text so far, the different types of breakdown structures, identified in the APM BoK 5ed as PBS, WBS, OBS and CBS, have been shown separately. This section will show how they can be interlinked.

Starting at the project level, the project is subdivided into its deliverables (deliverable 1 and deliverable 2). These deliverables are subdivided into the work packages required to make the deliverables (WP 1 to 6).

At the work package level, the work packages are linked to the OBS (OBS 1 to 3) which are the departments or the people performing the work. The work packages are also linked to the CBS (CBS 1 to 3) which are the budgets to pay for the work.

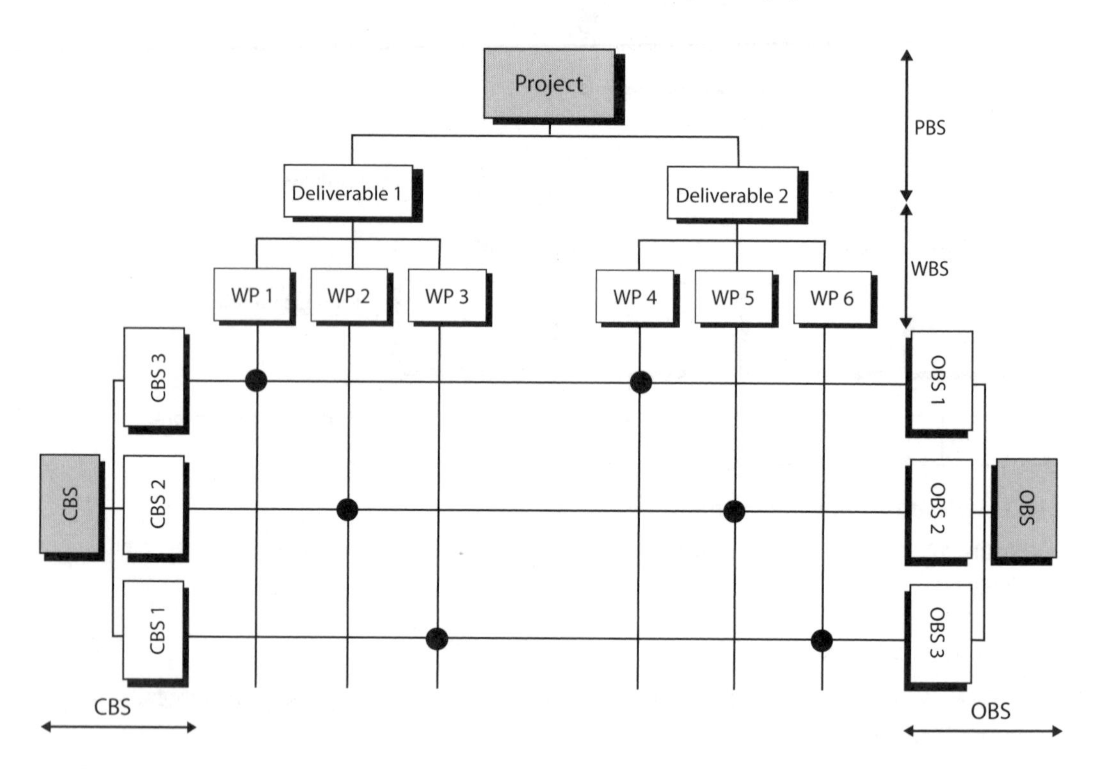

Figure 10.12: PBS/WBS/OBS/CBS Interface – shows how the four different breakdown structures can be interlinked

11. WBS Templates

Although 'product' and 'department' are popular criteria for subdivision, project managers sometimes have difficulty thinking of suitable methods to subdivide their projects. In practice companies tend to standardise their WBS using a standard WBS template for their projects. Instead of starting each project with a blank sheet of paper they might consider using a WBS format from a previous project as a template.

A standard WBS template ensures consistency and completeness as it also becomes a planning checklist with all the components that the company's particular type of projects would contain. Having a structured checklist also reduces the risk of omitting obvious items of work. It also enables policies, procedures, and lessons learnt from previous projects to be captured.

Even if the complete WBS template cannot be used, there could be portions of the template that are similar and can be copied across. Using proven structures will greatly speed up the planning process and help to structure company thinking. The figure 10.13 below outlines a WBS where the first level is subdivided by location, the second level by department and the third level by expense. This format could be used as a standard WBS template for the company. The only changes per project would be to the description rather than the structure.

Figure 10.13: WBS Template

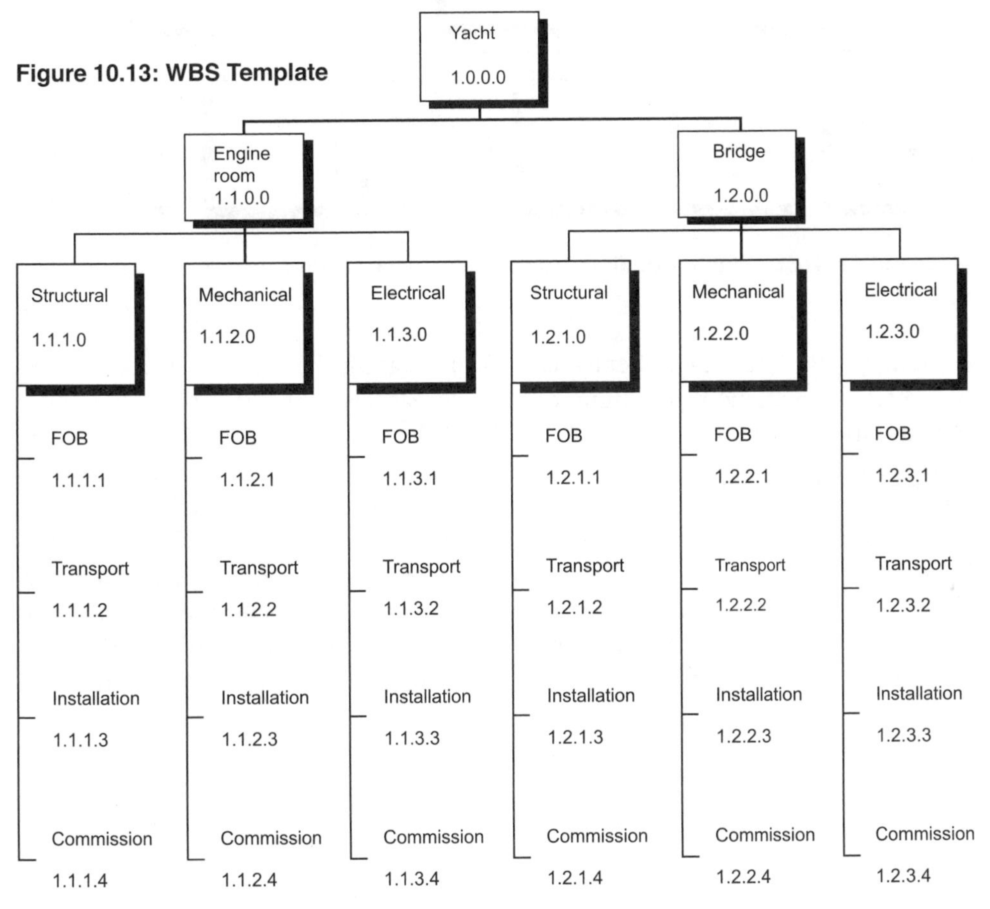

12. Spreadsheet Presentation

There are two basic methods of presenting the WBS:

- Graphically in boxes
- As a structured list within a spreadsheet.

The WBS is a hierarchical structure which, for training purposes, is usually presented by a graphical subdivision of the scope of work into boxes. This logical subdivision of all the work elements is easy to visually understand and easy to assimilate, thus helping the project participants to determine their responsibility and gain their commitment and support (see figure 10.14).

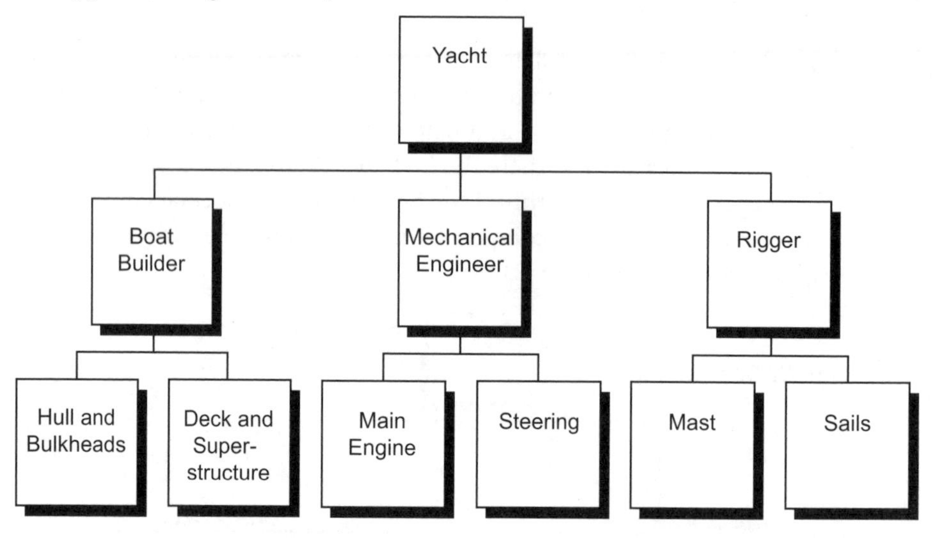

Figure 10.14: WBS – Yacht Building Project – shows the subdivision by discipline and then by product

Although boxes are an excellent means of presenting the scope of work, it soon becomes cumbersome as the number of work packages increase. The other, more practical method of presenting the WBS shows the scope of work as a structured list in a spreadsheet, where each level is tabbed to represent its level in the hierarchy. This format is similar to that used by computers to show folders and sub-folders.

The spreadsheet presentation is the most practical and widely used document to build up a complete list of work packages, jobs or checklists. Once this structure is established, additional columns can be inserted for other parameters, such as, budgets, man hours, materials, equipment, responsibility, quality, risk and communication. The more detailed the list, the greater the accuracy of the estimate and the greater the level of planning and control.

	A	B	C	D	E	F	G	H	I	J
1	WBS		Budget / Work Package	Budget / Department	Budget / Project	Manhours	Materials	Equipment	Responsibility	
2	1.0.0	Yacht Building Project			$					
3	1.1.0	Boat Builder		$						
4	1.1.1	Hull and Bulkheads	$							
5	1.1.2	Deck and Superstructure	$							
6	1.2.0	Mechanical Engineer		$						
7	1.2.1	Main Engine	$							
8	1.2.2	Steering	$							
9	1.3.0	Rigger		$						
10	1.3.1	Mast	$							
11	1.3.2	Sails	$							
12										

Figure 10.15: Project Control Sheet – shows the project's key parameters linked to the work packages

The project control sheet figure 10.15 is a summary sheet that contains information on all the key topics of the project on one sheet of paper. This enables the project manager to see an overview of the project and, therefore, be in a better position to make informed trade-offs between competing parameters. The project control sheet will be developed in the *Execution, Monitoring and Control* chapter.

This document is used to build-up a complete list of work packages, jobs or checklists. Once the project's structure has been established more columns can be added for the other parameters (budgets, man hours, material, equipment, responsibility, risk, quality and communication). The more detailed the list the greater the accuracy of the estimate and the greater the level of planning and control.

Once the WBS structure and numbering system have been established this format can now be used to structure a project control sheet. Figure 10.5 shows how the numbering system and the work package's description are the two left hand columns and form the backbone of the project. The other columns can be used to include all the important parameters of; budgets, man hours, material, responsibility etc.

Exercises:

1. Develop a shopping checklist which is structured by location, vendor and to line up with the layout of your local supermarket.
2. Show how the PBS can be used to subdivide a product you are familiar with into a number of deliverables, include a numbering system to uniquely identify each deliverable.
3. Show how a WBS template could be applied to your projects.
4. Show how the breakdown concept can be used to interlink the PBS, WBS, CBS (cost) and OBS (organization) and present it as a control sheet similar to figure 10.15.
5. Complete the numbering exercise in section 9 (see Appendix 1 for the solution).

Further Reading:

Burke, R., *Project Management Techniques*, will develop the following topics which relate to this knowledge area:

- Risk and uncertainty
- WBS roll-up.

11

Time Management (Estimating Time)

Learning Outcomes
After reading this chapter, you should be able to:
Develop a list of activities
Define an activity
Calculate an activity's level of effort
Estimate an activity's duration

The *Time Management* knowledge area includes the techniques required to complete the project on time. The time component, along with scope, cost and quality, is one of the key project deliverables and, therefore, a cornerstone of project management techniques.

This chapter will explain how to develop a list of activities and estimate their time durations as a prerequisite for a number of scheduling calculations that are included in the following chapters:

- The *CPM* chapter will explain how to present the logical sequence of the activities and calculate their start and finish dates.
- The *Gantt Chart* chapter will explain how to present schedule information in an easy to understand barchart format.
- The *Procurement Schedule* chapter will explain how to schedule the procurement requirement.
- The *Resource Planning* chapter will explain how to schedule the resource requirement.
- The *Project Cashflow* chapter will explain how to schedule the cost requirement.
- The *Execution, Monitoring and Control* chapter will explain how to measure the project's progress.
- The *Earned Value* chapter will explain how to integrate the project's schedule with a variable parameter (usually man hours).

Project Management Plan Flowchart

The project management plan flowchart (figure 11.1) shows the relative position of the time management technique (estimating time) with respect to the WBS and the CPM. It is logical that estimating time follows the *WBS* chapter because the WBS subdivides the scope of work into a list of work packages and a list of activities, and it is for these activities that a time estimate is developed.

It is also logical that estimating time should precede the CPM and Gantt chart calculations where the activity durations are an input into their calculations.

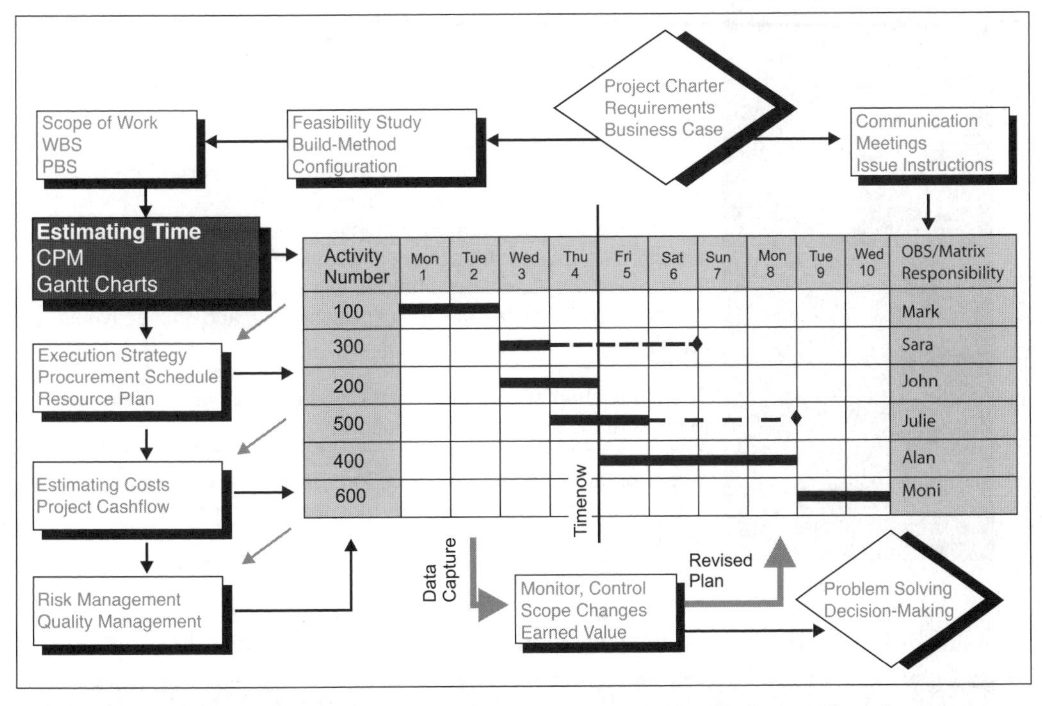

Figure 11.1: Project Management Plan Flowchart – shows the relative position of estimating time with respect to the WBS and the CPM

Body of Knowledge Mapping

PMBOK 4ed: The *Time Management* knowledge area includes the following:

PMBOK 4ed	Mapping
6.1: Define Activities	This chapter will explain how to develop a list of activities from the WBS work packages. It will also explain how the characteristics of an activity can be quantified.
6.2: Activity Sequence	The *CPM* chapter will explain how the sequence of activities can be presented as a network diagram.
6.3: Estimate Activity Resources	This chapter will explain how to estimate the type and quantities of material, people, equipment and supplies required to perform each scheduled activity.
6.4: Estimate Activity Durations	1. This chapter will explain how to develop a trade-off between level of work (effort) and resource availability. 2. The *Procurement Schedule* chapter will explain how to develop a trade-off between the project schedule and long lead items. 3. The *Project Cashflow* chapter will explain how to develop the trade-off between the project schedule and the activities' cashflow.
6.5: Develop Schedule	The *CPM* chapter and the *Gantt Chart* chapter will explain how to present the schedule information.
6.6: Control Schedule	The *Execution, Monitoring and Control* chapter will explain how to issue instructions, expedite, monitor and control the project schedule.

APM BoK 5ed: The *Time Management* knowledge area includes the following:

APM BoK 5ed	Mapping
3.2: Determine the overall project duration	The *CPM* chapter will explain how to calculate the overall project duration.
3.2: When activities and events are planned to happen	The *CPM* chapter will explain how to calculate when activities and events are planned to happen.
3.2: Identification of activities	This chapter will explain how to develop a list of activities from the WBS.
3.2: Logical dependencies	The *CPM* chapter will explain how to use the network diagram to show logical dependencies.
3.2: Estimation of activity durations	This chapter will explain how to estimate activity durations.

Unit Standard 50080 (Level 4): The unit standard for the *Time Management* knowledge area includes the following:

Unit 120384: *Develop a simple schedule to facilitate effective project execution*

Unit 120387: *Monitor, evaluate and communicate simple project schedules*

Specific Outcomes	Mapping
120384/SO1: Demonstrate an understanding of the purpose and process of scheduling project activities.	The *CPM* chapter will explain the purpose of scheduling.
120384/SO2: Define and gather information about project activities from technical (subject matter) experts and within own field of expertise.	This chapter will explain how to gather information about project activities from the technical experts.
120384/SO3: Develop a simple schedule for a project or part thereof.	The *CPM* chapter and *Gantt Chart* chapter will explain how to develop a simple schedule.
120387/SO1: Describe and explain a range of project schedule control processes and techniques.	The *Project Execution, Monitoring and Control* chapter will explain how to plan, monitor and control project activities.
120387/SO2: Monitor actual project work versus planned work (baseline).	The *Project Execution, Monitoring and Control* chapter will explain how to monitor and compare actual work against the baseline plan.
120387/SO3: Record and communicate schedule changes.	The *Project Communications Management* chapter will explain how to record and communicate schedule changes.

The PMBOK 4ed **Project Time Management** knowledge area defines time management as, *the processes required to accomplish timely completion of the project.*

The APM BoK 5ed defines **Scheduling** as, *the process used to determine the overall project duration and when activities and events are planned to happen. This includes identification of activities and their logical dependencies, and estimation of activity durations, taking into account requirements and availability of resources.*

1. WBS / Activity List

The *WBS* chapter explained how to subdivide the scope of work into a number of work packages which are a logical grouping of work for the purpose of cost estimating, budgeting, and management control. This section will take the WBS structure a step further and subdivide the work packages into a number of activities which are required for project scheduling and time control. It should be noted that there is usually a sufficient level of detail to manage the project budgets at the work package level but, for time planning and control, there needs to be a higher level of detail – hence the work packages need to be subdivided further into a list of activities as shown below, figure 11.2.

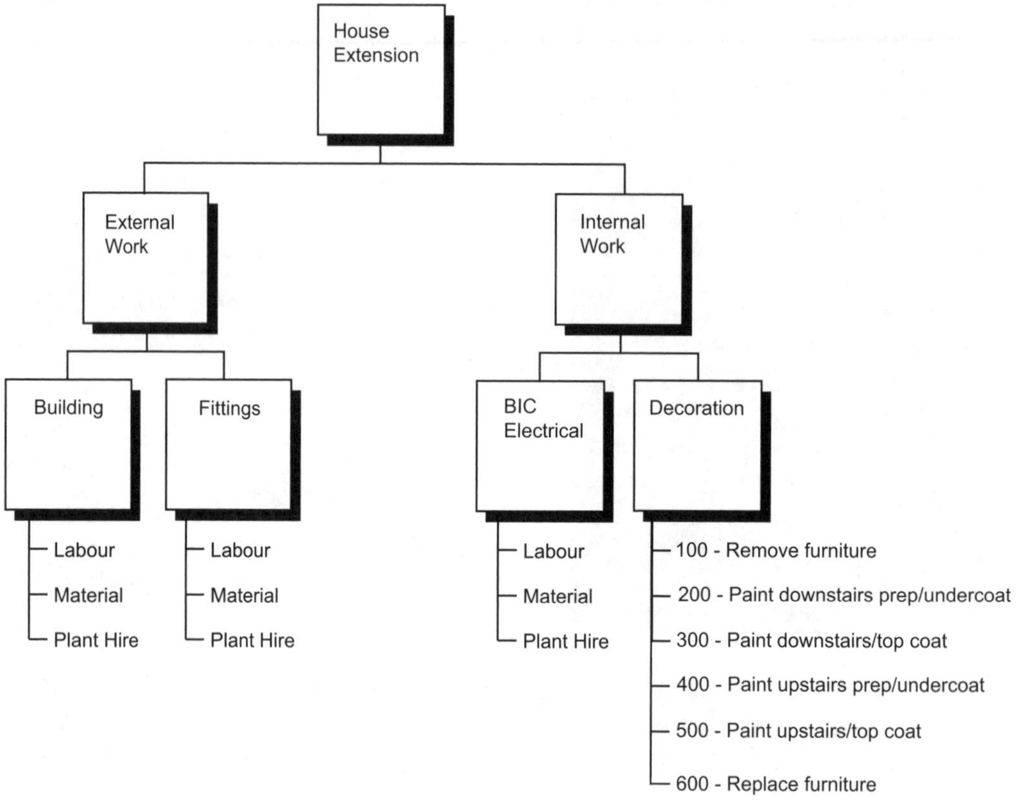

Figure 11.2: Activity List – shows how the WBS work packages can be subdivided into a list of activities

Decoration Example: Figure 11.2 above, is a good example of the different levels of breakdown required for budgeting and time management. Subdividing the scope of work to the work package level gives sufficient detail for cost estimating, budget assignment and cost control. But scheduling the work requires a greater level of detail - the work packages are subdivided into a list of activities that are required to complete the work or produce the deliverables, and then used to produce the CPM and the Gantt chart.

2. Definition of an Activity

An activity may be defined as any task, job or operation that must be performed to complete the work package or project. A WBS work package can be subdivided into one or more activities as shown in figure 11.2.

With the introduction of project planning software, certain terms and norms have been established - these terms will be used wherever possible. The characteristics of an activity include the following:

Number	An activity must have a unique activity code or number (A, or 010, or ABC100). The code can be alpha, numeric or alphanumeric.
Description	An activity must have a description. The description should be as informative and clear as possible to distinguish the activity from other activities. It is the project team members' responsibility to define and gather information about the project activities from the technical (subject matter) experts.
Logic	There will be logical relationships between the activities, expressed as activities in series (one after the other) or activities in parallel (at the same time).
Lag	A lag before or after an activity is a time delay. For example, if activity A is finished on Monday and the following activity B had a lag of two days this means that activity B cannot start before Thursday.
Duration	All activities have a time duration for completing the task. If the duration is zero, then it is called an event (see next section).
Calendar	All activities have a calendar or work pattern to indicate when the work can be scheduled. A calendar needs to be defined even if the work can be scheduled seven days a week (continuous working).
Target Date	An activity can have a target start date and a target finish date assigned. These are often referred to as milestones, key dates or deadlines. Certainly a start date for the project is required for the first activity.
Procurement	If an activity requires materials and services to be procured these can be linked to an activity and the delivery date can be established to produce a procurement schedule.
Resource Requirements	If an activity requires resources these can be linked to an activity and scheduled to produce a resource histogram.
Cashflow Statement	If an activity incurs costs these can be linked to an activity and scheduled to produce a project cashflow statement.
S Curve	When resources and costs are linked to the activities they can be integrated to generate an S curve similar to the exercise in the *Project Lifecycle* chapter.

Table 11.1: Activity Characteristics

Activities should be documented at an appropriate level of detail to support further planning activities. Specific project activities should be identified and information gathered from the technical experts.

Activity Box: In the network diagram, which is developed in the *CPM* chapter, an activity is always presented in a box with an identity number and a description. The activity should be given a description to ensure the project team members understand the work content - this can be expanded later in a job card.

EARLY START		EARLY FINISH
FLOAT	ACTIVITY NUMBER DESCRIPTION	DURATION
LATE START		LATE FINISH

Figure 11.3: Activity Box – shows how activity information can be presented in a box

3. Definition of an Event

An event may be defined as an activity with zero duration – it is simply a point in time. An event, also called a key date, a milestone, or a deadline, represents a happening on a particular day. This could be when an order is placed, the start of the project, the day the materials are received, or the end of the project.

Many project managers prefer to manage their projects by setting milestones. This effectively pushes the planning and control down a level and empowers the project team members to manage their own work to achieve the target milestones. For example, the project manager might issue an instruction that a certain activity must be finished by Friday the 10th June. The project team members must now do their own scoping and planning at their level, working back to achieve their scope of work by the assigned completion date.

The completion of the project itself might be leading up to an event, for example, a wedding, a concert, a sporting fixture – in these cases the project is to prepare for the event (see *Event Management* chapter).

There are three clear advantages of using a management-by-milestones approach which are:

- It simplifies the planning by focusing on a number of intermediate targets.
- It makes assigning work easier to focus on a number of achievable targets.
- It makes progress measurement easier to monitor and control, because the event's scope of work has either started or not, or it has either finished or not. It is a yes or no situation. This approach removes the need to interpolate with its inherent inaccuracy.

4. Activity Duration

Estimating an activity's duration is a trade-off between the amount of work required to complete the activity and the number of resources available. To estimate an activity's time duration the following information is required:

- Estimating data base
- Activity list
- Scope of work
- Level of effort
- Level of resources.

Estimating Data Base: A company's estimating data base of tariffs or rates outlines how much the company charges for its goods and services and is based empirically on past performance, efficiency and productivity. This is part of a company's terms and conditions of contract which is owned by the production manager or finance director. A project's estimate is based on these tariffs which might be expressed as per the table 11.2 below.

Type of Work	Rate
Painting	10 m^2/hour
Welding	10 m/day
Bricklayer	500 bricks/day

Table 11.2: Estimating Data Base – this table assumes an average productivity

Activity List: The activity list outlines the jobs to be performed to complete the project's scope of work and achieve the project's objectives. The project management approach is to subdivide the scope of work into work packages using a WBS, and then subdivide the work packages into activities so that they can be time planned and controlled (see figure 11.2).

Scope of Work: The scope of work is a measure of the amount of work each activity has to perform to complete the activity. This might be quantified by a bill of materials (BOM) or an appropriate measure of the work. For example, a painting job might be quantified as 'x' square meters, a report might be quantified as 'x' pages, or a software project quantified as 'x' lines of code (see table 11.3).

Type of Work	Quantified as:
Paint wall	20,000 m^2
Weld yacht	2,000 m
Build wall	10,000 bricks

Table 11.3: Scope of Work Quantified

Level of Effort (Man hours): Once an activity's scope of work has been quantified, this can now be converted into a level of effort (man hours). From a time perspective the level of effort can be expressed as so many man hours, or so many man days, or even so many man weeks depending on the size of the project.

Level of Resources: The level of resources quantifies the number of resources available to perform the work. Table 11.4 shows the trade-off between the resources available and the activity's duration. If the level of effort has been quantified as, say, 250 man days and there are 10 workers available then the activity will be completed in 25 days. However, as the number of resources changes so the duration also changes.

Decorating Example: One of the activities on a decorating project is to paint a surface area of 20,000 square meters. If the level of effort is quantified as 1 man-day per 10 square meters, then the level of effort is:

$$\frac{(20{,}000\ m^2)}{10\ \text{man hrs per}\ m^2 \times 8\ \text{hrs per day}} = 250\ \text{man days}$$

The resource / duration trade-off would then be as follows:

Level of Effort Man days	Resources Available	Duration (days)
250	10 men	25
250	11 men	22.7
250	12 men	20.8
250	13 men	19.2
250	14 men	17.9
250	15 men	16.6

Table 11.4: Resource Duration Trade-Off – shows the number of days to complete the work with the men available

Table 11.4 shows the trade-off between the number of resources and the activity's duration. It is the project manager's responsibility to manage the resource/duration trade-off and assign resources to achieve the desired activity duration. This decision will be influenced by an activity's dependencies – an activity could be time dependent (it has to finish at a certain time), or the activity could be resource dependent (there are only so many resources available or limited space).

At the outset the estimator will probably assume a certain resource availability to give preliminary time duration. During the detailed planning phase the estimator needs to check the resource availability and any other constraints and firm up on the figures.

Activity List	Scope of Work	Level of Effort Man days	Resources Available	Duration

Table 11.5: Activity Duration Table

Exercises:

1. Using a project you are familiar with, develop the WBS to the work package level and then subdivide one of the work pages into a number of activities for time planning (similar to figure 11.2). Use an activity control sheet similar to table 11.6 below, fill in the activity column and the description column.

Activity	Description	Man hours	Labour	Duration
100				
200				
300				
400				
500				
600				
Total				

Table 11.6: Activity List Control Sheet

2. Continuing with table 11.6 use your company's estimating data base to quantify the level of effort for each activity. Complete the man hours column in the activity control sheet.
3. From your company's estimating data base develop the labour / duration trade-off for:

   a). Resource dependent activities

   b). Milestone dependent activities.

Further Reading:

Burke, R., *Project Management Techniques*

12

Critical Path Method (CPM)

The *Time Management* knowledge area includes all the techniques required to plan and control the project's schedule. This chapter will explain how to use one of the key time management techniques – *Critical Path Method* (CPM).

The reason the CPM's iconic features are so important is because they identify and present the logical relationship between the project's activities and, further, identify the critical activities that determine the duration of the project. These critical activities are so named because, if any of the critical activities are delayed, this will extend the end date of the project. It is these critical activities that give this technique its name – the Critical Path Method.

Project Management Plan Flowchart

The project management plan flowchart (figure 12.1) shows where the CPM technique is positioned with respect to the WBS, estimating time, and the Gantt chart. It is logical that the CPM follows the WBS output because the CPM requires a list of activities which is developed from the WBS work packages. It is also logical that the CPM follows estimating time as the CPM calculation requires time durations for the activities. Using the output from the CPM analysis the Gantt chart and other schedule dependant documents can be developed, notably the procurement schedule, the resource histogram and the project cashflow statement.

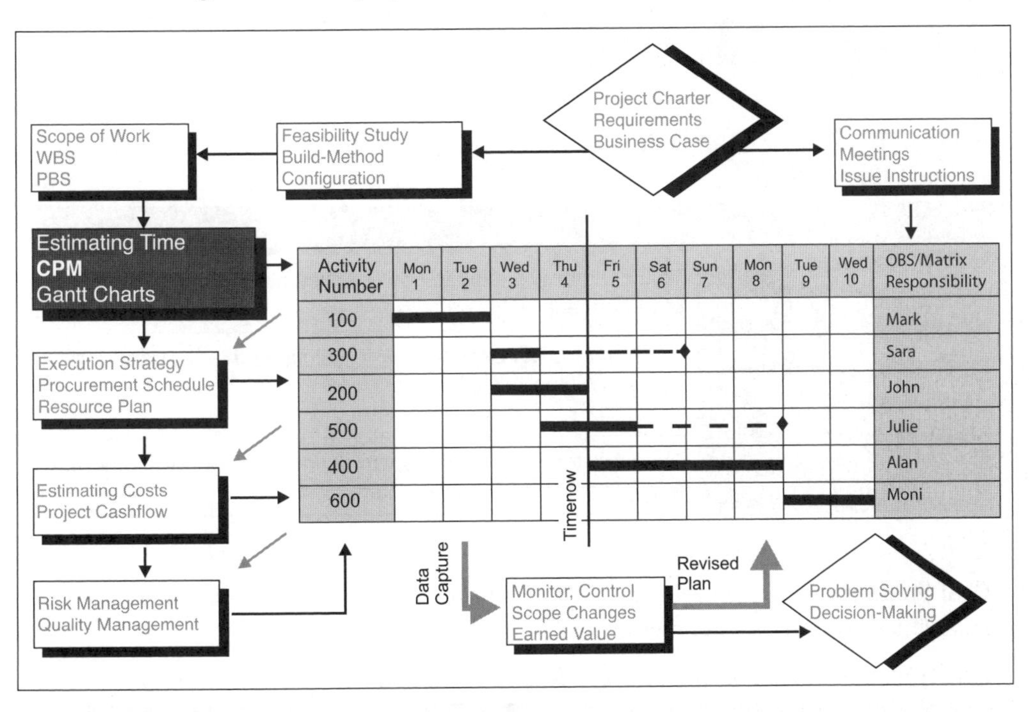

Figure 12.1: Project Management Plan Flowchart – shows the position of the CPM technique with respect to the other topics

Body of Knowledge Mapping

PMBOK 4ed: The CPM technique is part of the *Time Management* knowledge area which includes the following:

PMBOK 4ed	Mapping
6.2: Activity Sequence	This chapter will explain how to develop the network diagram to present the activities' sequence.
6.5: Develop Schedule	This chapter will explain how to calculate the activities' start and finish dates. The CPM iteration section will explain how to balance the trade-off between the parameters to reach an agreed schedule.

APM BoK 5ed: The CPM technique is part of the *Scheduling* knowledge area which includes the following:

APM BoK 5ed	Mapping
3.2: Determine the overall project duration	This chapter will explain how to calculate the overall duration of the project.
3.2: When activities and events are planned to happen	This chapter will explain how to calculate the start and finish dates of all the activities.
3.2: Logical dependencies	This chapter will explain how to develop the network diagram to present the activities' sequence.

Unit Standard 50080 (Level 4): The *CPM* is part of the *Time Management* knowledge area which includes the following:

Unit 120384: *Develop a simple schedule to facilitate effective project execution*

Specific Outcomes	Mapping
120384/SO1: Demonstrate an understanding of the purpose and process of scheduling project activities.	This chapter will explain the purpose of a project schedule.
120384/SO3: Develop a simple schedule for a project or part thereof.	This chapter will develop a simple schedule for a project.

1. Network Diagram

The network diagram, also referred to as a precedence diagram method (PDM), is a development of the activity-on-node (AON) technique.

The network diagram may be defined as a graphical representation of the project's activities showing the planned sequence of work. One of the powerful features of this presentation is that the project manager can see the logical sequence of work for the whole project on one diagram. To draw the network diagram and perform the CPM calculation the project manager needs the following items of information:

- List of activities (developed from the WBS)
- Activities' durations
- Work calendar
- Activity dependencies, also referred to as logic relationships
- Start date.

All these items of information should be recorded in an agreed format. If the project is simple and straight forward, then a Gantt chart with periodic updating should be sufficient but, as projects become more complex with a large number of participants and inter-dependencies, then a network diagram presentation is needed to outline the structure of the project.

Purpose of this Chapter: The purpose of this chapter is to help project managers and project team members acquire the competence and knowledge to be able to administer and perform the CPM calculation within projects of limited complexity. This chapter will explain:

- Step 1: How to draw the network diagram
- Step 2: How to add activity durations and the project start date
- Step 3: How to calculate a forward and backward pass to calculate the activity start and finish dates
- Step 4: How to calculate the activity float and identify the critical path.

Once the CPM structure has been established the other parameters of procurement, resources and costs can be included. This will almost certainly require a number of iterations with trade-offs and compromises in regard to resource availability, procurement supply and project cashflow.

2. Logical Relationships

The network diagram shows the logical sequence of the activities. There are two basic types of relationships:

- Activities in **series**
- Activities in **parallel**

Activities in Series: Activities in series are carried out one after the other. When the network diagram is first developed this will probably be the easiest type of logical relationship to use. An example of activities performed in series on a decoration project would be the surface preparation (activity 100), followed by painting the undercoat (activity 200), followed by painting the top coat (activity 300). The network diagram is typically read from top left to bottom right (see figure 12.2).

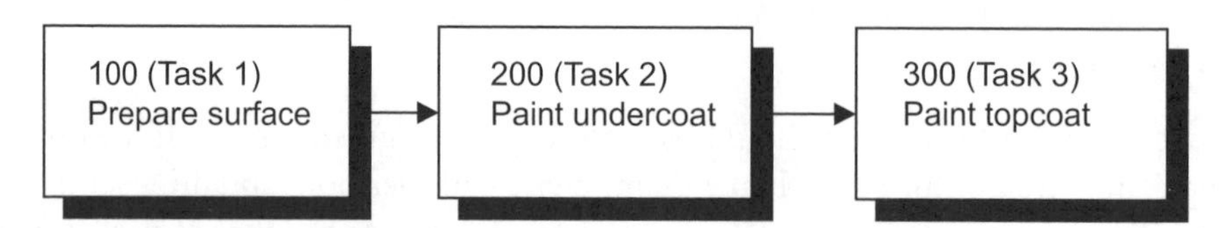

Figure 12.2: Activities in Series – shows activities planned to happen one after the other

Activities in Parallel: When activities are in parallel, the activities can be performed at the same time. This is a more efficient use of time compared to activities in series. An example on a decoration project would be the painting of the bedroom (activity 2000) and the painting of the dining room (activity 3000) simultaneously, after the surface preparation (activity 1000) is complete, and then followed by the replacement of the furniture (activity 4000) (see figure 12.3).

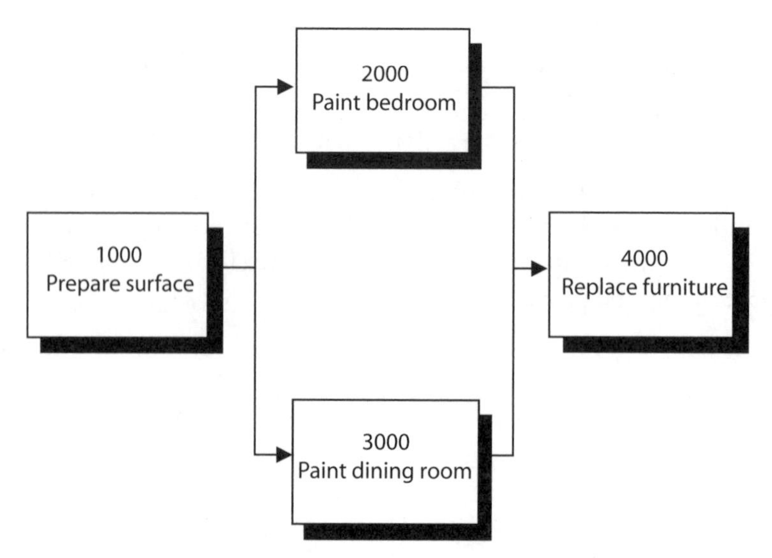

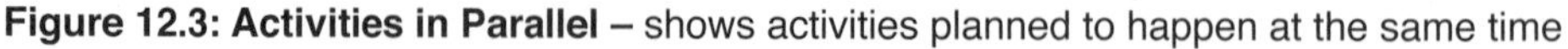
Figure 12.3: Activities in Parallel – shows activities planned to happen at the same time

3. How to Draw the Logical Relationships

Logical relationships, also referred to as constraints, dependencies and links are all used interchangeably to represent the lines drawn between the activity boxes. The preferred presentation shows the constraint lines drawn from left to right, starting from the right side of one activity box into the left side of the following box. However, some software packages draw the lines from the top and bottom of the boxes. Initially, an arrow at the end of the constraint might help to follow the direction of work flow.

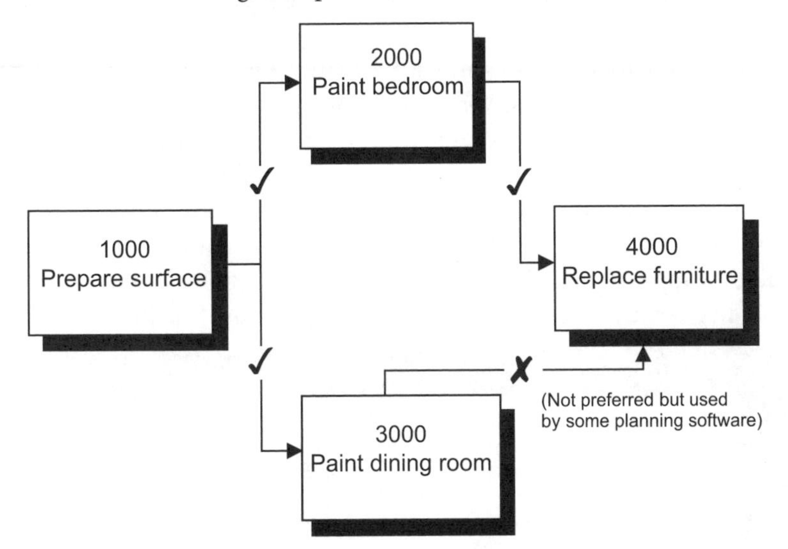

Figure 12.4: How to Draw Activity Constraints – shows the preferred presentation

Activity Box: The activity box's key indicates where to position the values in the activity box. This layout varies depending on the software package, but the format in figure 12.5 will be used throughout this book.

EARLY START		EARLY FINISH
FLOAT	ACTIVITY NUMBER DESCRIPTION	DURATION
LATE START		LATE FINISH

Figure 12.5: Activity Box – shows the position of the values in the activity box

4. Activity Logic – Tabular Reports

The activity relationships can be presented as either a network diagram or a tabular report. A tabular report would be a similar format to a data base used by the planning software, where each line defines a logical relationship. Planning software usually expresses the activity logic by stating the preceding activity.

Painting Example: This section will continue with the house decorating example. Taking the list of activities developed in the *WBS* chapter as the scope of work see figure 10.10.

The first task (activity 100) is to remove the furniture which is a start activity because there are no preceding activities. When the moving of the furniture is complete the painters can begin (activity 200) to prepare and paint the undercoat downstairs and, simultaneously, (activity 400) to prepare and paint the undercoat upstairs. When (activity 200) is finished the painters can begin (activity 300) to paint the top coat downstairs. Similarly when (activity 400) is finished the painters can begin (activity 500) to paint the top coat upstairs. When both (activities 300 and 500) are finished the painters can begin the last task (activity 600) to replace the furniture.

Activity	Description	Duration	Preceding Activity
100	Remove furniture	2	Start
200	Downstairs, prepare and paint undercoat	3	100
300	Downstairs, paint top coat	3	200
400	Upstairs, prepare and paint undercoat	2	100
500	Upstairs, paint top coat	2	400
600	Replace furniture	2	300, 500

Table 12.1: Activity Table – shows the network diagram information in a tabular format, activity list from *WBS* chapter, figure 10.10

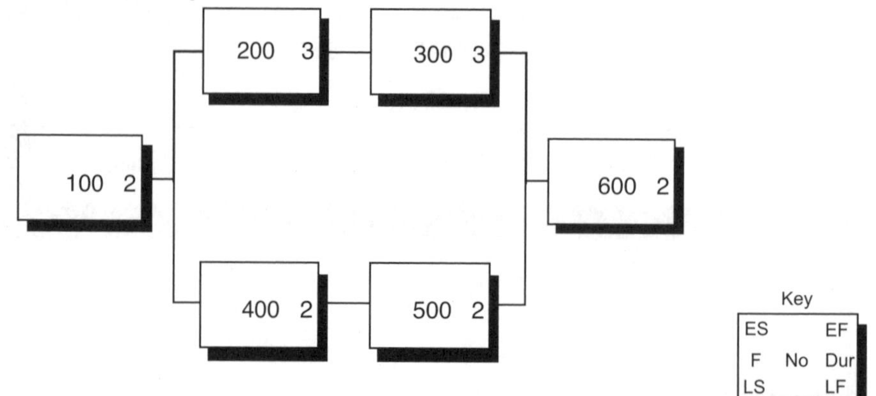

Figure 12.6: Network Diagram – shows the activity logic and durations and start date

5. Critical Path Method Steps

The second step is to impose activity dates and durations as per the following:

1. Start Date	The CPM analysis needs a start date from which to schedule the work. For this exercise the start date will be Monday 1st of the month. By setting the start date the first iteration will give the project team members a feel for the end date of the project using the given logic, activity durations and calendar. If a target completion date is given, the above parameters (logic, duration, calendar and start date), can be adjusted accordingly.
2. Activity Durations	The activity durations are a trade-off between the amount of work, the resources available, and the calendar. This is discussed in the *Time Management* chapter.
3. Calendar	For this exercise the work calendar will be continuous. This means the workers are working every day including the weekends.
4. Early Start	This is the earliest date by which an activity can start, assuming all the preceding activities are completed as planned.
5. Early Finish	This is the earliest date by which an activity can finish, assuming all the preceding activities are completed as planned.
6. Late Start	This is the latest date an activity can start to meet the planned completion date.
7. Late Finish	This is the latest date an activity can finish to meet the planned completion date.

Target Start and **Target Finish:** In addition to the calculated dates there could be a number of imposed dates, influenced by the delivery of materials, access to sub-contractors, or other milestones.

6. Forward Pass

The term forward pass is used to define the process of calculating the early start date (ES) and early finish date (EF) for all the activities. For clarity, this section will subdivide the forward pass calculation into four clouds as shown in figure 12.7.

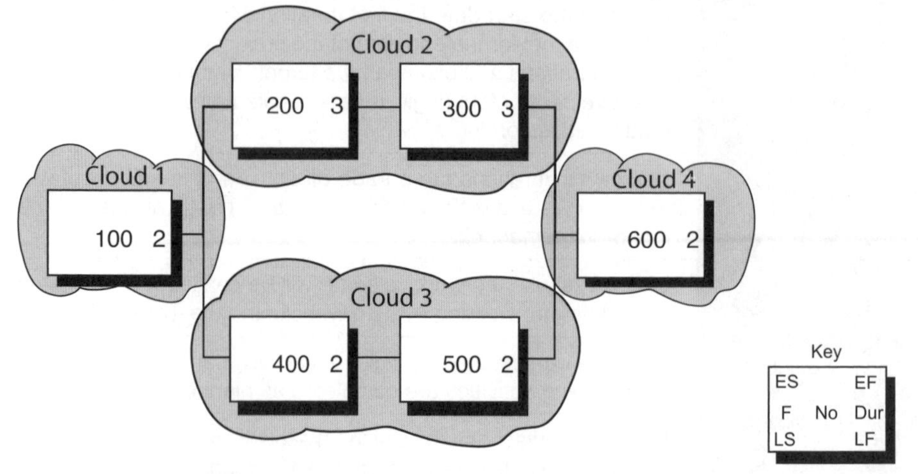

Figure 12.7: Forward Pass - shows the forward pass calculation subdivided into four clouds

Cloud 1: Cloud 1 calculates the activity 100 ES and EF. For convenience the early start date of the first activity in all the examples will be Monday the first day of the month (i.e. 1st May). For the painting project consider the activity 100, add the early start date (ES) of 1. The early finish (EF) date of activity 100 is calculated by adding the activity duration to the early start date (EF), using the following formula:

EF (100) = ES (100) + Duration (100) – 1

EF (100) = 1 + 2 – 1

= 2 (Tuesday)

In the equation the minus one is required to keep the mathematics correct. The Gantt chart in figure 12.12 will clarify this requirement. Shown as a Gantt chart it can be clearly seen that a two day activity that starts on day 1 (Monday) will finish on day 2 (Tuesday).

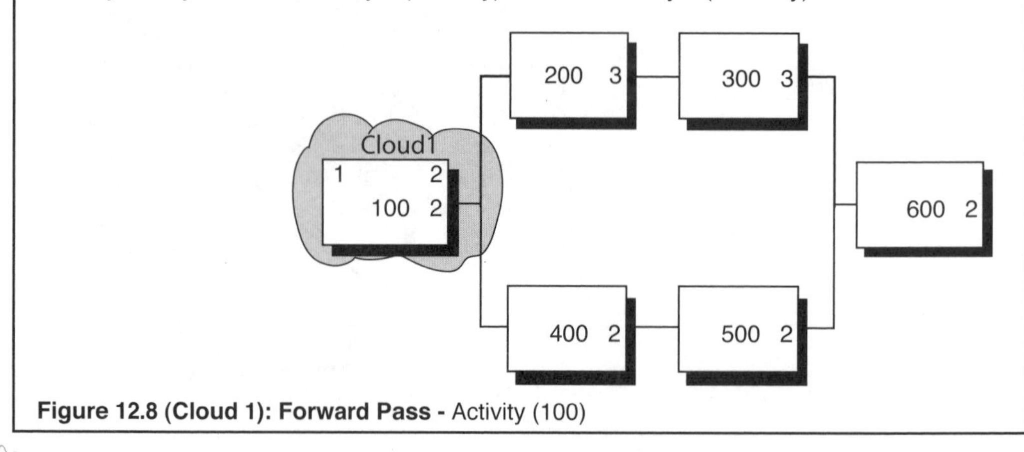

Figure 12.8 (Cloud 1): Forward Pass - Activity (100)

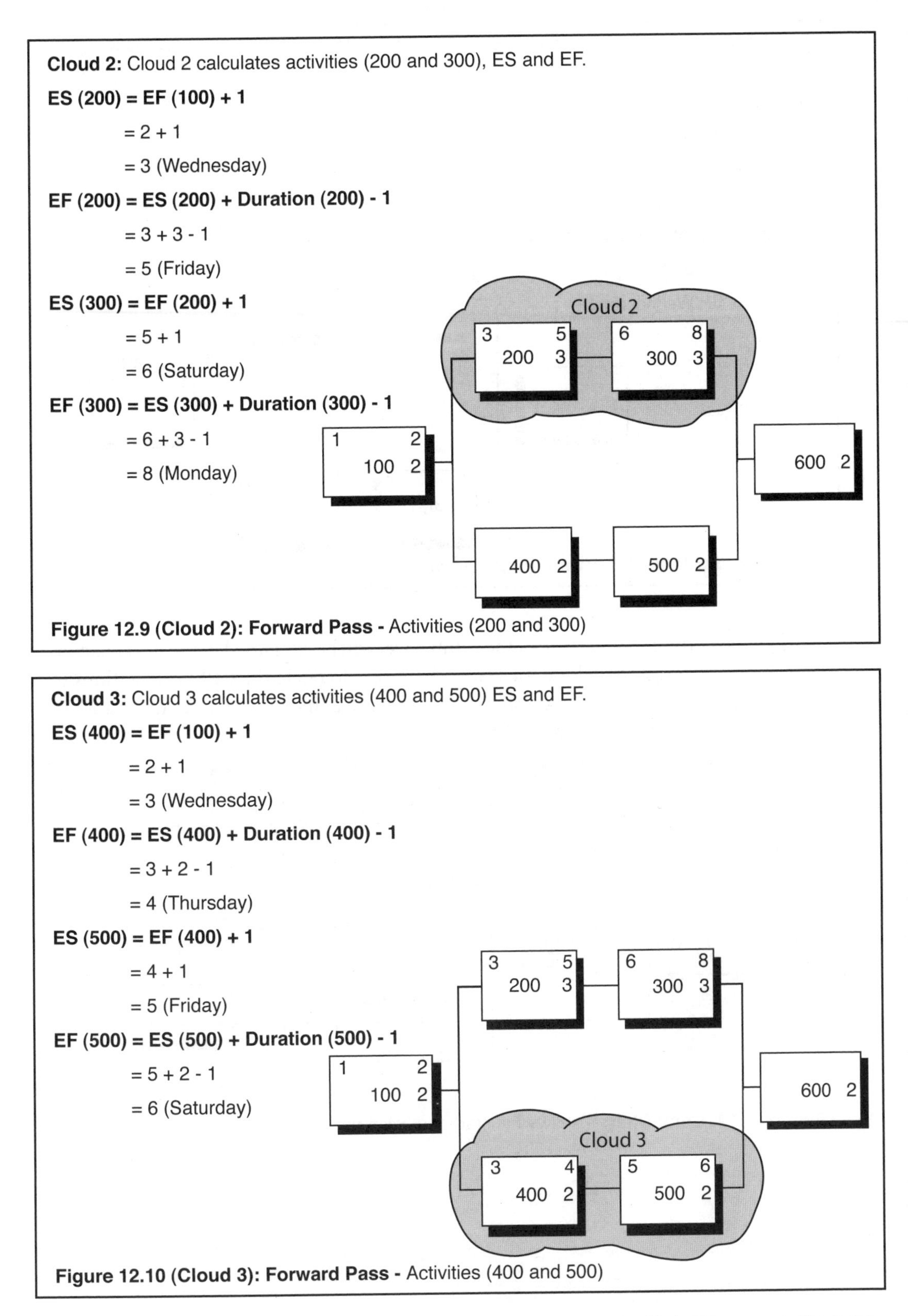

Cloud 2: Cloud 2 calculates activities (200 and 300), ES and EF.

ES (200) = EF (100) + 1

= 2 + 1

= 3 (Wednesday)

EF (200) = ES (200) + Duration (200) - 1

= 3 + 3 - 1

= 5 (Friday)

ES (300) = EF (200) + 1

= 5 + 1

= 6 (Saturday)

EF (300) = ES (300) + Duration (300) - 1

= 6 + 3 - 1

= 8 (Monday)

Figure 12.9 (Cloud 2): Forward Pass - Activities (200 and 300)

Cloud 3: Cloud 3 calculates activities (400 and 500) ES and EF.

ES (400) = EF (100) + 1

= 2 + 1

= 3 (Wednesday)

EF (400) = ES (400) + Duration (400) - 1

= 3 + 2 - 1

= 4 (Thursday)

ES (500) = EF (400) + 1

= 4 + 1

= 5 (Friday)

EF (500) = ES (500) + Duration (500) - 1

= 5 + 2 - 1

= 6 (Saturday)

Figure 12.10 (Cloud 3): Forward Pass - Activities (400 and 500)

Cloud 4: Cloud 4 calculates activity (600) ES and EF. When there is more than one activity leading into one activity, as activity (300) and (500) lead into activity (600), then select the highest EF to calculate the ES of the following activity.

ES (600) = EF (300) + 1

= 8 + 1

= 9 (Tuesday)

EF (600) = ES (600) + Duration (600) - 1

= 9 + 2 - 1

= 10 (Wednesday)

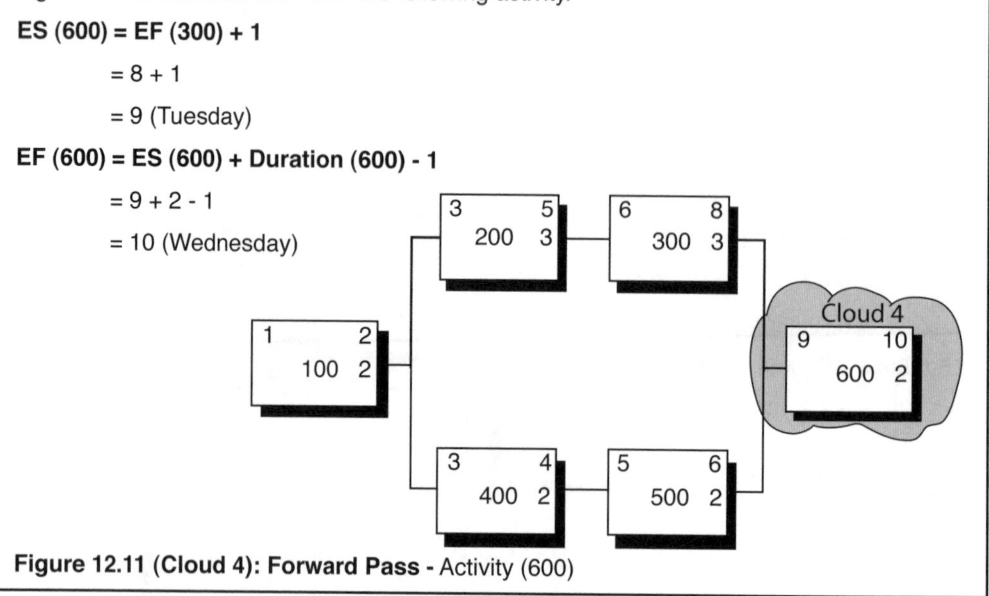

Figure 12.11 (Cloud 4): Forward Pass - Activity (600)

Gantt Chart: The characteristic of the Gantt chart will be explained in the next chapter. However, it is useful to include the Gantt chart here to relate the CPM calculations to a timeline. The Gantt chart is developed by drawing a line between the ES and EF of each activity, as show in figure 12.12 below.

Activity Number	Mon 1	Tue 2	Wed 3	Thu 4	Fri 5	Sat 6	Sun 7	Mon 8	Tue 9	Wed 10
100	ES	EF								
200			ES		EF					
300						ES		EF		
400			ES	EF						
500					ES	EF				
600									ES	EF

Figure 12.12: Gantt Chart – shows the start and finish dates in the Gantt chart format

7. Backward Pass

The next step is to perform a backward pass to calculate the late start date (LS) and late finish date (LF) of each activity. For clarity, this section will subdivide the backward pass calculation into four clouds as shown in figure 12.13.

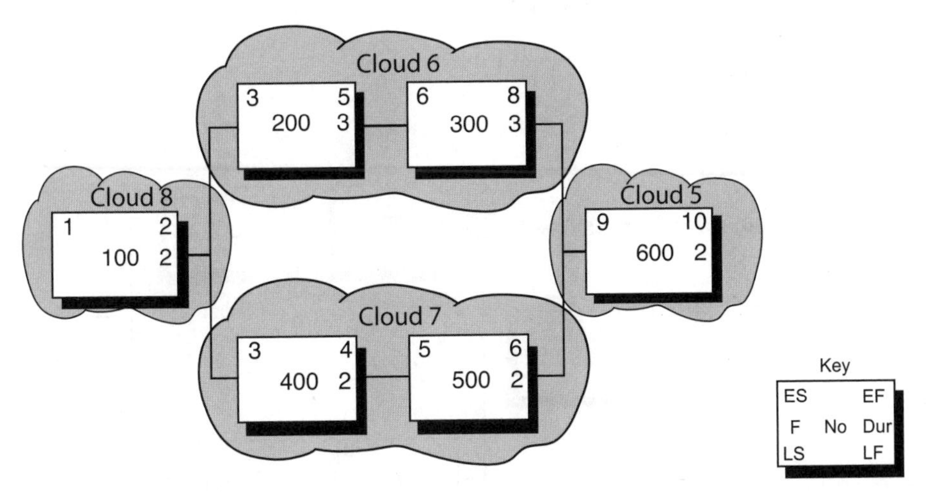

Figure 12.13: Backward Pass

Cloud 5: Cloud 5 calculates activity (600) LS and LF. The late finish (LF) date for the last activity might be assigned as a target finish date; if not, use the early finish (EF) date of the last activity (see figure 12.13).

LF (600) = EF (600)
= 10 (Wednesday)

LS (600) = LF (600) - Duration (600) + 1
= 10 - 2 + 1
= 9 (Tuesday)

Note the plus one in the formula to keep the mathematics correct.

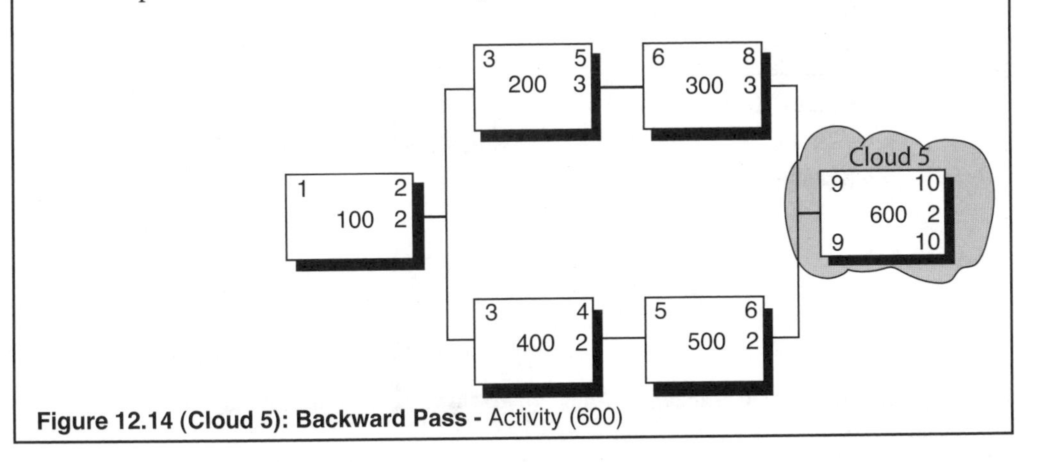

Figure 12.14 (Cloud 5): Backward Pass - Activity (600)

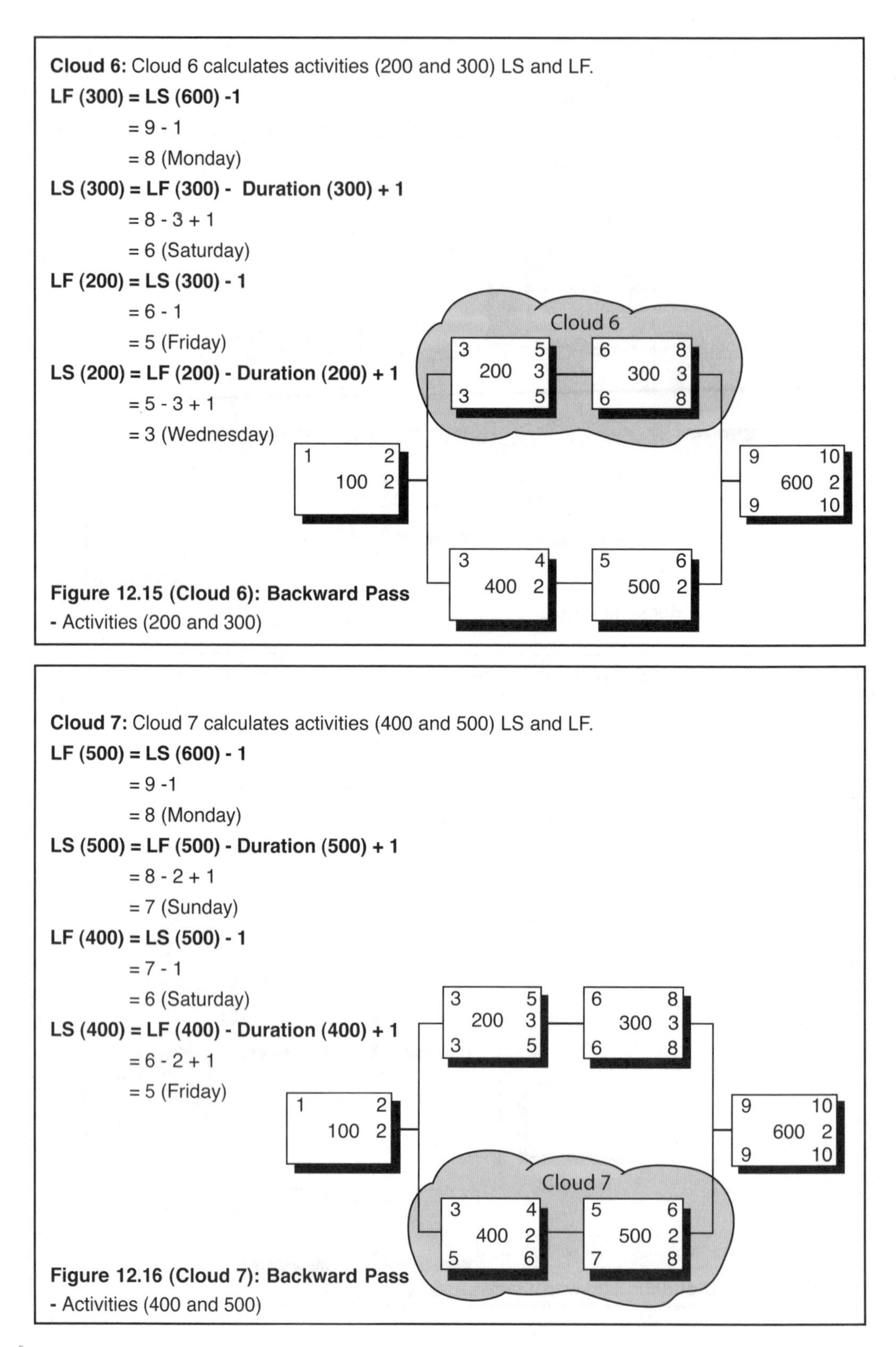

Cloud 6: Cloud 6 calculates activities (200 and 300) LS and LF.

LF (300) = LS (600) -1
= 9 - 1
= 8 (Monday)

LS (300) = LF (300) - Duration (300) + 1
= 8 - 3 + 1
= 6 (Saturday)

LF (200) = LS (300) - 1
= 6 - 1
= 5 (Friday)

LS (200) = LF (200) - Duration (200) + 1
= 5 - 3 + 1
= 3 (Wednesday)

Figure 12.15 (Cloud 6): Backward Pass
- Activities (200 and 300)

Cloud 7: Cloud 7 calculates activities (400 and 500) LS and LF.

LF (500) = LS (600) - 1
= 9 -1
= 8 (Monday)

LS (500) = LF (500) - Duration (500) + 1
= 8 - 2 + 1
= 7 (Sunday)

LF (400) = LS (500) - 1
= 7 - 1
= 6 (Saturday)

LS (400) = LF (400) - Duration (400) + 1
= 6 - 2 + 1
= 5 (Friday)

Figure 12.16 (Cloud 7): Backward Pass
- Activities (400 and 500)

Cloud 8: Cloud 8 calculates activity (100) LS and LF. On the backward pass, when a number of activities lead into one, select the LS with the earliest date.

LF (100) = LS (200) - 1

= 3 - 1

= 2 (Tuesday)

LS (100) = LF (100) - Duration (100) + 1

= 2 - 2

= 1 (Monday)

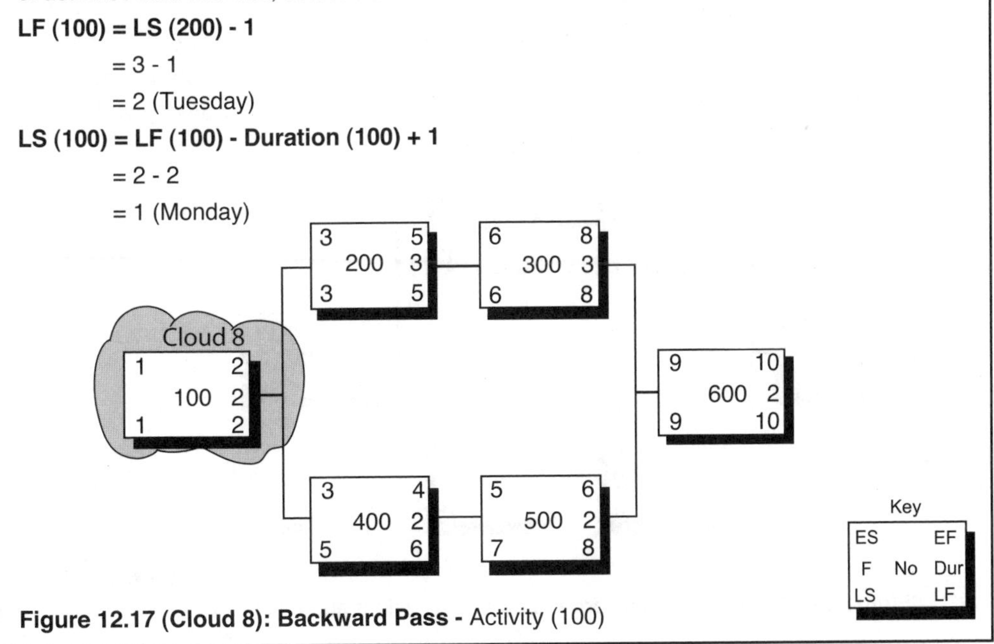

Figure 12.17 (Cloud 8): Backward Pass - Activity (100)

8. Activity Float

Activity float, also referred to as **slack**, is a measure of a project's flexibility, or inherent surplus time in an activity's scheduling. This indicates how many working days the activity can be delayed before it will extend the completion date of the project or any intermediate target finish dates (milestones). Float is calculated as:

Float = Late Start - Early Start

From the painting project:

Float (100) = LS (100) - ES (100)
= 1 - 1 = 0

Float (200) = LS (200) – ES (200)
= 3 – 3 = 0

Float (300) = LS (300) - ES (300)
= 6 - 6 = 0

Float (400) = LS (400) – ES (400)
= 5 – 3 = 2

Float (500) = LS (500) - ES (500)
= 7 - 5 = 2

Float (600) = LS (600) - ES (600)
= 9 - 9 = 0

The next step is to transfer the float values into the float slots as shown on the network diagram key. On the network diagram, where the float value is zero, the activity boxes need to be joined with a bold line (a red line if you have a colour printer). Note activities (100, 200, 300 and 600) all have zero float, this means they are on the critical path. If any of these activities are delayed it will delay the whole project. Meanwhile the other activities (400 and 500) have 2 days float, so these activities could be delayed 2 days before impacting on the end date of the project.

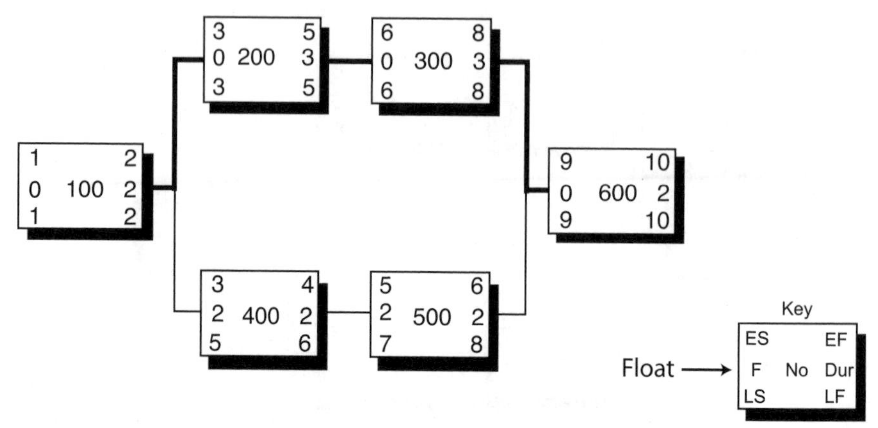

Figure 12.18: Network Diagram - shows activity float and critical path

Activity Table: The CPM calculation can be presented in a tabular format. Using the headings in table 12.2, the values are transferred across from figure 12.18. Table 12.2 also includes a responsibility column to indicate who is going to carry out the work.

Activity Number	Duration	Early Start	Early Finish	Late Start	Late Finish	Float	Responsibility
100	2	1 May	2 May	1 May	2 May	0	Glynn
200	3	3 May	5 May	3 May	5 May	0	Warren
300	3	6 May	8 May	6 May	8 May	0	Jan
400	2	3 May	4 May	5 May	6 May	2	Jim
500	2	5 May	6 May	7 May	8 May	2	Tony
600	2	9 May	10 May	9 May	10 May	0	Trish

Table 12.2: Tabular Report - shows the CPM values in a tabular report format

Worked Example CPM 1: Given the activity table (table 12.3) draw the network diagram and carry out the CPM analysis (figure 12.19) and produce the tabular report (table 12.4).

Activity	Preceding	Duration
100	Start	1
200	100	1
300	100	2
400	100	3
500	200	2
600	300	2
700	500	3
800	700, 600, 400	2

Table 12.3: CPM 1 Activity Table – for worked example CPM 1

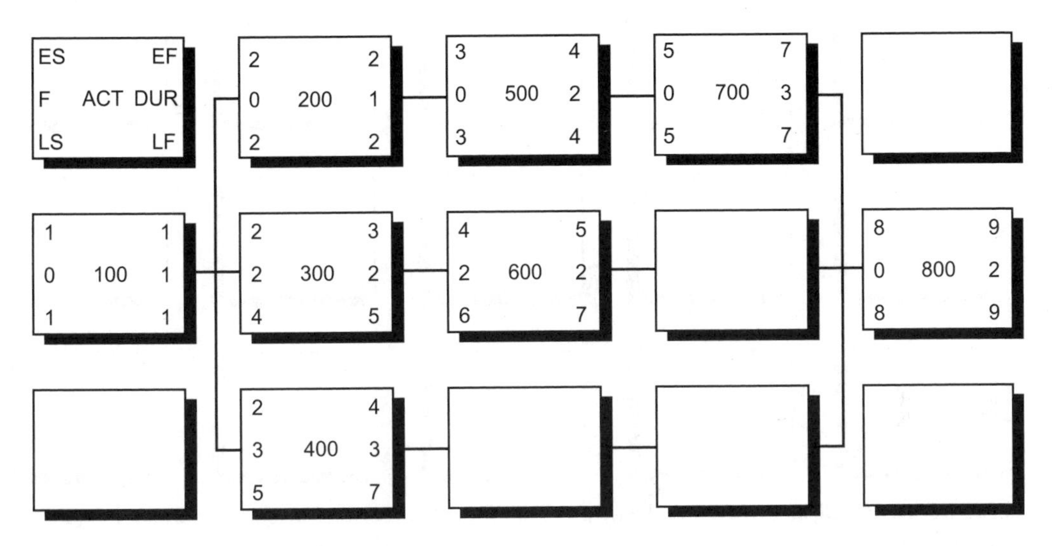

Figure 12.19: CPM 1 Network Diagram – for worked example CPM 1

Activity	Duration	Early Start	Early Finish	Late Start	Late Finish	Float
100	1	1	1	1	1	0
200	1	2	2	2	2	0
300	2	2	3	4	5	2
400	3	2	4	5	7	3
500	2	3	4	3	4	0
600	2	4	5	6	7	2
700	3	5	7	5	7	0
800	2	8	9	8	9	0

Table 12.4: CPM 1 Tabular Report – for worked example CPM 1

Exercise CPM 2: Given the activity table (table 12.5) draw the network diagram and carry out the CPM analysis (figure 12.20) and produce the tabular report (table 12.6).

Activity	Preceding	Duration
100	Start	1
200	100	2
300	200	3
400	100	2
500	400	3
600	500	3
700	100	3
800	700	3
900	300, 600, 800	1

Table 12.5: CPM 2 Activity Table – for exercise CPM 2

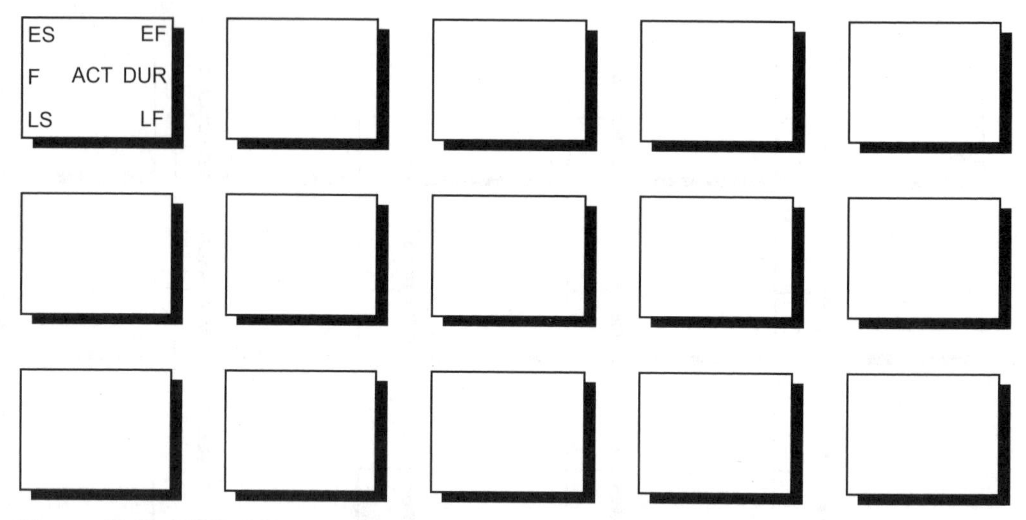

Figure 12.20: CPM 2 Network Diagram Template – (for solution see appendix 1 figure 12.22)

Activity	Duration	Early Start	Early Finish	Late Start	Late Finish	Float
100	1					
200	2					
300	3					
400	2					
500	3					
600	3					
700	3					
800	3					
900	1					

Table 12.6: CPM 2 Tabular Report Template – (for solution see appendix 1 table 12.9)

Exercise CPM 3: Given the activity table (table 12.7) draw the network diagram and carry out the CPM analysis (figure 12.21) and produce the tabular report (table 12.8).

Activity	Preceding	Duration
100	Start	1
200	Start	2
300	100, 200	3
400	300	3
500	400	4
600	100, 200	4
700	100, 200	2
800	700	3
900	500	2
1000	600, 800	2

Table 12.7: CPM 3 Activity Table – for exercise CPM 3

Figure 12.21: CPM 3 Network Diagram Template – (for solution see appendix 1 figure 12.23)

Activity	Duration	Early Start	Early Finish	Late Start	Late Finish	Float
100	1					
200	2					
300	3					
400	3					
500	4					
600	4					
700	2					
800	3					
900	2					
1000	2					

Table 12.8: CPM 3 Tabular Report Template – (for solution see appendix 1 table 12.10)

13

Gantt Charts

Learning Outcomes
After reading this chapter, you should be able to:
Draw a Gantt chart
Draw a Gantt chart from a CPM tabular report
Produce a hammock barchart
Produce a milestone chart
Produce a rolling horizon Gantt chart
Produce a revised Gantt chart

The *Time Management* knowledge area includes all the techniques required to plan and control the project's schedule. This chapter will explain how to produce one of the best known techniques – the Gantt chart.

The Gantt chart or barchart is one of the most widely used planning and control documents for communicating schedule information. It is ideal for small projects or sub-projects because it is easy for most people to understand and assimilate, and it also conveys the planning and scheduling information accurately and precisely.

It was originally designed before the First World War by an American, Henry Gantt, who used it as a visual aid for planning and controlling his shipbuilding projects. In recognition, planning barcharts often bear his name (Gantt chart).

Project Management Plan Flowchart

The project management plan flowchart shows the relative position of the Gantt chart with respect to the CPM and the WBS techniques. On projects of limited complexity the Gantt chart can be drawn straight from the WBS, but as projects become larger and more complex the CPM calculation should be carried out first and then the Gantt chart drawn from the CPM's tabular report.

The Gantt chart must be developed as a time structure for other techniques which include; procurement schedule, resource histogram and the project cashflow statement.

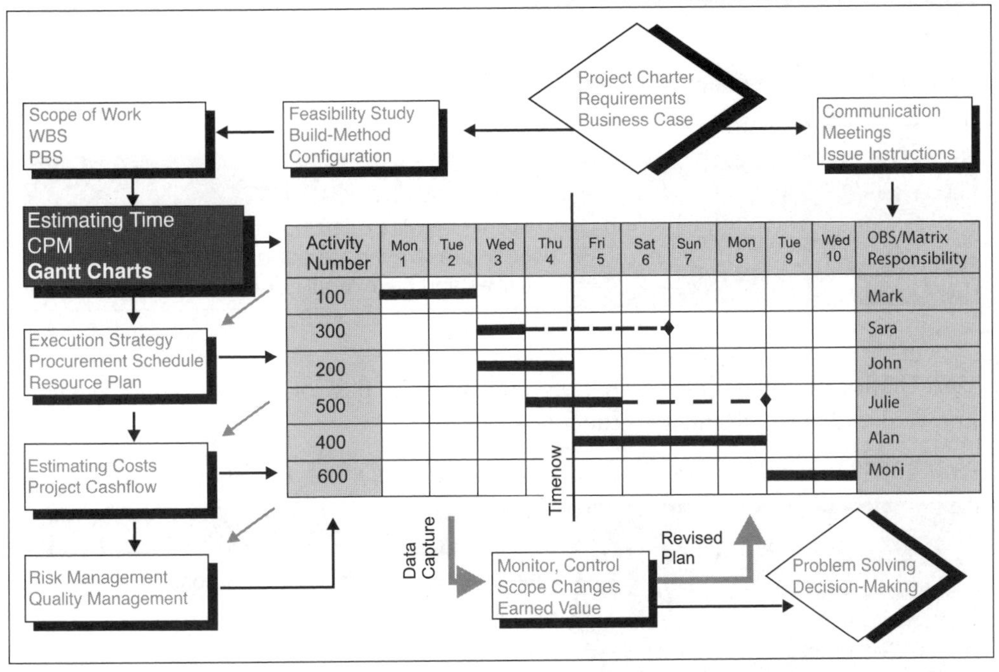

Figure 13.1: Project Management Plan Flowchart – shows the relative position of the Gantt chart with respect to the other tools and formats

Purpose of this Chapter: The purpose of this chapter is to help project managers and project team members acquire the competence and knowledge to be able to produce the Gantt chart for projects of limited complexity. This chapter will explain:

- How to draw the Gantt chart
- How to link the Gantt chart to tabular reports
- How to draw activity float.

The project manager and team leader are responsible for producing the Gantt chart as a planning and instruction document. It is essential that the team members understand the Gantt chart's content and purpose. The team members can also use the Gantt chart to record their progress and give short term planning by using a rolling horizon Gantt chart.

Body of Knowledge Mapping

PMBOK 4ed: The Gantt chart technique is part of the *Project Time Management* knowledge area and includes the following:

PMBOK 4ed	Mapping
6.5: Develop Schedule	This chapter will explain how to use the Gantt chart techniques to develop a project schedule.
6.6: Control Schedule	This chapter will explain how to use the revised Gantt chart for controlling the schedule.

APM BoK 5ed: The Gantt chart technique is part of the *Schedule* knowledge area and includes the following:

APM BoK 5ed	Mapping
3.2: Determine the overall duration of the project and when activities are planned.	This chapter will explain how the Gantt chart can be used to present the start and finish dates of all the activities and the overall duration of the project.

Unit Standard 50080 (Level 4): The Gantt chart is part of the *Time Management* knowledge area which includes the following:

Unit 120384: *Develop a simple schedule to facilitate effective project execution*

Specific Outcomes	Mapping
120384/SO1: Demonstrate an understanding of the purpose and process of scheduling project activities.	This chapter will explain the purpose of a project schedule.
120384/SO3: Develop a simple schedule for a project or part thereof.	This chapter will develop a simple schedule for a project.

Unit 120387: *Monitor, evaluate and communicate simple project schedules*

Learning Outcomes	Mapping
120387/SO1: Describe and explain a range of project schedule control processes and techniques.	This chapter will explain how to produce a revised Gantt chart.

1. How to Draw a Gantt Chart

The Gantt chart in figure 13.2 lists the scope of work as WBS work packages or activities in the left hand column [1], against a time scale along the top of the chart [2]. The scheduling of each activity is represented by a horizontal line or bar, from the activity's start date [3] to the activity's finish date [4]. The length of the activity line is proportional to its estimated duration. Consider the following list of activity data from a garage building project (table 13.1):

Activity Number	Activity Description	Duration	Start Date	Finish Date
1000	Lay Garage Foundations	4 Days	1 May	4 May
2000	Build Garage Walls	5 Days	5 May	9 May
3000	Fit Garage Roof	4 Days	10 May	13 May
4000	Install Garage Doors	2 Days	14 May	15 May

Table 13.1: Activity Data – shows a garage building project

The calendar time scale is usually presented in days or weeks, but hours, months and years are also possible. The examples here will use days (see figure 13.2).

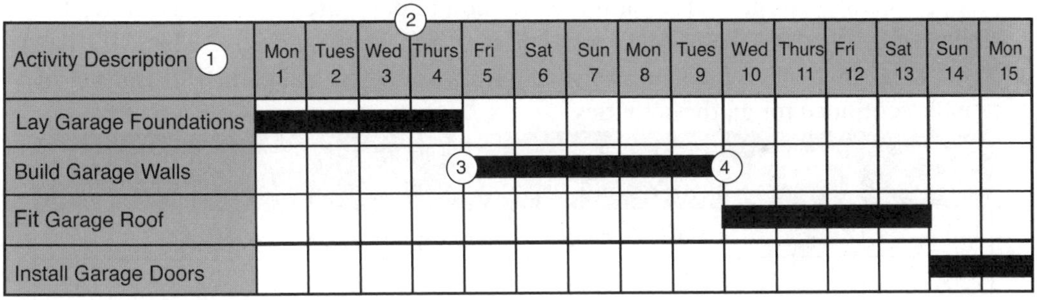

Activity Description (1)	Mon 1	Tues 2	Wed 3	Thurs 4	Fri 5	Sat 6	Sun 7	Mon 8	Tues 9	Wed 10	Thurs 11	Fri 12	Sat 13	Sun 14	Mon 15
Lay Garage Foundations															
Build Garage Walls				(3)					(4)						
Fit Garage Roof															
Install Garage Doors															

Figure 13.2: Simple Gantt Chart – shows a garage building project

Although the Gantt chart looks like a simple document, it does contain a wealth of information. For example, the list of activities and sequence (logic) will have been defined and gathered from the technical experts and the project team members' own field of expertise. The activities' start and finish dates imply knowledge or assumption of procurement delivery dates, resource loading, resource availability, funds availability and cashflow.

2. Tabular Reports

Tabular reports provide an excellent structure to store and present project information; and provide an important link between the CPM time analysis and the Gantt chart. On complex projects it is essential to develop the CPM before developing the Gantt chart, as the CPM is the best document to establish the logical sequence of work. Figure 13.3 shows the network diagram developed in the *CPM* chapter as figure 12.18. The data is transferred from figure 13.3 to the tabular report table 13.2.

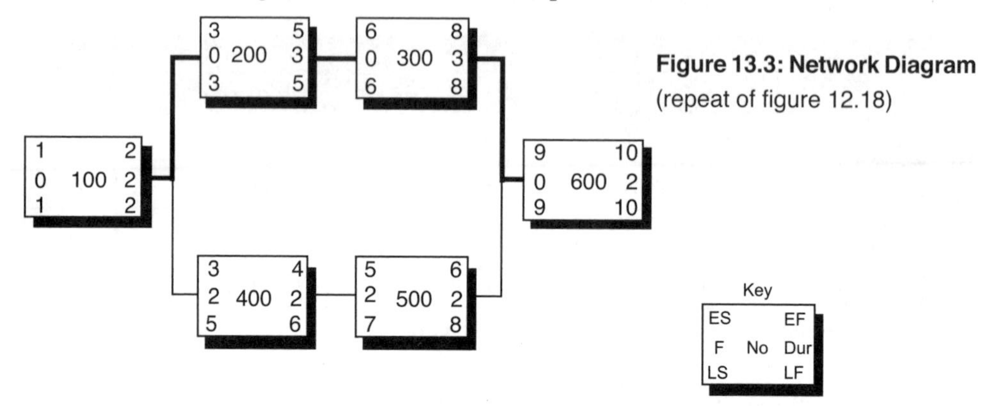

Figure 13.3: Network Diagram (repeat of figure 12.18)

Consider the key in figure 13.3, the seven parameters within the key become the headings in table 13.2, to which responsibility has been added. Starting with activity 100, the activity number (100) and the duration (2 days) are transferred to the table followed by the dates, ES (1 May), EF (2 May), LS (1 May), and LF (2 May). The last item is the float which indicates if the activity is on the critical path. This transfer of data is continued for all the activities.

Activity Number	Duration	Early Start	Early Finish	Late Start	Late Finish	Float	Responsibility
100	2	1 May	2 May	1 May	2 May	0	Glynn
200	3	3 May	5 May	3 May	5 May	0	Warren
300	3	6 May	8 May	6 May	8 May	0	Jan
400	2	3 May	4 May	5 May	6 May	2	Jim
500	2	5 May	6 May	7 May	8 May	2	Tony
600	2	9 May	10 May	9 May	10 May	0	Trish

Table 13.2: Tabular Report (the data was transferred from figure 13.3)

Tabular Report to Gantt Chart: The tabular data (table 13.2) is transferred on to a Gantt chart (figure 13.4). Starting with activity 100, the bar is drawn from the ES date (1 May) to the EF date (2 May). This is continued for all the activities.

Figure 13.4: Scheduled Gantt Chart (drawn from table 13.2)

Activity Number	Mon 1	Tue 2	Wed 3	Thu 4	Fri 5	Sat 6	Sun 7	Mon 8	Tue 9	Wed 10
100	■	■								
200			■	■	■					
300						■	■	■		
400			■	■						
500					■	■				
600									■	■

3. Activity Float

The Gantt chart presentation can also show the activity float. Figure 13.5 shows the accepted presentation with the float at the end of the activity from early finish (EF) to late finish (LF), and is denoted as a dotted line with a symbol at the end (usually a diamond or an upturned triangle).

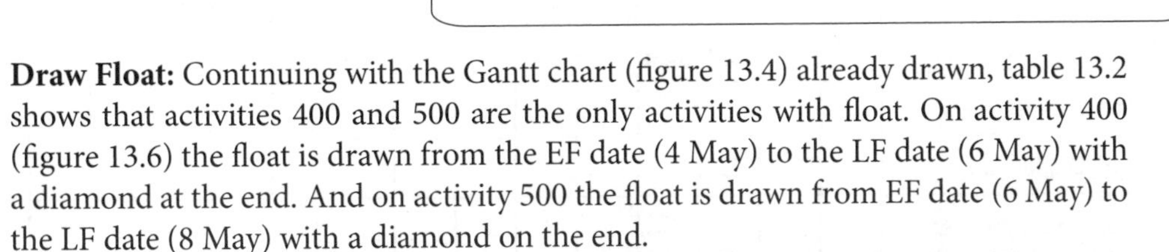

Figure 13.5: Activity Float

Draw Float: Continuing with the Gantt chart (figure 13.4) already drawn, table 13.2 shows that activities 400 and 500 are the only activities with float. On activity 400 (figure 13.6) the float is drawn from the EF date (4 May) to the LF date (6 May) with a diamond at the end. And on activity 500 the float is drawn from EF date (6 May) to the LF date (8 May) with a diamond on the end.

Figure 13.6: Scheduled Gantt Chart - shows activity float

Activity Number	Mon 1	Tue 2	Wed 3	Thu 4	Fri 5	Sat 6	Sun 7	Mon 8	Tue 9	Wed 10
100	■	■								
200			■	■	■					
300						■	■	■		
400			■	■	---	---◆				
500					■	■	---	---◆		
600									■	■

By implication it may be assumed that any activity without float is on the critical path. However, in practice, planners are always reluctant to show float as it is only human nature for people to delay work to the last minute which makes all their activities critical. It is better to try and make everyone start at the early start date (ES) and keep the float as a contingency.

4. Hammock Activities

A hammock or summary activity is used to gather together a number of sub-activities into one master activity. This is a useful technique for layering or rolling-up the project's schedule. Consider the garage building project from the beginning of this chapter. Table 13.3 and figure 13.7 show how the first activity 1000 (lay the garage foundation) can be subdivided or broken down into three tasks; mark out the foundations, dig the foundations and throw the concrete.

Activity Number	Activity Description	Duration	Start Date	Finish date
1000	Lay Garage Foundation	4 days	1 May	4 May
1001	Mark Out Garage Foundation	1 day	1 May	1 May
1002	Dig Garage Foundation	2 days	2 May	3 May
1003	Throw concrete	1 day	3 May	4 May

Table 13.3: Hammock Activities – shows the included sub-activities

Activity Description	Mon 1	Tues 2	Wed 3	Thurs 4	Fri 5	Sat 6	Sun 7	Mon 8	Tues 9	Wed 10	Thurs 11	Fri 12	Sat 13	Sun 14	Mon 15
1000 Lay Garage Foundations															
1001 Mark out garage															
1002 Dig foundation															
1003 Throw concrete															
2000 Build Garage Walls															
3000 Fit Garage Roof															
4000 Install Garage															

Figure 13.7: Hammock Gantt Chart – shows the hammock bar drawn from the start of the earliest activity to the end of the latest included activity - the sub activities are shown as a white bar

With hammock activities the Gantt chart can now be presented at the appropriate level of detail. Less detail is required by the senior managers who only need to see the overall picture of the project and key dates. But more detail is required by the project team members who are carrying out the work. The ability to vary the level of detail is a fundamental feature of project planning and control.

5. Events, Key dates, Milestones and Deadlines

The principle difference between an activity and an event is that an event has zero duration - it is a point in time. An event, also referred to as key date, milestone or deadline, represents a happening on a particular day. This could be when the order is placed, the design plans are approved, goods are received, or even the start and finish dates of an activity. An event has the following characteristics:

Zero Duration	An event has zero duration, it is a point in time. In Microsoft Project, for example, an event is an activity with zero duration and would appear on the screen as a diamond symbol.
Start or Finish	An event could be the start or finish of an activity, the WBS work package, the project phase or the project itself.
Checkpoint	An event focuses the project on a checkpoint, a major accomplishment, a deliverable result, a stage payment or an approval to proceed.
Interface	An event could be the interface between trades or contractors as one hands over to the other.
Data Capture	Data capture will be more accurate if the scope of work is subdivided into a number of milestones.

Managing events, or management-by-events, gives a clear focus on when work must be completed and hence a clear measure of progress. Using plenty of milestones pushes the planning down a level and empowers the project team members. Consider the painting example shown as an event Gantt chart (figure 13.4).

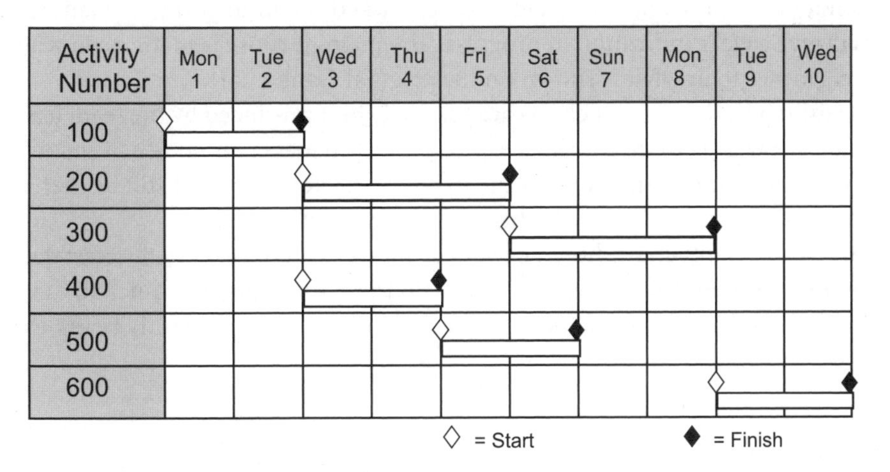

Figure 13.8: Event Gantt Chart – shows events at the start and finish of all the activities

The milestone schedule offers another type of planning presentation which can be used on its own or in conjunction with a scheduled Gantt chart.

6. Rolling Horizon Gantt Chart

The rolling horizon Gantt chart, or rolling wave Gantt chart is a simplified version of the Gantt chart which focuses on the activities that are currently working or will be working in the next few weeks. This short period or time window may be two or three weeks ahead for the activities being worked on, and four weeks ahead for pre-planning (making sure all the drawings, procedures, job cards, equipment, materials etc. are going to be in place). This type of Gantt chart lends itself to a manual presentation as the scope of work is limited to just the activities that are currently being worked on.

The rolling horizon Gantt chart can be partly computerized by preparing a planning template, as shown below (figure 13.9), which includes the scope of work (scheduling is optional). The production manager or foreman can then draw up the project schedule by hand.

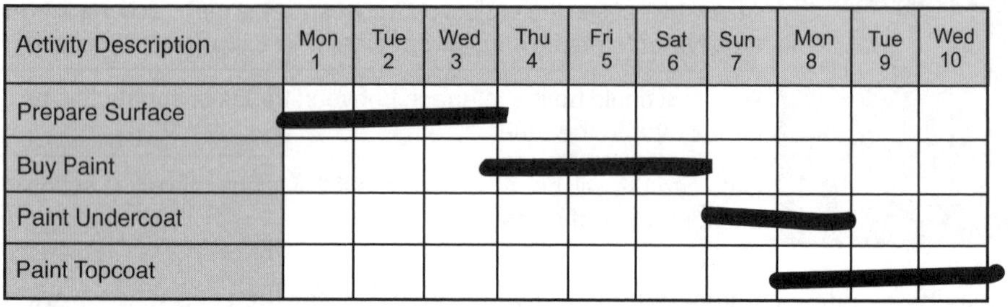

Activity Description	Mon 1	Tue 2	Wed 3	Thu 4	Fri 5	Sat 6	Sun 7	Mon 8	Tue 9	Wed 10
Prepare Surface										
Buy Paint										
Paint Undercoat										
Paint Topcoat										

Figure 13.9: Rolling Horizon Gantt Chart – shows how a Gantt chart can be marked up by hand for the next two or three weeks

The rolling horizon Gantt chart can also be marked up on the original Gantt chart - this will give a clear indication of progress. The main purpose is to focus on what can be done, rather than what is shown on the original Gantt chart.

The rolling horizon Gantt charts are generally best produced by the team leader or site manager who is at the work face. The project manager should then incorporate the information into a master schedule to check the key dates will still be met and, if they have changed, communicate this to the interested parties.

This type of Gantt chart should be very accurate as it is based on the latest data and drawn up by someone who is working close to the action. It is very quick to draw and only includes relevant information on the activities that are currently being worked on.

7. Revised Gantt Chart

The Gantt chart was originally designed as a planning and control tool where the actual progress was marked up against the original plan. This progress bar can be drawn above, inside or underneath the original bar. This way the project manager can see at a glance how each activity is progressing and where control might be required to guide the project to completion.

Baseline Plan: For effective control the project's baseline plan must be frozen, with the only changes being those approved by the scope change control system. Without a baseline plan it would be difficult to calculate progress variances and control would be lost.

Data Capture: To draw the revised Gantt chart the project manager needs to capture progress information. This can be achieved by setting up a progress reporting sheet. Consider the following (table 13.4):

Project:		Report Date:	2nd May			
WBS	**Activity Number**	**Description**	**Start Date**	**Finish Date**	**Percentage Complete**	**Remaining Duration**
	1000		1 May		50%	2 days
	2000				0%	
	3000		1 May		75%	1 day
	4000				0%	
	5000		1 May		25%	3 days
	6000				0%	

Table 13.4: Data Capture Template

The progress report should start with the reporting date or timenow. The progress should be reported against a WBS number or an activity number. The start and finish dates are important milestones which clearly state if the activity has started and if the activity has finished.

To draw the revised Gantt chart the project manager needs to know the remaining duration as the Gantt chart scale is a time scale not a percentage complete scale. Percentage complete and remaining duration are often used interchangeably, but this can be misleading or inaccurate. Consider the following example:

When the project manager reports that a 12 day activity will take another 6 days to complete, he has implicitly stated that the material, resources and funds are available to complete the work. However, if the project manager reports the progress is 50% complete, this will confirm what has been done, but say nothing about when he intends to complete the activity.

If an activity has a long duration it may be more convenient to initially report percentage complete and assume the remaining duration. But as the activity nears completion the project manager will have a more accurate feel for the remaining duration as opposed to the percentage complete. For example, if painting a house is one activity, then as the work progresses through the house it may be easier to say each room is 5% of the work. But as the painter starts painting the last few rooms it should be more accurate to state how many days it will take to finish.

Activity No	Mon 1	Tue 2	Wed 3	Thu 4	Fri 5	Sat 6	Sun 7	Remaining duration	Percentage complete	Comments
1000								2	50%	On time
2000								2	0%	Not started
3000								1	75%	1 day ahead
4000								2	0%	Not started
5000								3	25%	1 day behind
6000			Timenow					2	0%	Not started

Figure 13.10: Revised Gantt Chart - shows the progress at timenow 2nd May

Figure 13.10 shows the progress drawn relative to timenow. The progress analysis shows that activity 1000 has a remaining duration of two days, which means the activity is on time and no control is required. Activity 3000 has a remaining duration of one day, which means it is one day ahead and no control is required. Activity 5000 has a remaining duration or three days, which means it is one day behind and control is required to speed up the progress to finish the activity on time. The revised Gantt chart is easy to mark up by hand and clearly indicates progress against the original baseline plan and, therefore, where control is required. If the progress over the successive weeks is marked up on the same document, progress trends can also be established.

If the project manager is intending to apply control to bring the project back on track the knock-on impact to the end date of the project is of little interest. However, if any key dates are delayed this could disrupt the project. The disruption on the key dates, in the short term, can be identified by using a three week rolling horizon Gantt chart (see previous section). As the work progresses there might be changes to the project's activity logic. These should be reported so the CPM's network diagram can be updated.

Exercises:

1. Discuss the benefits of using Gantt charts to plan and control your projects.
2. Given figure 13.11, draw the Gantt chart (see Appendix 1 for the solution).

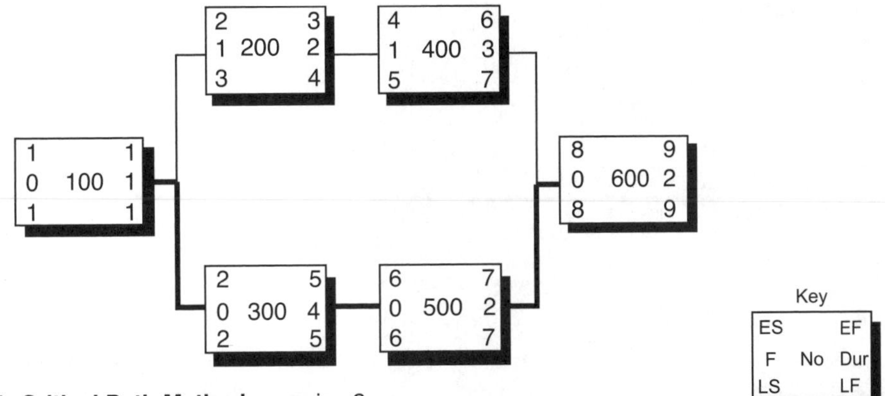

Figure 13.11: Critical Path Method, exercise 2

3. Given figure 13.12, draw the Gantt chart (see Appendix 1 for the solution).

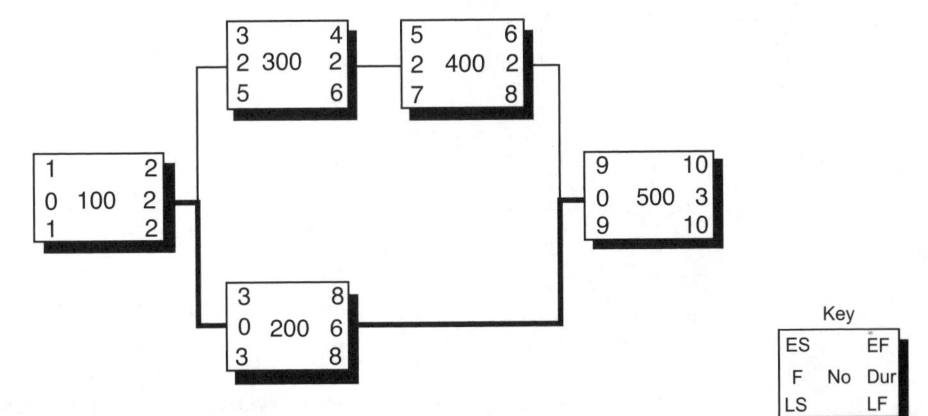

Figure 13.12: Critical Path Method, exercise 3

4. Discuss the benefits of using milestone planning and give examples based on your experience in a project environment.
5. Discuss the benefits of using hammock activities and give examples based on your experience in a project environment.

Further Reading:

Burke, R., *Project Management Techniques*, will develop the following topic which relates to this knowledge area:

- Select and sort activities.

Note: Microsoft conducted a survey recently of their Microsoft Project software users - the findings indicated that 90% of the users preferred the Gantt view in preference to the network diagram.

14

Procurement Schedule

Learning Outcomes

After reading this chapter, you should be able to:

- Develop a procurement list
- Administer the procurement process
- Develop the procurement schedule
- Expedite procured orders

The *Procurement Management* knowledge area includes the techniques to acquire the goods and services from contractors and suppliers, outside of the project organization. This could be drawings, materials, components, equipment or professional services required to perform and complete the project's scope of work. The procurement schedule will link the procurement bill of materials (BOM) with the CPM project schedule.

This chapter will explain how to manage the procurement cycle and calculate the procurement schedule.

The PMBOK 4ed defines **Project Procurement Management** as, *the processes required to purchase or acquire products, services, or results needed from outside the project team to perform the work.*

The APM BoK 5ed defines **Procurement** as, *the process by which the resources (goods and services) required by the project are acquired. It includes development of the procurement strategy, preparation of contracts, selection and acquisition of suppliers, and management of the contracts.*

Project Management Plan Flowchart

The project management plan flowchart (figure 14.1) shows the relative position of the procurement schedule with respect to the other topics. The procurement schedule should be considered after the Gantt chart has been established, but before the resource histogram and project cashflow statement are calculated. This is because it is important to identify the long lead items (particularly those items on the critical path), as the work cannot start until the materials are available.

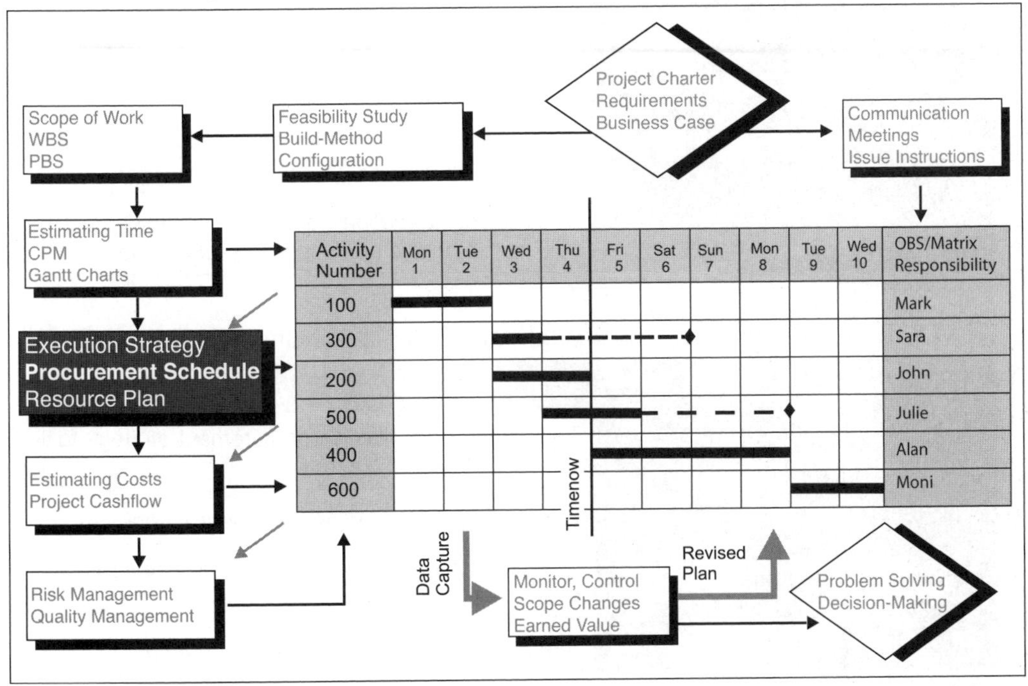

Figure 14.1: Project Management Plan Flowchart – shows the relative position of the procurement schedule with respect to the other topics

Purpose of this Chapter: The purpose of this chapter is to help project managers and project team members acquire the competence and knowledge to be able to produce, co-ordinate and administer the procurement schedule within projects of limited complexity. This chapter will explain:

- How to develop the procurement list
- How to link the procurement list with the activities schedule
- How to identify any long lead items
- How to administer and expedite the procurement process.

Body of Knowledge Mapping

PMBOK 4ed: The *Procurement Management* knowledge area includes the following:

PMBOK 4ed	Mapping
12.1: Plan Procurements	This chapter will explain how to develop the procurement list from the WBS and develop the procurement schedule from the project schedule.
12.2 Conduct Procurements	This chapter will explain how to invite the suppliers to tender, adjudicate the tenders and award the contracts.
12.3: Administer Procurements	This chapter will explain how to expedite the procurement process.
12.4: Close Procurements	See *Project Integration Management* chapter which will explain how to produce a closeout report.

APM BoK 5ed: The *Procurement Schedule* knowledge area includes the following:

APM BoK 5ed	Mapping
5.4: Development of the procurement strategy	This chapter will discuss procurement strategy to buy, or make, or to contract.
5.4: Preparation of contracts	See **Burke**, R., *Project Management Techniques.*
5.4: Selection and acquisition of suppliers	This chapter will discuss selection and acquisition of suppliers.
5.4: Management of the contracts	See **Burke**, R., *Project Management Techniques.*

Unit Standard 50080 (Level 4): The unit standard for the *Procurement* knowledge area includes the following:

Unit 120386: *Provide procurement administration support to a project*

Specific Outcomes	Mapping
120386/SO1: Compile and process procurement requests to required standards and needs.	This chapter will explain how to manage the procurement process.
120386/SO2: Source suppliers/ sellers to meet procurement requirements.	This chapter will explain how to identify and pre-qualify potential suppliers in accordance with the project's quality plan.
120386/SO3: Receive and evaluate proposals and make recommendations.	This chapter will explain how to adjudicate the tenders in accordance with the accepted company criteria and the risk management plan.
120386/SO4: Maintain and administer procurement records.	This chapter will explain how to maintain and administer procurement records.

1. Procurement Process

The procurement process can be effectively presented as a flowchart outlining a series of steps. In practice, there might be a number of iterations between certain steps and, also, the steps could be carried out in a different order.

Scope of Work: In figure 14.2 the scope of work is the starting point of the procurement process, as it is the WBS work packages that identify the complete list of jobs together with their associated materials, components and equipment requirements.

Procurement Planning: The procurement planning will be viewed from the perspective of the project team's buyer within the project office. Procurement planning is the process of identifying what products and services are best procured from outside the project organization. This is the **buy-or-make** decision which is a key component of the execution strategy that considers:

- What to procure
- How much to procure
- When it is required
- When to procure
- Where to procure
- How to procure (type of contract).

From this material list the project manager must decide on the execution strategy to buy-or-make. This decision might require input from external experts and suppliers to assess market conditions together with input from internal experts to assess company expertise and resource loading. The buy-or-make decision is based on:

Buy the Components	When the company resources lack the expertise and machinery, or when the resources are overloaded, or when an outside sub-contractor makes an offer the company cannot refuse.
Make the Components	When the company resources, expertise and machinery are available and under utilized, and the internal costs are less than when using outside contractors or outsourcing.

Table 14.1: Buy-or-Make Decision

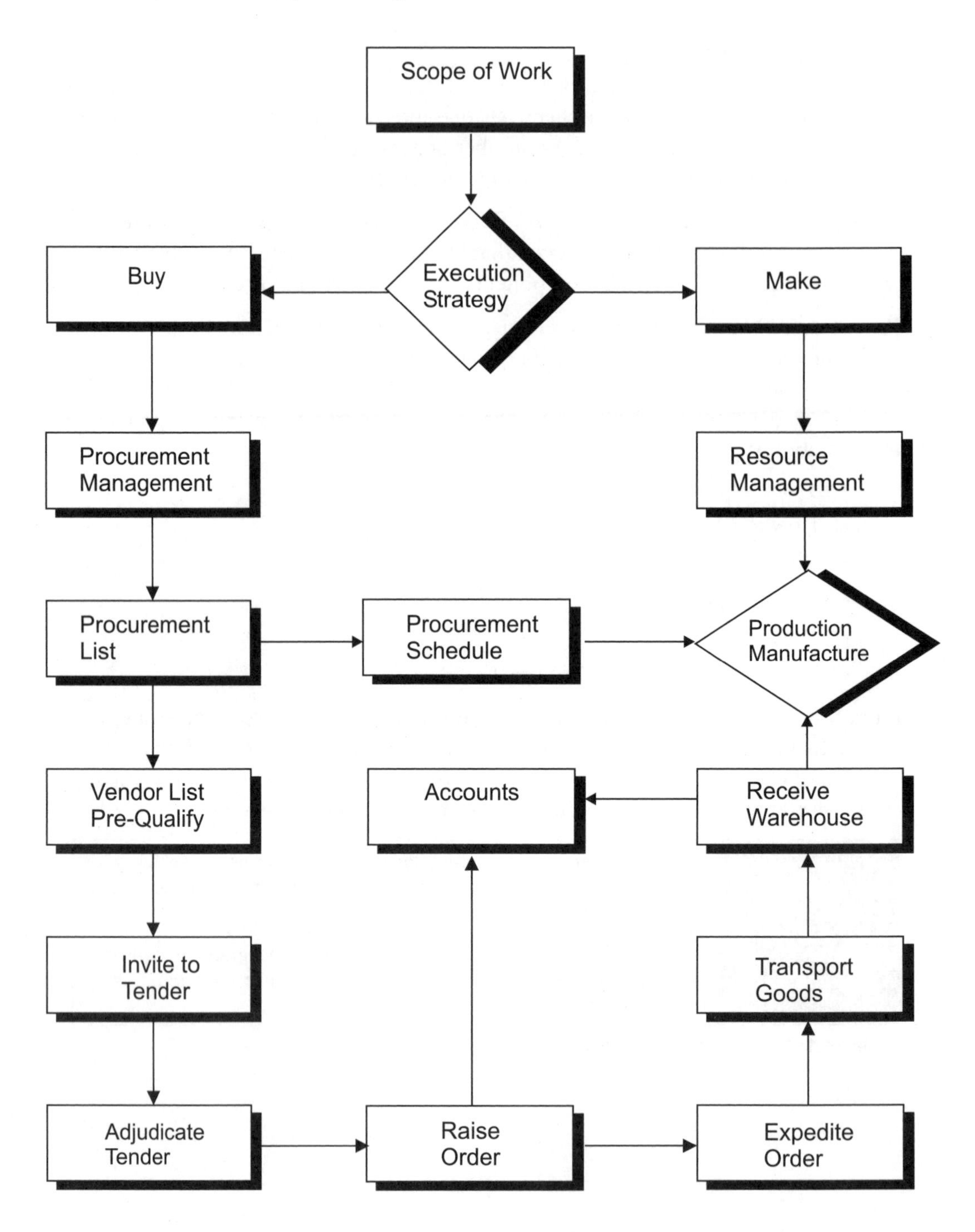

Figure 14.2: Procurement Process – shows the relationship between the procurement topics

Procurement List: The procurement list is developed from the project's scope of work and compiled in the agreed format and to the agreed time frame. On a small project this could be an expansion of the WBS (see table 14.2 below). Against each work package or activity, the material list should give all the product details; manufacturer, model number, specification, type, colour, rating, level of inspection, etc.

WBS	Description Specification	Material Components	Services
100			
200			
300			

Table 14.2: Procurement List – shows the link between the WBS and the procurement topics

Procurement Schedule: The procurement schedule integrates the procurement list with the schedule Gantt chart. This will highlight the '**order-by-date**' to meet the project schedule. This will be discussed in the next section.

Supplier and Vendor List: All potential suppliers need to be identified and pre-qualified according to criteria set out in the project quality plan. The project manager needs to be satisfied that the suppliers have the production and quality management systems to deliver the product to the required specification, quality standards and schedule. The reputation and financial stability of the company should also be considered. For example, it would be sensible to check the quality of the products the company has previously made.

Vendor List	Date Pre-Qualified	Comments
Vendor 1		
Vendor 2		
Vendor 3		

Table 14.3: Vendor List

Invitation to Tender: The tender documents or bid package should be sent to the suppliers (on the short list) within the stipulated time frame. As these initial enquiry documents are the basis for the contract, they need to be progressively adjusted and marked up as more information becomes available.

Tender Adjudication: The tender adjudication process scrutinizes the quotations (tenders) by compiling a technical and commercial bid tabulation (spreadsheet) of the quotations. This ensures the buyer is comparing 'apples with apples' and meeting the selection criteria established by the company. The adjudication process should consider the suppliers' suggestions and always be prepared to negotiate to achieve the best price and conditions, while striving for a win-win arrangement. This process should take advantage of market conditions and accommodate any extension requests.

The adjudication process should identify the risks associated with the supply of the goods or services, and document the risks. These risks should be communicated to the relevant parties. Procurement documentation is processed within confidentiality requirements and in accordance with company policies and procedures.

Raising the Order: The purchase order should be raised using the company's standard terms and conditions of contract, where the terms and conditions of contract are signed off by the company lawyer. The purchase order should be a stand-alone document, superseding all previous documents and correspondence, and must be formally accepted by the supplier.

Where the client imposes contractual requirements on the project these need to be passed on with back-to-back agreements with the suppliers. For example, the project manager does not want to be held liable for a 'schedule penalty' which cannot be passed on to the supplier who has caused the delay in the first place. For this reason, it is risky to finalise the contracts with the sub-contractors before signing a contract with the client.

Expedite: The expediting process follows up on the purchase orders to ensure and encourage the suppliers to meet their contractual requirements (particularly quality and delivery). Expediting makes the order happen (see section later in this chapter). On-going monitoring of supplier delivery should be conducted in terms of procurement agreement.

Transport: The transport process considers the different methods of moving the materials and components from the supplier to the project. Where possible the shipment's quality should be checked and confirmed before leaving the factory - this particularly applies to items being imported or exported.

'Sorry for the late delivery. I hope you haven't been waiting too long!'

Receiving: The receiving process checks the procured goods into the company. This involves checking the goods against the delivery note, checking the delivery note against the purchase order, and checking the quality of the products against the required condition. Any non-conformances should be identified, recorded and reported in accordance with company policies and procedures. The reported non-conformances should then be resolved in accordance with company policies and procedures.

Date	Receiving and Inspection	Purchase Order	Specification	Delivery	Inspection
		PO 1			
		PO 2			
		PO 3			

Table 14.4: Receiving Inspection Control Sheet

Warehousing: The warehouse facility stores the delivered goods for safe keeping and prompt retrieval. The procurement process should confirm if the warehousing requires any special handling equipment and storage facilities. For example, the goods might be delivered on a pallet and require a forklift truck to off-load, or the goods might be perishable and require refrigeration.

Warehouse No.	Purchase Order	Handling Receiving	Storage
100	PO 1	Fork lift	Refrigeration
200	PO 2	Craneage	Ventilation
300	PO 3	Manually (by hand)	Temporary container

Table 14.5: Warehouse Control Sheet

Accounts: The accounts department's job in the procurement process is to check the budget, the purchase order, the invoice and the delivery note for variances before making payment. Table 14.6 shows how the variances can be highlighted; in this case, WBS 2000 with a ($500) variance would attract attention.

1. WBS	2. Description	3. Budget	4. Purchase Order	5. Actual Cost	6. Variance
1000	Buy Computer	$10,000	PO 1001	$8,500	$1,500
2000	Buy Materials	$5,000	PO 2001	$5,500	($500)
3000	Website Design	$3,000	PO 3001	$3,000	$0

Table 14.6: Purchase Order Control Sheet – shows the list of items to be procured linked to the WBS. The actual costs are compared to the budget and any variance is reported to the responsible person

Document Storage: All procurement requests, records and documents should be collated, filed and maintained according to project office policies and procedures.

2. Procurement Schedule

The procurement schedule is developed after the CPM and Gantt charts have established a logical sequence of work and a schedule but, in most cases, before the resource histograms and project cashflow statements have been considered. The sequence of the procurement schedule, resource histogram and project cashflow might be different on different projects for the following reasons:

- Consider procurement first when there are long lead items that could delay critical activities.
- Consider the resource histogram first when there are resource constraints that could delay critical activities.
- Consider project cashflow first when there are funding constraints that could delay critical activities.

Procurement Control Sheet: In figure 14.7, the procurement control sheet integrates the following parameters:

- Column 1: The WBS scope of work
- Column 2: The procurement list/purchase order (PO) (goods to buy)
- Column 3: The procurement lead time (delivery time)
- Column 4: The warehousing just-in-time (JIT) stock control
- Column 5: The project schedule (Gantt chart, when work is scheduled to start)
- Column 6: The order-by-date to meet the schedule
- Column 7: The actual delivery date
- Column 8: The schedule variance.

Procurement Schedule Example: Consider the following example of three sets of two sequential activities (100 and 200, 300 and 400, 500 and 600) to show three possible situations. Table 14.7 shows the procurement schedule in a tabular format, and figure 14.3 shows the procurement schedule in a Gantt chart format.

1. WBS	2. PO	3. Lead Time (LT)	4. JIT	5. Early Start (ES)	6. Order-by-Date (ES - LT - JIT - 1)	7. Delivery Date (DD)	8. Variance (ES - DD)
100	PO 1	3	2	7	1 May		
200				9			
300	PO 2	3	3	7	30 April		
400				9			
500	PO 3	2	2	7	2 May		
600				9			

Table 14.7: Procurement Schedule Table - shown in a tabular format

Activity Number	Procurement Status	Mon 1	Tues 2	Wed 3	Thur 4	Fri 5	Sat 6	Sun 7	Mon 8	Tues 9	Wed 10
100	Delivered on time	OBD ◆		Lead Time		JIT		ES			
200											
300	Delivered late	OBD ◆		Lead Time		JIT		ES			
400											
500	Delivered early	OBD ◆	Lead Time		JIT			ES			
600											

Figure 14.3: Procurement Schedule - shown in a Gantt chart format

Where **'lead time'** is the time from the 'order date' (or order-by-date) to the 'delivery date', this could be an issue for goods coming from overseas. **JIT** is an internal requirement - how long does the project manager want to have the goods on site before they are required? For example, the warehouse and quality department might require a few days to unpack and test the goods.

Analysis: The analysis from table 14.7 and figure 14.3 shows that activity 100's procurement profile has a lead time of 3 days and a JIT of 2 days. This means that if the order is placed by 1st May then the goods will be delivered by 6th May and be available for the activity's early start on the 7th May.

Activity 300 shows what happens when the goods are delivered late. Activity 300's procurement profile has a lead time of 3 days and a JIT of 3 days. This means that if the order is placed on the 1st May then the goods will be delivered on the 7th May the day activity 300 is scheduled to start. This means activity 300 cannot start until the 8th May. If activity 300 is on the critical path, this will delay the project's completion date.

Activity 500's procurement profile has a lead time 2 days and a JIT of 2 days. This means that if the order is placed on the 1st May then the goods will be delivered on the 5th May and available on the 6th May, which gives the procurement one day float.

Order-By-Date: In practise, the order-by-date is calculated first to ensure the materials are delivered before the activity's early start by using the following equation:

Order-by-date = ES - LT - JIT - 1

Activity 100 = 7 - 3 - 2 - 1 = 1 May

Activity 300 = 7 - 3 - 3 - 1 = 30 April

Activity 500 = 7 - 2 - 2 - 1 = 2 May

By setting up a procurement table similar to table 14.7, this gives the project manager a simple summary document to plan and control the procurement process.

House Extension Example: Consider the Gantt chart for a house extension project where figure 14.4 (copied from figure 13.6) shows when the work is planned, and table 14.8 shows when the materials are required with respect to the early start of each activity.

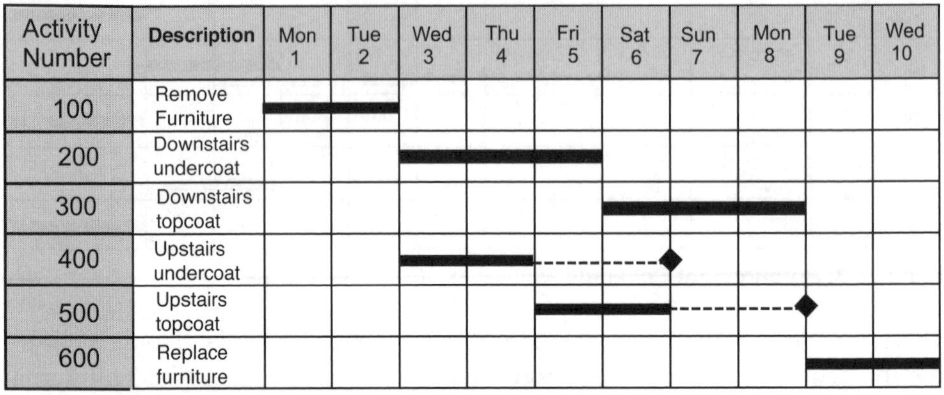

Activity Number	Description	Mon 1	Tue 2	Wed 3	Thu 4	Fri 5	Sat 6	Sun 7	Mon 8	Tue 9	Wed 10
100	Remove Furniture										
200	Downstairs undercoat										
300	Downstairs topcoat										
400	Upstairs undercoat										
500	Upstairs topcoat										
600	Replace furniture										

Figure 14.4: Gantt Chart - House Extension Schedule

Procurement Table: In figure 14.8, the procurement table integrates the following parameters:

- Column 1: The WBS list of work packages or activities.
- Column 2: The description of the scope of work.
- Column 3: The description of the goods to be procured – what needs to be bought-in from outside the project organization.
- Column 4: The date the materials are required to be delivered to meet the project's schedule. In this case, it is one day before the early start date (ES) of the activity.
- Column 5: The forecasted delivery date (given by the procurement buyer which is based on the lead time and JIT not shown in this example).
- Column 6: The variance between delivery date and required date. Any negative variances means the goods will be delivered after the planned start.

1. WBS	2. Description	3. Material	4. Required Date (ES – JIT)	5. Delivery Date	6. Variance
100	Remove furniture				
200	Downstairs undercoat	Paint 1	2 May (Tuesday)	2 May	0
300	Downstairs top coat	Paint 2	5 May (Friday)	6 May	-1
400	Upstairs undercoat	Paint 3	2 May (Tuesday)	3 May	-1
500	Upstairs top coat	Paint 4	4 May (Thursday)	4 May	0
600	Replace furniture				

Table 14.8: Procurement Table – shows the difference between the delivery date and the required date

From the variance column [6], table 14.8, the deliveries of the paint for activities 200 and 500 are on time, but the deliveries for activities 300 and 400 are going to be 1 day late. The schedule Gantt chart (figure 14.4) shows activity 300 is on the critical path, so if it is not possible to organize a faster delivery or shorter duration this will extend the completion date of the project. On the other hand, activity 400 has 2 days of float, which means this activity can accommodate a delivery 1 day late.

Trade-Off: This is where the procurement schedule trade-offs and compromises start. There are basically three options:

- Speed up the delivery of the goods to meet the schedule
- Reduce the duration of a critical activity to meet the completion date
- Change the logic to meet the completion date.

Revising the schedule to meet the procurement delivery dates impacts on the resource requirements and project cashflow. This means the resource histogram and project cashflow will also have to be recalculated and rescheduled.

Procurement Expeditor!

3. Expediting (Control)

The duties of the progress expeditor, also referred to as a progress chaser, are to follow up on all the purchase orders and instructions to the suppliers. On a large project the expeditor is an important team member who becomes the project manager's eyes and ears. On a small project the expeditor and project manager might be the same person. Consider the following expediting checklist:

Received Order	Confirm the contractor has received the purchase order and instructions.
Understand Order	Confirm the purchase order and instructions are understood.
Job Number	Confirm the supplier's job number.
Project Manager	Confirm who is the supplier's project manager or foreman.
Planned	Confirm the job has been planned into the supplier's production system. Confirm this by sighting their planning documents.
Drawings	Confirm the product has been designed. Confirm this by sighting the construction drawings and specifications.
Materials	Confirm the supplier has ordered the materials and components they require. Check the materials have been received and inspected.
Instructions	Confirm the instructions have been given to the supplier's foreman to perform the work.
Resources	Confirm the company has sufficient suitably qualified resources available. If necessary, confirm names and sight their qualifications.
Progress	Confirm the work is progressing as planned.
Problems	Confirm the supplier has all the relevant information and instructions, and there are no problems that could impact on progress or lead to a claim. Note the response in the minutes.
Delivery	Confirm the supplier is on track to meet the contracted delivery date.

By asking all these questions the expeditor becomes an invaluable source of progress information giving early warning of any supply problems so that the project manager has time to take effective action.

The APM BoK defines **Expediting** as, *facilitation and acceleration of progress by the removal of obstacles.*

Exercise:

1. Discuss how you would set up a project procurement system in a project environment.
2. Discuss how you would identify long lead items. Give examples of long lead times and how they have been addressed on your projects.
3. Discuss how you would expedite project procurement in a project environment.
4. Given the information in table 14.9, complete figure 14.5 to show the order-by-date, lead time and JIT for all the activities (see Appendix 1 for solution).

1. WBS	2. PO	3. Lead Time (LT)	4. JIT	5. Early Start (ES)	6. Order-by-Date (ES - LT - JIT - 1)	7. Delivery Date (DD)	8. Variance (ES - DD)
100	PO 1	3	2	9		8	
200	PO 2	4	0	9		8	
300	PO 3	3	3	9		8	
400	PO 4	4	2	9		8	
500	PO 5	4	4	9	1	9	
600	PO 6	2	2	9	1	5	

Table 14.9: Procurement Table

Activity Number	Procurement Status	Mon 1	Tues 2	Wed 3	Thur 4	Fri 5	Sat 6	Sun 7	Mon 8	Tues 9	Wed 10
100										ES	
200										ES	
300										ES	
400										ES	
500										ES	
600										ES	

Figure 14.5: Procurement Schedule Question Template

Further Reading:

Burke, R., *Project Management Techniques*, will develop the following topics which relate to this knowledge area:

- Type of contracts
- B2B procurement
- Just in time (JIT).

15

Resource Planning

Learning Outcomes After reading this chapter, you should be able to:
Draw the resource histogram
Smooth resource overloads

This chapter will explain how to produce the *Resource Histogram*. Resource planning integrates the requirements of the *Human Resource Management (HRM)* knowledge area and the *Time Management* knowledge area, where the HRM identifies the human resource requirements, and the time management links the resource requirements with the project schedule.

Project resources may be defined (in its widest sense) as the machinery and people who perform the scope of work. Resource planning is, therefore, forecasting the number of resources required to perform the scope of work to achieve the project schedule.

Project Management Plan Flowchart

The project management plan flowchart (figure 15.1) shows the relative position of the resource planning with respect to the time estimate, CPM, Gantt chart and the procurement schedule. The resource planning should be considered after the CPM, the Gantt chart and the procurement schedule have been developed, but before the cashflow statement. However, for projects where the procurement is mostly off-the-shelf, the resource planning might become the key parameter determining the projects schedule.

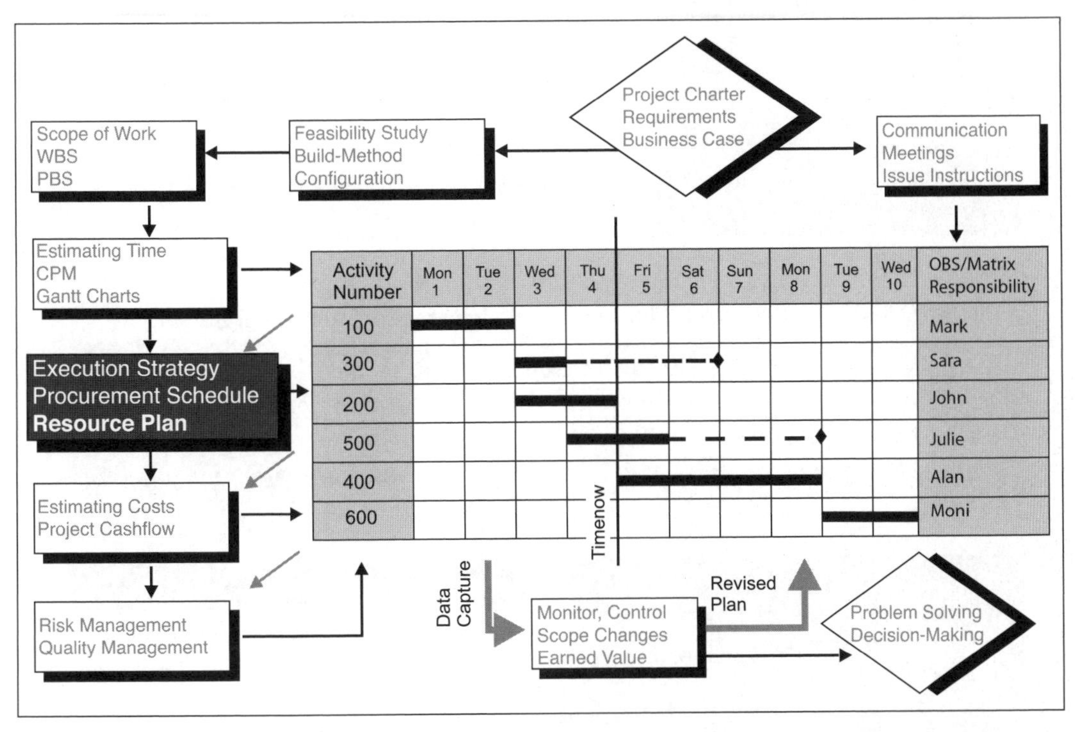

Figure 15.1: Project Management Plan Flowchart – shows the relative position of resource management with respect to the other topics

Body of Knowledge Mapping

PMBOK 4ed: The *Resource Planning* technique falls under the *Human Resource Management* knowledge area and the *Time Management* knowledge area which includes the following sections:

PMBOK 4ed	Mapping
9.1: Develop Human Resource Plan	This chapter will explain how to develop the resource histogram.
9.2: Acquire Project Team	See **Burke,** R., **Barron, S.,** *Project Management Leadership.*
9.3: Develop Project Team	See *Human Resource Management* (Project Teams) chapter.
9.4: Manage Project Team	See *Human Resource Management* (Project Teams) chapter.

APM BoK 5ed: The *Resource Planning* knowledge area includes the following:

> The APM BoK 5ed defines **Resource Management** as, *identifying and assigning resources to activities so that the project is undertaken using appropriate levels of resources and within an acceptable duration. Resource allocation, smoothing, leveling and scheduling are techniques used to determine and manage appropriate levels of resources.*

APM BoK 5ed	Mapping
3.3: Identifying and assigning resources to activities	This chapter will explain how to identify and assign resources to activities as a first step to produce the resource histogram.
3.3: Resource allocation	This chapter will explain how to allocate resources to activities and produce the resource histogram.
3.3: Resource smoothing, leveling	This chapter will explain how to smooth and level the resource profile to match resource availability with resource demand.
3.3: Resource scheduling	This chapter will explain how to adjust the project schedule to include resource requirements and resource availability.

Unit Standard 50080 (Level 4): There are no units for this subject area.

1. How to Draw the Resource Histogram

In the previous chapters on developing the CPM, the Gantt charts and the procurement schedule, the calculations assumed an unlimited supply of resources to be available as and when required. In reality, this is obviously not the case. This section will explain how to use the resource histogram to integrate the resource requirements with the project's schedule.

Estimating an activity's duration has already been explained in the *Time Management* chapter, where the activity's duration is determined by the trade-off between the amount of work and the resources available.

The resource histogram is a popular planning tool because it gives a powerful visual presentation of the resource loading which is easy to understand and easy to communicate. The prerequisites are:

- An early start Gantt chart (this considers the procurement requirements to ensure the materials and equipment are available for the work/job)
- A resource forecast (estimate) per activity per day.

By using the early start Gantt chart it is assumed that the project manager wishes to start all activities as soon as possible and keep the activity float for flexibility. Consider the following steps explaining how to draw the resource histogram:

Step 1	Collect data from the resource table which links the resources required per day to the activities (table 15.1)
Step 2	Draw the Gantt chart (see figure 15.2) (This example is copied from the *Gantt Chart* chapter figure 13.6)
Step 3	Transfer the resources per day from the table 15.1, to the Gantt chart figure 15.2 and insert the numbers into the activity bar.
Step 4	Add the resources per day vertically to give a total daily requirement.
Step 5	Plot the resource histogram (see figure 15.2).

Activity No	Description	Start Date	Finish Date	Resource Per Day
100	Remove furniture	1	2	4, 4
200	Paint downstairs, prep / undercoat	3	5	3, 1, 1
300	Paint downstairs, top coat	6	8	3, 1, 1
400	Paint upstairs, prep / undercoat	3	4	3, 1
500	Paint upstairs, top coat	5	6	3, 1
600	Replace furniture	9	10	4, 4

Table 15.1: Resource Table – shows the resources required per day linked to the activities

Activity Number	Mon 1	Tue 2	Wed 3	Thu 4	Fri 5	Sat 6	Sun 7	Mon 8	Tue 9	Wed 10	Thu 11
100	4	4									
200			3	1	1						
300						3	1	1			
400			3	1							
500					3	1					
600									4	4	
Total	4	4	6	2	4	4	1	1	4	4	

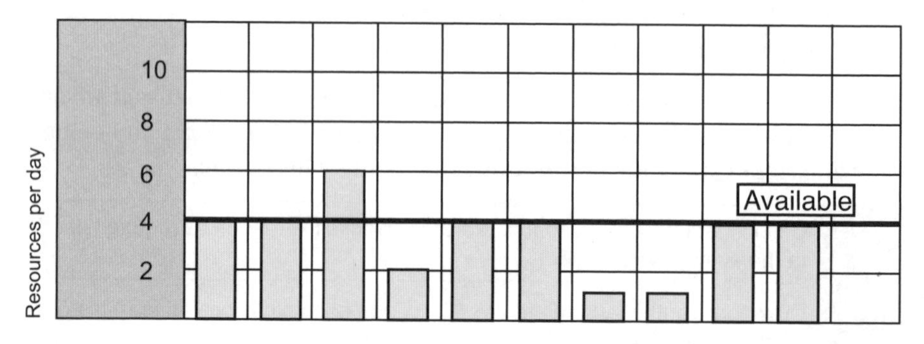

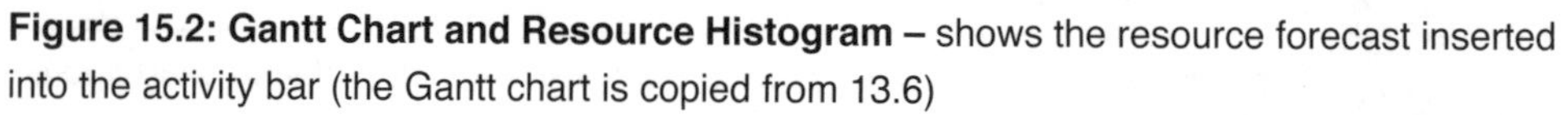

Figure 15.2: Gantt Chart and Resource Histogram – shows the resource forecast inserted into the activity bar (the Gantt chart is copied from 13.6)

Once the resource requirements have been added to the early start Gantt chart, the daily requirements are summed by moving forward through the Gantt chart one day at a time to give the total resource required per day.

The total daily resource requirements are then plotted vertically to give the iconic resource histogram. It is important to note that separate resource histograms are required for each resource type (painter, computer operator etc.).

2. Resource Loading

The resource forecast or requirement is now compared with the resources available. The ideal situation is achieved when the resource requirement equals the resources available. Unfortunately, in the real world this seldom happens, because it is not always possible to adjust supply to suit the demand, so the project manager either has to reschedule and/or adjust the number of resources.

A resource overload is when the resource forecast requirement exceeds the available resources; for example, if the planning requires 10 men, but only 8 men are available. While a resource underload is when the resource forecast is lower than the available resources; for example, if the planning requires only 10 men, but 12 men are available. A resource overload will lead to some activities being delayed, while a resource underload will under utilize the company's resources, which could have a negative impact on the company's profitability.

In figure 15.2 the resource histogram shows the total resource required per day and compares it with the resources available (4 in this case). On Wednesday 3rd May there is an overload (6 painters are required, but only 4 painters are available). However, on Thursday 4th May, only 2 painters are required, and on Sunday and Monday, only 1 painter is required. This situation would seem to offer the opportunity of moving the activities to balance the supply to the demand.

3. Resource Smoothing

Resource smoothing is the process of moving activities and/or adjusting the availability of the resources to improve the resource histogram profile. The project manager has a number of resource smoothing options - consider the following:

Resource Smoothing	Assign resources to critical activities first, because any delays to a critical activity will delay the whole project.
Time-Limited Resource Scheduling	If the end date of the project is fixed the resources must be increased to address any overloads.
Resource-Limited Resource Scheduling	If the maximum number of resources is fixed the end date may need to be extended to address any overload.
Increase Resources	Increase resources to address an overload.
Reduce Resources	Reduce resources to address an underload (under utilized).

Consider the painting example in figure 15.2, when there are only 4 resources available the resource histogram will be overloaded on 3rd May (Wednesday). This resource overload can be addressed by simply moving activity 400 by two days within the float available. However, as the network diagram, figure 15.3 (copied from 12.18) shows, activity 400 has a finish-to-start relationship with activity 500, therefore, activity 500 will also have to move forward by two days. The resulting Gantt chart and smoothed histogram are shown in figure 15.4.

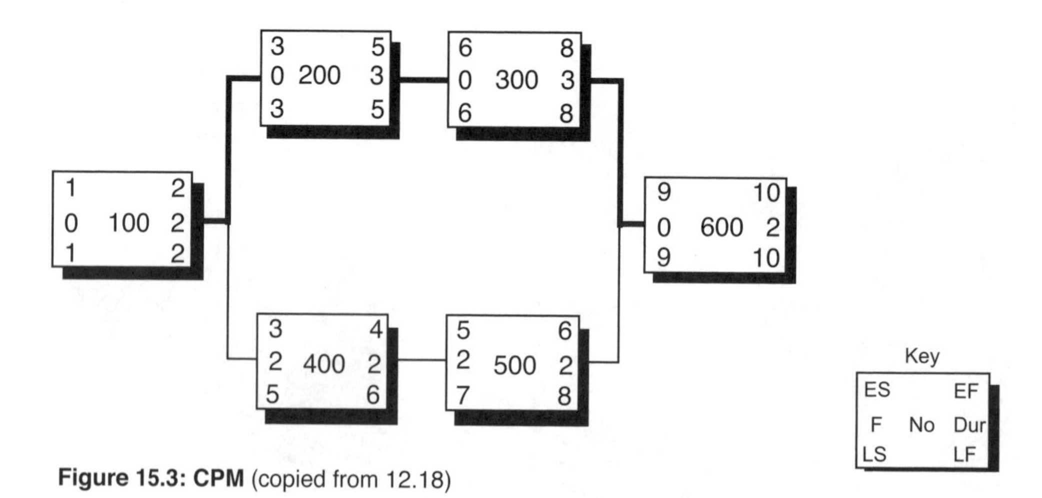

Figure 15.3: CPM (copied from 12.18)

	Activity Number	Mon 1	Tue 2	Wed 3	Thu 4	Fri 5	Sat 6	Sun 7	Mon 8	Tue 9	Wed 10	Thu 11
	100	4	4									
	200			3	1	1						
	300						3	1	1			
Move >>>>	400					3	1					
Move >>>>	500							3	1			
	600									4	4	
	Total	4	4	3	1	4	4	4	2	4	4	

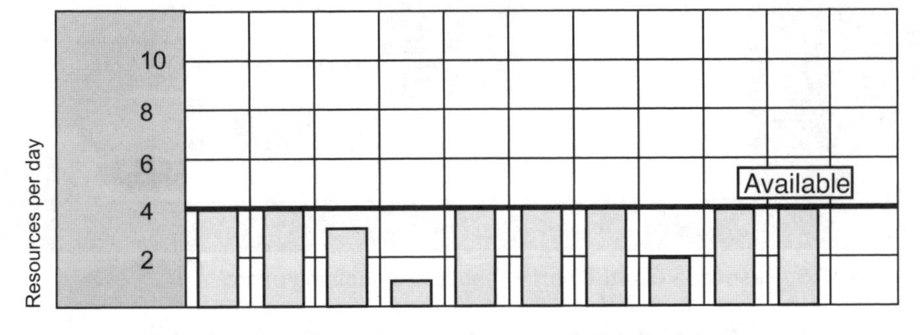

Figure 15.4: Smoothed Gantt Chart and Resource Histogram – shows the smoothed resource histogram after moving activities 400 and 500 within their float

Exercises:

Resource Smoothing Exercise 1: Given the resource Gantt chart (figure 15.5) draw the resource histogram and smooth the activities to give the best distribution. Assume all the activities can move within their float (see Appendix 1 for the solution).

	1	2	3	4	5	6	7	8	9	10	
100	7										
200		4	4	****	****	****	****	****			
300		4	4	****	****	****	****				
400		3	3	3							
500					3	3					
600				4	4	****	****	****			
700				4	4	****	****	****	****		
800							3	3	3		
900										7	
Totals	7	11	11	11	11	3	3	3	3	7	70

Figure 15.5: Resource Gantt Chart – resource smoothing exercise 1

Resource Smoothing Exercise 2: Given the resource Gantt chart (figure 15.6) draw the resource histogram and smooth the activities to give the best distribution. Assume all the activities can move within their float and the profile can change within the activity (see Appendix 1 for the solution).

	1	2	3	4	5	6	7	8	9	10	
100	5										
200		5	5	5	5	****	****	****	****		
300		4	4	****	****	****	****				
400		3	3	3	3	****	****	****	****		
500						5	5				
600						3	3				
700								3	3		
800										5	
900										3	
Totals	5	12	12	8	8	8	8	3	3	8	75

Figure 15.6: Resource Gantt Chart – resource smoothing exercise 2

Resource Smoothing Exercise 3: Given the resource Gantt chart (figure 15.7) draw the resource histogram. Assume the activity logic is the same as figure 15.3 (although the dates are different), smooth the activities for 5 resources (see Appendix 1 for the solution).

Activity Number	Mon 1	Tue 2	Wed 3	Thu 4	Fri 5	Sat 6	Sun 7	Mon 8	Tue 9	Wed 10	Thu 11
100	5	5									
200			5	5	2	2					
300							5	5	1	1	
400			3	3							
500					4	4					
600											6
Total	5	5	8	8	6	6	5	5	1	1	6

Figure 15.7 Resource Gantt Chart – resource smoothing exercise 3

4. Draw the resource histogram for your project and comment on the type of resource which is most likely to be overloaded and underloaded.
5. As project manager you have a number of options on how to address resource overloads. Discuss the options you use on your projects.

Further Reading:

Burke, R., *Project Management Techniques*, will develop the following topics which relate to this knowledge area:

- How to increase resources
- How to reduce resources
- Resource S curve.

Burke, R., **Barron**, S., *Project Management Leadership*

16

Project Cost Management

(Estimating Costs)

Learning Outcomes

After reading this chapter, you should be able to:

- Identify and calculate the different types of project costs
- Calculate labour costs for your project
- Calculate procurement costs
- Prepare a project budget

The *Project Cost Management* knowledge area includes the techniques to estimate the project costs, to assign budgets, and to monitor and control the expenditure to ensure the project can be completed within the approved budget.

Estimating costs uses a portfolio of special project management techniques to predict what will happen in the future. Most of the components of the baseline plan are underpinned by the cost estimate. The accuracy of the planning and control is, therefore, directly dependent on the accuracy of the cost estimate. Project cost estimating also underpins problem solving, decision-making and risk management.

Estimating in the project management context typically means predicting and forecasting what could happen in the future. Although there is an element of estimating in all the techniques, this book focuses on two estimating methods;

- Estimating time explains how to estimate the time durations of project activities (see *Time Management* chapter).
- Estimating costs and the different types of costs incurred in making the project (see this chapter).

Project Management Plan Flowchart

The project management plan flowchart (figure 16.1) shows the relative position of cost management with respect to the other topics. It is logical that estimating costs follows the WBS, because the WBS techniques subdivide the scope of work into work packages, and it is at the work package level that the costs are estimated and budgets assigned.

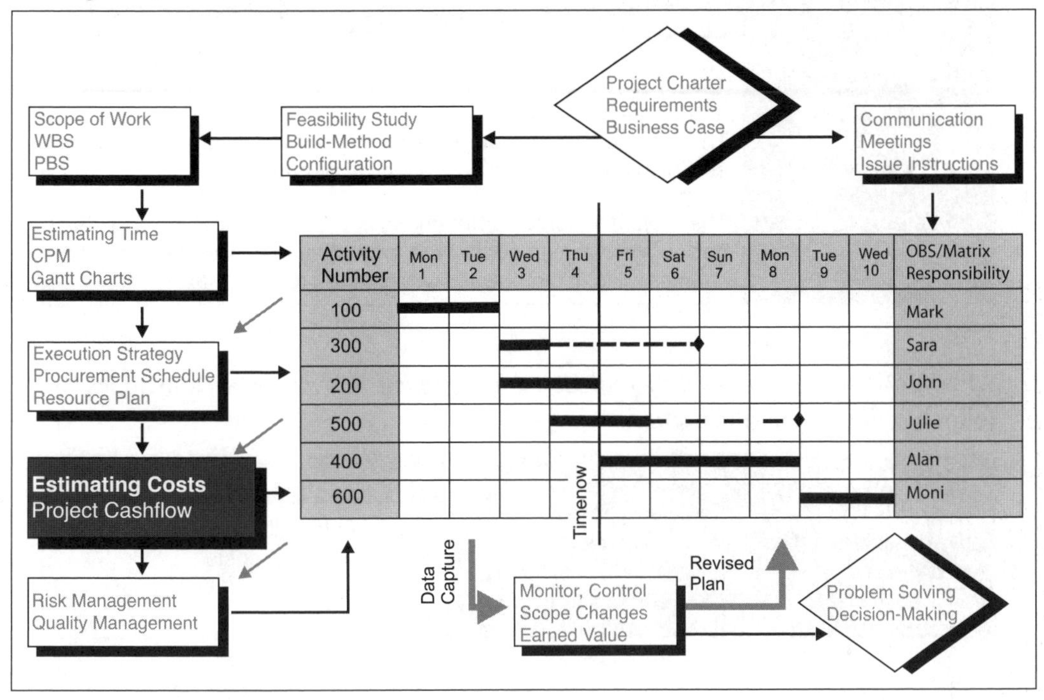

Figure 16.1: Project Management Plan Flowchart – shows the relative position of cost estimating to the other topics

Purpose of this Chapter: The purpose of this chapter is to help project managers and project team members acquire the competence and knowledge to be able to produce the cost estimates, assign budgets and control expenditure within projects of limited complexity. This chapter will explain:

- The different types of costs
- How to estimate labour costs
- How to estimate material costs (or procurement costs)
- How to use unit rates
- How to present the estimate (estimate format)
- How to assign a budget.

Body of Knowledge Mapping

PMBOK 4ed: The *Project Cost Management* knowledge area includes the following:

The PMBOK defines **Estimating Costs** as, *the process of developing an approximation of the monetary resources needed to complete project activities.*

PMBOK 4ed	Mapping
7.1: Estimate costs	This chapter will explain how to estimate the cost to complete the project work packages.
7.2: Assign budgets	This chapter will explain how to aggregate the project costs to establish an authorized cost baseline or budget.
7.3: Control costs	The *Project Execution, Monitoring and Control* chapter will explain how to monitor and control the project budget. The *Earned Value* chapter will explain how to forecast the project costs.

APM BoK 5ed: The *Budgeting and Cost Management* knowledge area includes the following definitions and sections:

APM BoK 5ed	Mapping
The APM BoK 5ed defines Cost Management as, the estimating of costs and the setting of an agreed budget, and the management of actual and forecast costs against that budget.	This chapter will explain how to estimate project costs.

Unit Standard 50080 (Level 4): The unit standard for the *Project Cost Management* knowledge area includes the following:

Unit 120375: *Participate in the estimation and preparation of cost budget for a project or sub project and monitor and control actual cost against budget*

Specific Outcomes	Mapping
120375/SO1: Identify elements and resources to be costed through interpreting the project scope statement, work breakdown structure and other project data.	This chapter will explain how to identify elements and resources to be costed.
120375/SO2: Participate in the preparation and production of a cost budget.	This chapter will explain how to produce a budget.
120375/SO3: Contribute to the monitoring and controlling of cost budget performance by maintaining records and communicating.	The *Project Execution, Monitoring and Control* chapter will explain how to monitor and control cost budgets.

1. Estimating Costs

The WBS technique enables the project to be estimated at different levels of detail. It can either be a high level, top down estimate for the whole project, or a more accurate bottom up estimate at the work package level. The work package can then be further subdivided into the different types of costs, see figure 16.2.

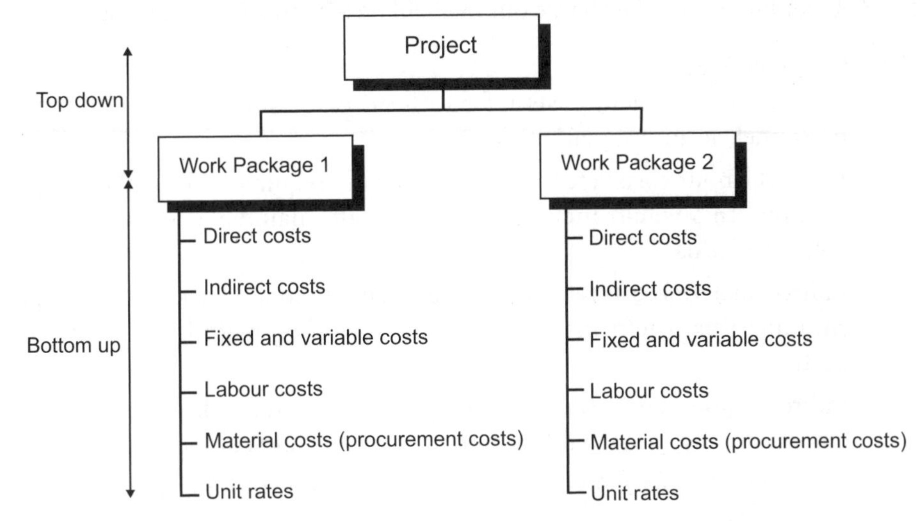

Figure 16.2: Top Down, Bottom Up Estimates - shows how each work package can be subdivided into different types of costs

This chapter will show how all these different types of costs can be collated and presented in a structured format.

2. Direct Costs

For ease of estimating, project costs can be subdivided into a number of different categories. Direct costs, as the term implies, are costs that can be clearly identified and assigned to a job. The project management philosophy is to assign as much work as possible to direct costs as direct costs can be individually budgeted and controlled far more effectively and accurately than indirect costs. Consider the following:

- Direct management costs refer to the project manager, project team members and other people working on the project
- Direct labour costs refer to the people working on an activity
- Direct material costs refer to the material used to complete an activity
- Direct equipment costs refer to the cost of machinery and equipment used to complete an activity
- Direct bought-in expenses refer to the cost of services used directly to complete an activity.

The distinctive nature of a direct cost means that the total expense can be charged to a job or project - this is the basis of the project management approach.

3. Indirect Costs

Indirect costs are overhead type costs that cannot be directly attributed to any particular job or project, but are required to keep the company functioning. Indirect costs are usually financed by an overhead recovery charge which is generally included in or added to the labour rate. For example, if the overhead recovery charge is 20% and the labour rate is $50 per hour, then the charge out rate will be $50 plus 20% = $60 per hour.

Consider the following:

- Indirect management costs refer to head office staff, this would include; personnel, accounts, marketing.
- Indirect labour costs refer to the staff who are required to keep the company running; this would include reception staff, maintenance engineers and security guards
- Indirect material costs refer to materials that are required to keep the company running; this would include stationery, cleaning materials, maintenance parts.
- Indirect equipment costs refer to equipment required to keep the company's offices running; this would include photocopier and computers.
- Indirect costs refer to expenses; this would include training and insurance.

Indirect costs are notoriously difficult to manage as they are difficult to estimate (other than from historical costs) and difficult to identify and pin down. They are usually financed through an overhead recovery charge added to the labour rate, but if not properly managed they will eat into the company's profits. The acid test is to check if it is more economical to use a service from an outside contractor. If it is cheaper to outsource say a cleaning or maintenance service, this means the internal indirect cost is either too high or the workers are inefficient.

4. Fixed and Variable Costs

Fixed and variable costs relate to how the costs change with the number of products made or services provided. For example, a hotel has:

- Fixed costs of rent, labour and certain overheads which are independent of the number of people staying at the hotel.
- Variable costs that increase with each, additional guest, for example, there will be an increase in materials (food), equipment (electricity), transport (to and from airport).

It is, for this reason, that during the off-season a hotel will often reduce their rates (specials), because any income above the hotel's variable costs is a contribution towards their fixed costs.

5. Labour Costs

Labour costs are a key component of most jobs. If a labour rate can be assigned to the estimated number of hours to perform a task, this will simplify the estimating process. Labour costs are generally expressed as so much per hour, or a fixed cost based on an estimated number of hours.

This section will explain how to determine the labour charge-out rate for the project. The labour costs considered here (table 16.1) are for the workforce and are, therefore, a direct cost. Although the salaries of the project's workforce can be clearly identified, there are also a number of other associated costs which form part of the labour rate. The labour rate is calculated by aggregating the various costs and dividing them by the number of hours worked. This process is explained in the following worked example. Here the costs have been subdivided into four main headings:

A.	Employee's salary	$4,000
B.	Employee's associated labour costs	See below
C.	Employee's contribution to overheads	25% of salary = $1,000
D.	Employee's contribution to company profit.	25% of salary = $1,000

		Costs per month	Days lost per month
B1	Medical insurance	$200	
B2	Sickness benefits		1
B3	Annual holiday		2
B4	Training course	$50	1
B5	Protective clothing	$50	
B6	Car allowance	$400	
B7	Housing allowance	$100	
B8	Subsistence allowance	$100	
B9	Pension	$200	
B10	Tool allowance	$100	
B11	Standing time		1
B12	Inclement weather		1
	Total	$1,200	6 days

Table 16.1: Labour Rate

$$\text{Labour rate} = \frac{\text{Total monthly costs (A + B + C + D)}}{\text{Total number of normal working hours per month}}$$

Where the average working month is 21 days, the average days lost per month are six days and eight hours are worked per day.

$$\text{Labour rate} = \frac{4000+1200+1000+1000}{(21 - 6) \times 8}$$

$$= \$\ 60/\ \text{hr.}$$

Some of the above items may be difficult to quantify without access to statistical analysis; for example, days lost due to sickness and standing time, or waiting time. They should, however, be recognized as potential costs and a figure assigned, if only as a contribution to an unknown amount. The end product of this analysis should be a labour rate per hour. As a rule of thumb, the labour rate can be calculated approximately from the worker's wages (which are known). The costs can be subdivided into thirds as below:

- 1/3 labour wages (package)
- 1/3 contribution to overheads
- 1/3 company profit.

An even easier way to determine the labour rate is to see what other companies are charging! The project manager will have to do this anyway to ensure the project's labour rates are competitive.

6. Material Costs

This section will outline how to determine the cost to procure all the bought-in materials, goods and outsourced services. The procurement costs can be broken down into a number of separate costs. Consider the following table:

Department	Scope of Work	Cost
Design Office	The design office and planning office incur costs developing the procurement list from the bill of materials and the specifications.	$1,800
Buying Office	The buying office incurs costs managing the tender process (see procurement cycle in the *Procurement Schedule* chapter).	$500
Quality Department	The quality department incurs costs pre-qualifying the suppliers to make sure they are capable of supplying the goods or services to the required condition. Further, quality costs are incurred inspecting and receiving delivered goods.	$500
Warehouse	The warehouse incurs costs handling, moving, storing, and dispatching the goods together with inventory and stock control.	$1,500
Accounts	The accounts department incurs costs approving suppliers' credit worthiness and paying invoices.	$500
Total Costs		$4,800

Table 16.2: Procurement Costs

The procurement costs are usually added as a percentage to the buying price to cover all the procurement costs as an indirect cost.

$$\text{Procurement percentage} = \frac{\text{Procurement costs}}{\text{Total cost of materials}} \times 100$$

For example, if the total cost of all the procured items is $50,000 and the cost of running the procurement office is $4,800 as per table 16.2, then the percentage will be as below:

$$\frac{\$4{,}800}{\$50{,}000} \times 100 = 9.6\% \text{ (call it 10\%)}$$

The above costs will need to be checked to ensure they are not being covered by another budget; for example, the inspection and pre-qualifying of suppliers could fall under the buying department or QA budget. The procurement costs would generally be developed at the company level and apply to all the company's projects.

7. Unit Rates

Although projects tend to be a unique undertaking (by definition), the scope of work is usually similar to previous jobs performed by the company in line with their field of expertise. It is, therefore, expedient to use unit rates for common items of work based on the company's performance on previous projects. This technique estimates a job's cost from an empirically developed tariff of unit rates. Consider the following parameters in the table below:

Type of Rate	Type of Work	Unit Rate
Per linear metre	Piping, wiring, welding, textiles	$
Per square metre	Decorating, painting, house building	$
Per cubic metre	Concrete, water supply	$
Per tonne	Ship building, cargo freight	$
Per HP or KW	Power, electrical supply, installing a generator	$
Per mile or Km	Transport	$
Per day	Plant hire, car hire	$
Per hour	Labour	$
Per minute	Fashion garment construction	$
Per GB	Internet download	$

Table 16.3: Unit Rates – shows the relationship between the type of work and the unit rate

Unit rates are probably the most commonly used estimating technique and will form the basis of most estimates because they are easy to measure, easy to budget for, and easy to control. Even fixed price contracts usually contain a unit price clause for charging out additional work.

The automotive industry uses unit rates - if a car is taken to a garage for a repair or maintenance, the garage should be able to provide a quote based on a national tariff of charges which have been compiled by the motor industry in association with labour unions and employers.

8. Assign Budgets

It is important to distinguish between a cost estimate and a budget. A cost estimate is a forecast of what the estimator calculates a work package will cost to complete. Whereas, a budget is the allocation of funds to complete the work package. To develop an estimate into an assigned budget might require a certain amount of trade-offs and compromises with the stakeholders.

The BoKs present the following definitions:

The PMBOK 4ed defines the **Budget** as, *the process of aggregating the estimated costs of individual activities or work packages to establish an authorized cost baseline.*

The APM BoK 5ed defines a **Budget** as, *the agreed cost of the project.*

Budget Trade-Off: Determining the project budget at a work package level or a project level is an iterative process. The cost estimate is developed based on a certain scope of work, build-method and time scale. If these estimates are in conflict with the funds available then alternative options will need to be discussed and result in trade-offs and compromises.

Compiling a budget is, therefore, an iterative process of aggregating the estimated costs and discussing options with the stakeholders, technical experts, experienced managers and consultants, until there is an agreement on the scope of work and the funds available. These funds allocated will then become the agreed budget available to complete the baseline scope of work in an agreed time frame.

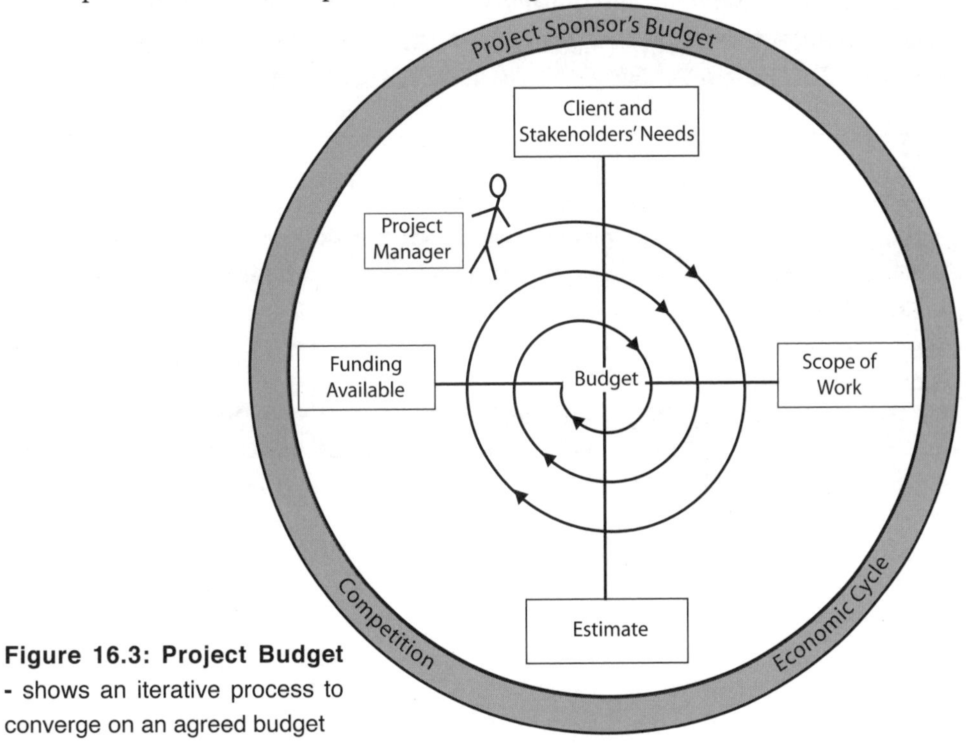

Figure 16.3: Project Budget - shows an iterative process to converge on an agreed budget

9. Budget Format

The WBS structure of work packages can be used to format the budget's presentation; where column [1] lists the scope of work as work packages to capture all the project's budgets, column [2] is a description of the work packages, and column [3] is the budget in monetary terms.

1. WBS (work package)	2. Description	3. Budget
1000		$12,000
2000		$11,000
3000		$17,000
Sub-Total		$40,000
Project Management fee	Project Management fee (10%) of sub total	$4,000
Profit	Quasi cost (20%) of subtotal	$8,000
Total		$52,000

Table 16.4: Estimating Format – shows the budget per work package

Each job can be further subdivided into labour, material, equipment and transport. By setting up the budget on a spreadsheet, the amounts can be added horizontally and vertically to give the total budget per work package (job), the total budget for all the labour, material, equipment and transport, and the total budget of the new venture.

WBS	Labour	Material	Machinery	Transport	Total
1000	$5,000	$4,000	$2,000	$1,000	$12,000
2000	$6,000	$3,000	$1,000	$1,000	$11,000
3000	$10,000	$3,000	$2,000	$2,000	$17,000
Sub-Total	$21,000	$10,000	$5,000	$4,000	$40,000
Project Office costs (10%)	$	$	$	$	$4,000
Profit (20%)	$	$	$	$	$8,000
Total	$	$	$	$	$52,000

Table 16.5: Estimating Format – shows the WBS budgets subdivided into labour, material, machinery and transport

10. Cost Control

Once the project is executed it is the project manager's responsibility to control the project costs to within the agreed budget. The BoKs define cost control as the following:

The PMBOK 4ed defines **Cost Control** as, *the process of monitoring the status of the project to update the project budget and managing changes to the cost baseline.*

The APM BoK 5ed defines **Cost Control System** as, *any system of keeping costs within the bounds of budgets or standards based upon work actually performed.*

Cost Control Sheet: A cost control sheet will capture all the project budgets and actual costs, and highlight any variances. The control of these variances will determine the financial success of the project.

WBS	Budget	Actual Cost	Variance	Variance %
100	$100	$100	0	0
200	$100	$120	($20)	(20%)
300	$100	$90	$10	10%
400				
Totals				

Table 16.6: Cost Control Sheet

Exercises:

1. Using a similar labour rate template as outlined in this chapter (table 16.1), calculate the labour rate for your project.
2. Show how unit rates can be used on your projects.
3. Using the WBS technique developed in the *WBS* chapter show how top-down and bottom-up estimating apply to your type of projects (The example in the *WBS* chapter referred to a house building project where it is common to refer to building costs as $\$x/m^2$ at the top level, and unit rates $x/m at the lower level).

Further Reading:

Burke, R., *Project Management Techniques*, will develop the following topics which relate to this knowledge area:

- Estimating to costing continuum
- Economy of scale
- Bidder's dilemma.

17

Project Cashflow

Learning Outcomes

After reading this chapter, you should be able to:

- Understand how the invoice and payment timing affects the project cashflow
- Produce a project cashflow statement
- Produce a cost 'S' curve

The project cashflow statement technique integrates the needs of the *Project Cost Management* knowledge area to determine and control the budget with the needs of the *Time Management* knowledge area, which is to plan and control the schedule.

The financial success of a project depends not only on the project finishing on time and making a profit but, also, on being able to finance the project's expenses through the project lifecycle. Statistics clearly indicate that more companies go into liquidation because of cashflow problems than for any other reason. It is, therefore, essential for the project manager and the project team members to closely monitor the project's cashflow.

Project accounting techniques should not be confused with financial accounting or management accounting which are both used within the corporate environment.

Financial Accounting: Financial accounts keep a record of all the financial transactions, payments in and payments out, together with a list of creditors and debtors. This information gives the financial status of a company using the generally accepted accounting principles. The three main reports are; the balance sheet, the income statement and the cashflow statement.

Management Accounting: Management accounting, also referred to as cost accounting, uses the above financial information, together with the profit and loss account, to analyse company performance. This analysis will assist management decision-making with respect to estimating, planning, budgeting, implementation and control.

Project Accounting: Project accounting uses a combination of both financial accounting and management accounting, together with a number of special project management tools (*WBS and CPM*), to integrate the project accounts with the other project parameters to produce the project cashflow statement and the earned value forecast.

Project Management Plan Flowchart

The project management plan flowchart (figure 17.1) shows the relative position of the project cashflow statement with respect to cost estimating and the Gantt chart. It is logical that the project cashflow statement follows the project schedule and the budget, as these two items of information are required in the cashflow calculation.

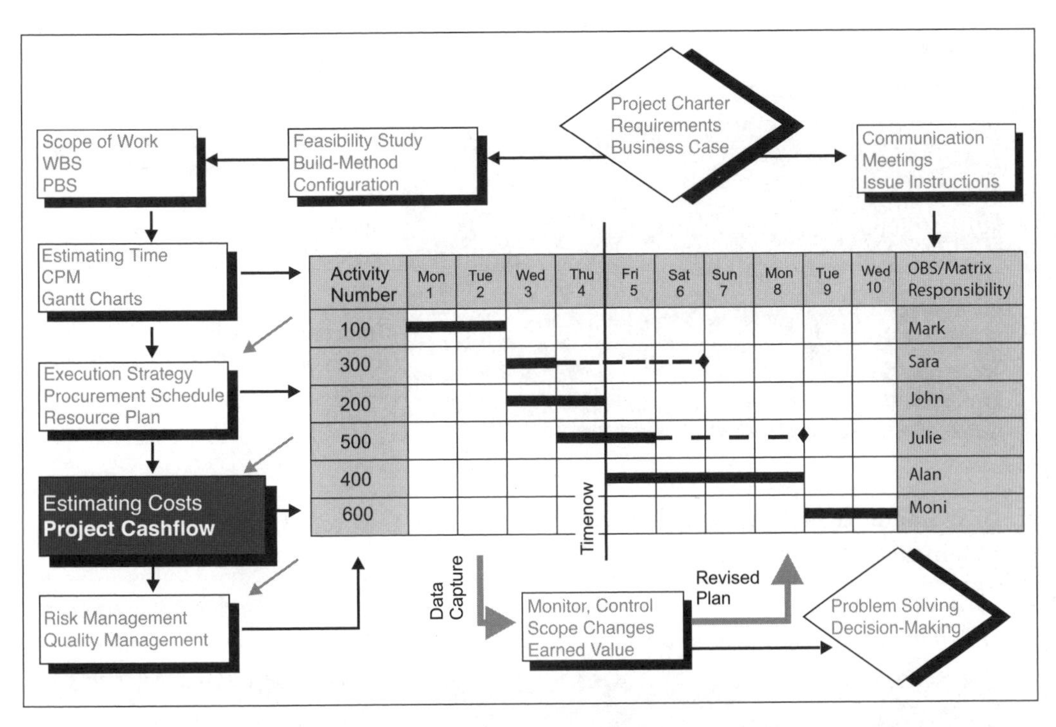

Figure 17.1: Project Management Plan Flowchart – shows the relative position of the project cashflow statement with respect to the other techniques

Up to now the schedule calculations for the CPM, the procurement schedule and the resource planning have all assumed the funds are available as and when required. By developing the project cashflow statement the project manager and project team members will be able to show the actual funding requirements.

Body of Knowledge Mapping

PMBOK 4ed: The *Project Cashflow Statement* technique is part of the *Project Cost Management* knowledge area and the *Time Management* knowledge area which includes the following:

PMBOK 4ed	Mapping
6.5 Develop schedule	This chapter will explain how to balance the project cashflow with respect to the timing of the project costs and incomes.
6.6: Control schedule	This chapter will explain how to control the timing of the project cashflow.
7.1: Determine budget	This chapter will explain how to determine the budget's cashflow.
7.2: Control costs	This chapter will explain how to monitor and control the project cashflow.

APM BoK 5ed: The *Project Cashflow Statement* technique is part of the *Budgeting and Cost Management* knowledge area which includes the following:

APM BoK 5ed	Mapping
3.4: The budget is phased over time to give a profile of expenditure. This is an important part of the budgeting process as the profile of expenditure is used in project financing and funding.	This chapter will explain how to develop a project cashflow which highlights the funding profile.
3.4: The budget is phased over time to give a profile of expenditure. This will allow a cash flow forecast for the project to be developed, and a drawdown of funds to be agreed with the organization.	This chapter will explain how to develop a project cashflow statement.

Unit Standard 50080 (Level 4): The unit standard for the *Project Cashflow* knowledge area includes the following:

Unit 120375: *Participate in the estimation and preparation of cost budget for a project or sub project and monitor and control actual cost against budget*

Specific Outcomes	Mapping
120375/SO3: Contribute to the monitoring and controlling of cost budget performance by maintaining records and communicating.	This chapter will explain how to monitor and control project cashflow records.

1. Project Cashflow Statement

The project cashflow statement is a document that models the flow of money in and out of the project's account. The time frame is usually monthly, to coincide with the normal business accounting cycle. The project cashflow statement is based on the same information used in a typical bank statement except that here the income (cash inflow) and expenditure (cash outflow) are grouped together and totalled to simplify the presentation.

Contractor's Perspective: The project cashflow statement developed in this chapter will be presented from the contractor's perspective. During a project, the contractor's income will come from the monthly progress payments (invoices), and the expenses will be wages, materials, overheads, bought-in services and finance interest charges.

Client's Perspective: The client and project sponsor have a different perspective of the project cashflow. The client will have no income from the facility during the project's implementation period. The client's income will come from the operation of the facility after the project has been completed. The client's expenses will be the invoices from the contractors, suppliers and other professional services required to make the facility or product.

Project Cashflow Example 1 (Incurred Costs): The first project cashflow example focuses on the incurred costs during a four-month project from February to May.

The first step considers how the project costs are incurred and how they will appear on a project cashflow statement. This project cashflow statement does not look at the timing of the costs (exercise 2 will consider the timing of the costs).

This is a simple example of a four-month project (from February to May) which looks at the incurred costs from the contractor's position.

Client	The client pays four monthly payments of $5,000 from February to May.
Labour	The contractor employs labour at $2,000 per month from February to May.
Materials	The contractor procures materials at $500 per month from February to May.
Equipment	The contractor hires equipment at $1,500 per month from February to May.

The following steps can be used as a guideline to produce the project cashflow statement (incurred costs):

Step 1	The project cashflow statement headings are set up as per table 17.1, using monthly headings (fields or columns) to cover the duration of the project - six months in this case.
Step 2	The brought forward (B / F) for February is zero.
Step 3	The incurred income is $5,000 per month from February to May.
Step 4	There are three items of expenditure from February to May; Labour at $2,000 per month Materials at $500 per month Hire of equipment at $1,500 per month.
Step 5	Starting with February, the total funds available are $5,000. The total expenses are $4,000. The closing balance is therefore, $1,000
Step 6	The closing balance for February of $1,000 becomes the brought forward for March.
Step 7	The calculations are repeated each month up to May.

	January	February	March	April	May	June
Brought Forward		$0	$1,000	$2,000	$3,000	
Income		$5,000	$5,000	$5,000	$5,000	
Total $ Available		**$5,000**	**$6,000**	**$7,000**	**$8,000**	
Expenses						
Labour		$2,000	$2,000	$2,000	$2,000	
Materials		$500	$500	$500	$500	
Equipment		$1,500	$1,500	$1,500	$1,500	
Total Expenses		**$4,000**	**$4,000**	**$4,000**	**$4,000**	
Closing Balance		$1,000	$2,000	$3,000	$4,000	

Table 17.1: Project Cashflow Statement (incurred costs) – shows the incurred costs between February and May

Note: The profit at the end of the project is $4,000.

2. Cashflow Timing

The previous section considered when the project costs were incurred; this section will consider when the cash actually flows in or out of the project's accounts.

The project cashflow statement, as the name suggests, is a measure of the cash in and cash out of the project's account. The catch is that the project cashflow is usually not the same as the sales figures or expenses for the month because of the different timing of the incomes and payments. Listed below are some typical examples of project cashflow timings:

Part payment	Part payment with placement of order - this is often used to cover the manufacturer's cost of materials and ensure the purchaser's commitment, particularly on imported goods.
Stage payments	Stage payments or progress payments for a project which could take many months to complete.
Cash On Delivery (COD)	Payment on purchase - this is normal practice with retailers.
Monthly	Monthly payments for labour, rent, telephone and other office expenses.
Credit	30, 60 or 90 days credit; normal terms for bought-in items.

It may help the learning process to look at the data presented the other way round.

Labour	Internal labour costs are usually paid in the month the work is performed, but contractors usually give one month credit.
Material	Material costs can vary from an up-front payment, cash on delivery (COD), to one to three months credit.
Services	Bought-in services and plant hire costs can be paid within one to three months after delivery.
Income	Income from the client can vary from up-front payments, stage payments or progress payments one month after invoice is issued.

These figures are usually compiled monthly on a creditors and debtors schedule. It is the project accountant's responsibility to chase up late payments.

Non-Project Cashflow Items: Company assets should not appear on a project cashflow statement as they do not represent a movement of cash. Although appreciation and depreciation may represent a flow of value, physically, they do not represent an inflow or outflow of cash. This also applies to the revaluation of property and the value of the company's shares.

3. Cashflow Distribution

The project cashflow statement is an integral part of the critical path method (CPM) which combines the WBS, the estimate (budget), the project schedule, the procurement schedule and the resource planning (resource histogram). At this point some assumptions need to made about the distribution and profile of the project cashflow with respect to the schedule of the activities. For ease of calculation, the project cashflow within each activity is usually assumed to be linear unless otherwise stated (see figure 17.2).

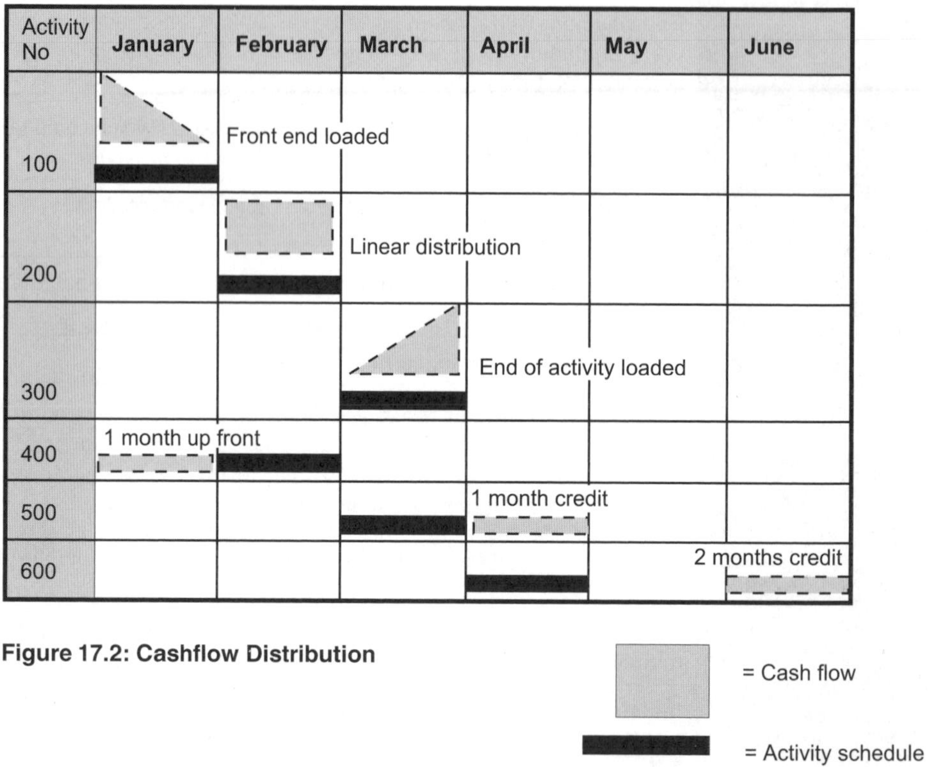

Figure 17.2: Cashflow Distribution

Labour costs are generally uniform over the duration of the activity, whereas, the cost of materials and other bought-in items might need to be qualified. As stated in the previous section, the cashflow can vary from up-front payments to one, two or three months later depending on the supplier.

Projects with many activities will tend to smooth out any distortions caused by non-linear project cashflows. However, if there are activities with disproportionately large material or equipment payments, these can be separated out to form new activities with appropriate durations to match the expense profile.

Project Cashflow Example 2: The second project cashflow example will apply the timing of the income and payments to the project cashflow example 1. The timing of the invoices and payments are as follows:

Client	The client pays four monthly payments of $5,000 one month after invoice.
Labour	The contractor employs labour at $2,000 per month paid in month of work.
Materials	The contractor procures materials at $500 per month; the supplier gives one month credit.
Equipment	The contractor hires equipment at $1,500 per month; payment is required one month in advance.

Step 1	The project cashflow statement headings are set up as per table 17.2. Use monthly headings (fields or columns) to cover the duration of the project.
Step 2	The brought forward (B / F) for January is zero.
Step 3	The client pays the invoices after 30 days, therefore, the $5,000 earned in February is paid from March and so on for the other payments.
Step 4	There are three items of expenditure from February to May; Hire of equipment - $1,500 per month Labour - $2,000 per month Materials - $500 per month.
Step 5	The equipment hire is paid one month in advance, therefore, insert the payments of $1,500 per month from January to April.
Step 6	The labour is paid for in the month the work is performed, therefore, insert the payments of $2,000 per month from February to May.
Step 7	The material supplier gives one month credit, therefore, insert the payments of $500 per month from March to June.
Step 8	Starting in January, there is zero balance brought forward and no income, but there is an expense of $1,500, which gives a closing balance of ($1,500) and, therefore, an opening balance for February of ($1,500).
Step 9	In February there is no income so there is ($1,500) available and the expenses are $3,500 which gives a closing balance of ($5,000). This calculation is continued through to June.

	January	February	March	April	May	June
Brought Forward	$0	($1,500)	($5,000)	($4,000)	($3,000)	($500)
Income	$0	$0	$5,000	$5,000	$5,000	$5,000
Total Available	$0	($1,500)	$0	$1,000	$2,000	$4,500
Expenses						
Equipment	$1,500	$1,500	$1,500	$1,500		
Labour		$2,000	$2,000	$2,000	$2,000	
Materials			$500	$500	$500	$500
Total Expenses	$1,500	$3,500	$4,000	$4,000	$2,500	$500
Closing Balance	**($1,500)**	**($5,000)**	**($4,000)**	**($3,000)**	**($500)**	**$4,000**

Table 17.2: Project Cashflow Statement

Both project cashflow example 1 and 2 end with a balance of $4,000, but when the project cashflow is considered in example 2 all the months end with a negative balance until June. This means outside finance would need to be organised in advance.

4. How to Draw an Expense S Curve

Another method for modelling the project cashflow is to use an **S curve** analysis, which provides the link between the budget and the timeline. Experience has shown that a project's accumulated costs tend to follow an S shape curve. To draw the S curve use the following procedure:

Step 1	Draw an early start Gantt chart for the project, assume figure 17.3 as given.
Step 2	Assign expenses evenly over the duration of the activity. For example, activity 100 is $100 over two days, giving $50 per day. Do this for all the activities.
Step 3	Add the cost values vertically to get daily totals. On 1st May the total is $50 as there is only one activity working; do the same for 2nd, 3rd and 4th, but on 5th May the total is $200 because there are four activities working (200, 300, 400 and 500). Do this for all the activities.
Step 4	Plot the daily total costs on a graph of costs against time to obtain the daily rate of expenditure curve, figure 17.4.
Step 5	The accumulative S curve is calculated by adding the daily values from left to right. This is done by adding 1st May total [$50] to 2nd May total [$50] giving [$100]. Then move to the next day [$50] + the running total [$100] giving $150. Do this for all the activities.
Step 6	Plot accumulated figures on a graph of cost against time (figure 17.4). This will produce the distinctive **S curve**.

Activity/Date	$	1	2	3	4	5	6	7	8	9	10
100	100	50	50								
200	150			50	50	50					
300	100					50	50				
400	100					50	50				
500	150					50	50	50			
600	100							50	50		
700	100							50	50		
800	100								50	50	
900	50									50	
1000	50										50
Daily Total		**50**	**50**	**50**	**50**	**200**	**150**	**150**	**150**	**100**	**50**
Accumulated		50	100	150	200	400	550	700	850	950	1000

Figure 17.3: Gantt Chart – shows how the expenditure can be presented as a S curve

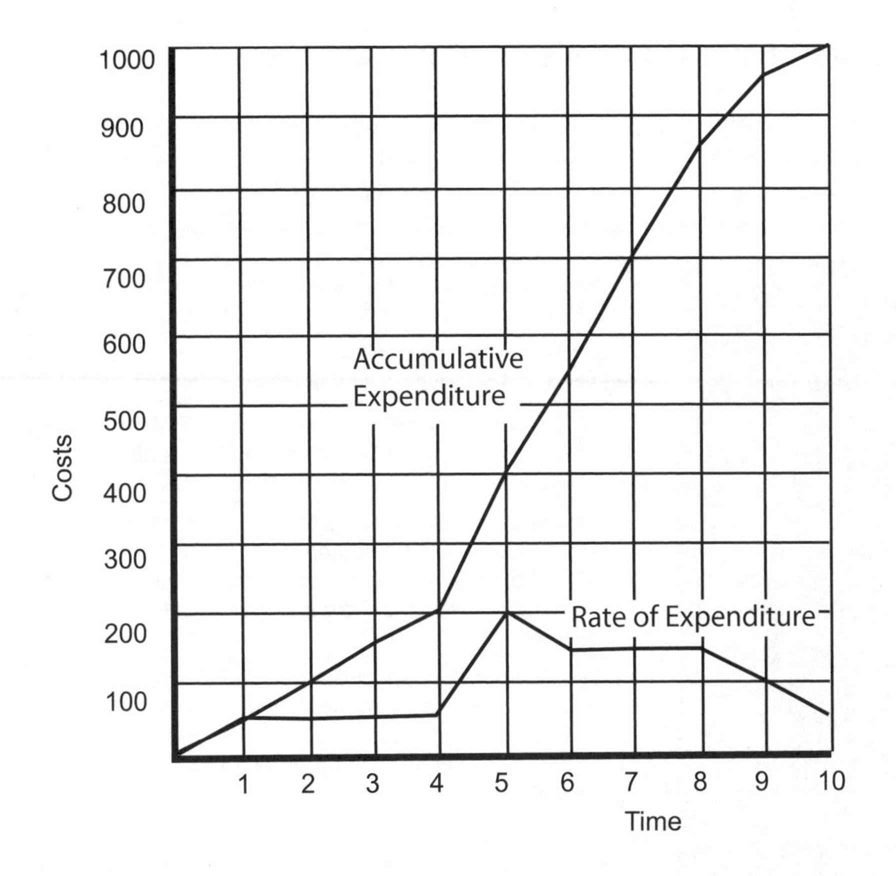

Figure 17.4: S Curve – shows the S curve of accumulated expenses and rate of expenditure

5. Benefits of Using a Project Cashflow Statement

Although statistics clearly indicate that more companies go into liquidation because of project cashflow problems than for any other reason, the full benefits of project cashflow management are not always appreciated. Listed below are some of the many benefits associated with using project cashflow-modelling techniques:

Plan Ahead	The project manager can plan ahead knowing what funds are required, when they are required and how much is required.
Warnings	It gives a timely warning of negative project cashflow which needs to be financed and positive project cashflow which should be invested.
Rate of Invoicing	It gives the client a forecast rate of invoicing (FRI) so that the client can produce his own cashflow statement. This is often a contractual requirement with some of the larger corporations.
Business Case	The project cashflow statement is one of the main items of a business case, as it will show the bank manager or lender how much money is needed, when it is needed and most importantly when the money will be paid back. It will also show that the project manager has done his homework.
Cheaper Loans	The project cashflow will establish the lending and repayment dates which usually make a loan cheaper than an overdraft. With an overdraft the bank has no idea when the company is going to borrow or pay back, so the bank has to build up an extra margin to cover its own funding costs.
Lending on Project Cashflow	'Secured' lending, even if it is based on the borrower's assets, still depends on the borrower's cashflow to repay the loan. Assets are only worth as much as someone is prepared to pay for them, and that valuation is likely to be based on the asset's ability to produce cash.
Expenditure Curves	The project cashflow statement can be developed into expenditure curves, rates of expenditure and accumulated expenditure, all of which are required for earned value project control.
What-if	The project cashflow statement can be used to perform a what-if simulation which will indicate where the project's sensitivity lies. This forms the basis of the sensitivity analysis.
Pay back Period	The project cashflow can be used as a data source to calculate an investment's payback period.
Discounted Project Cashflow	The discounted cashflow (DCF) introduces *the time value of money* (see **Burke,** R., *Project Management Techniques*).
Asset Register	The project cashflow statement can be the data source for the company's asset register, asset depreciation and company taxes.

These benefits clearly indicate why the project cashflow statement underpins effective project cost planning and control. This concludes the chapter on *Project Cashflow* and all the planning techniques outlined in the planning and control cycle. The following chapter will discuss how to execute, monitor and control the project.

Exercises:

Exercise 1: Given incurred costs (table 17.3), convert the incurred costs to a cashflow statement using a format similar to table 17.2 (see Appendix 1 for solution).

	Cashflow	Jan	Feb	Mar	Apr	May	Jun
B/F		1,000					
Income	1 month credit		5,000	5,000	5,000	5,000	
Total Income							
Equipment	1 month upfront		1,000	1,000	1,000	1,000	
Labour	Same month		2,000	2,000	2,000	2,000	
Material	1 month credit		1,000	1,000	1,000	1,000	
Total Expenses							
Closing							

Table 17.3: Incurred Costs - exercise 1

Exercise 2: Given incurred costs (table 17.4), convert the incurred costs to a cashflow statement using a format similar to table 17.2 (see Appendix 1 for solution).

	Cashflow	Jan	Feb	Mar	Apr	May	Jun
B/F		2,000					
Income	1 month credit		6,000	6,000	6,000	6,000	
Total Income							
Equipment	1 month upfront		1,000	1,000	0	0	
Labour	Same month		2,000	2,000	2,000	2,000	
Material	1 month credit		2,000	2,000	1,000	0	
Total Expenses							
Closing							

Table 17.4: Incurred Costs - exercise 2

Exercise 3: Draw the rate of expenditure curve and the accumulative S curve from the Gantt chart of expenditure below (see Appendix 1 for solution).

Activity/Date	$	1	2	3	4	5	6	7	8	9	10
100	100	50	50								
200	150		50	50	50						
300	100			50	50						
400	100				50	50					
500	150					50	50	50			
600	100						50	50			
700	100							50	50		
800	100								50	50	
900	50									50	
1000	50										50
Daily Total											
Accumulated											

Figure 17.5: Gantt Chart - exercise 3

18

Project Execution, Monitoring and Control

Learning Outcomes

After reading this chapter, you should be able to:

- Identify what can be planned and what can be controlled
- Understand the project control cycle
- Execute processes and standards to support project change control

This chapter will explain how to execute, monitor and control the project's scope of work. Project execution, monitoring and control are key techniques within the *Project Management Process*, which in turn is a key topic within the *Project Integration Management* knowledge area.

Project control is the flipside of planning, because planning is a pointless exercise unless the execution of the plans are tracked, monitored and controlled through accurate reporting on performance.

The project's baseline plan is the course to steer, with the tracking and monitoring functions ascertaining the project's position with respect to scope, time, procurement, resources and costs. If the project is off course, then control in the form of corrective action must be applied to bring the project back on course.

Project Management Process

In the previous chapters, the development of the project management plan or baseline plan completes the planning process within the project management process (see figure 18.1). This chapter will explain the next process; project execution, monitoring and control using the baseline plan as the means to achieving the project objectives and an outline of the required condition.

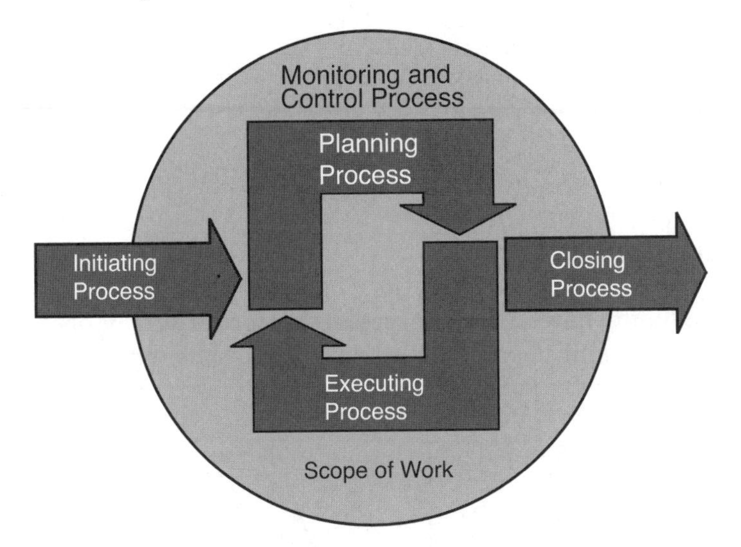

Figure 18.1: Project Management Process – shows the sub-processes of initiating, planning, execution, monitoring, controlling and closing

Body of Knowledge Mapping

PMBOK 4ed: The *Project Integration Management* knowledge area includes the following:

PMBOK 4ed	Mapping
4.1: Develop a project charter	See *Project Integration Management* chapter.
4.2: Develop a project management plan	See *Project Management Plan* chapter.
4.3: Direct and manage project execution	This chapter will explain how to direct and manage project execution.
4.4: Monitor and control project work	This chapter will explain how to monitor and control project work.
4.5: Perform integrated change control	See *Scope Management* chapter.
4.6: Close project or phase	See *Project Integration Management* chapter.

APM BoK 5ed: The *Project Execution, Monitoring and Control* technique falls under *Executing the Strategy* knowledge area which includes the following requirements:

APM BoK 5ed	Mapping
3: The ongoing measurement and management of the project's performance	This chapter will explain how to monitor and control project progress.

1. Planning and Control Spiral

The planning and control spiral shows a suggested sequence of discrete operations that are repeated iteratively until an optimum baseline plan is achieved.

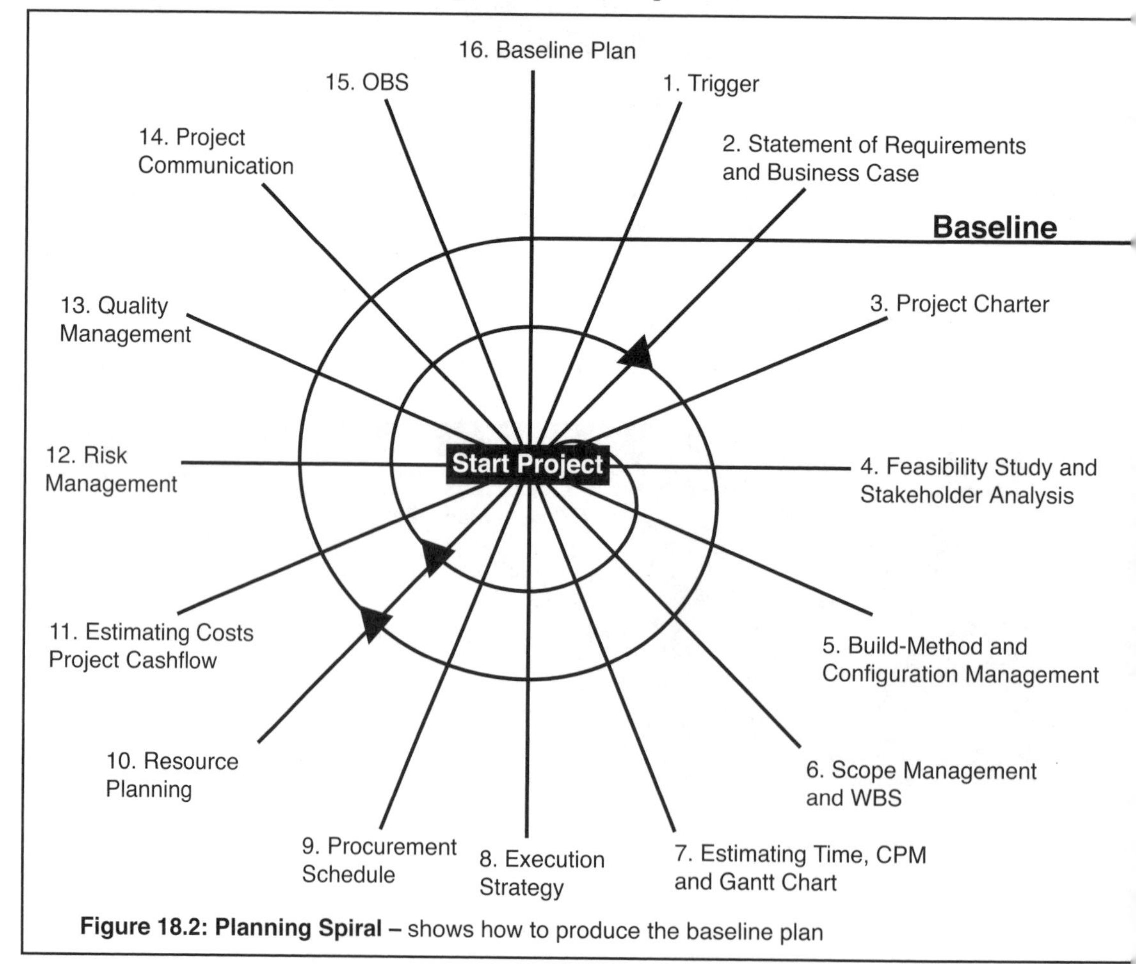

Figure 18.2: Planning Spiral – shows how to produce the baseline plan

Unit Standard 50080 (Level 4): *Project Execution, Monitoring and Control* is included in the following unit standards:

Unit 120372: *Explain fundamentals of project management*

Unit 120387: *Monitor, evaluate and communicate simple project schedules*

Specific Outcomes	Mapping
120372/SO5.3: The reasons for planning and controlling a project are explained.	This chapter will outline the reasons for planning and controlling a project.
120387/SO1: Describe and explain a range of project schedule control processes and techniques.	This chapter will explain a range of project schedule control processes.
120387/SO2: Monitor actual project work versus planned work (baseline).	This chapter will explain how to compare actual work with planned work.
120387/SO3: Record and communicate schedule changes.	This chapter will explain how to record schedule changes.

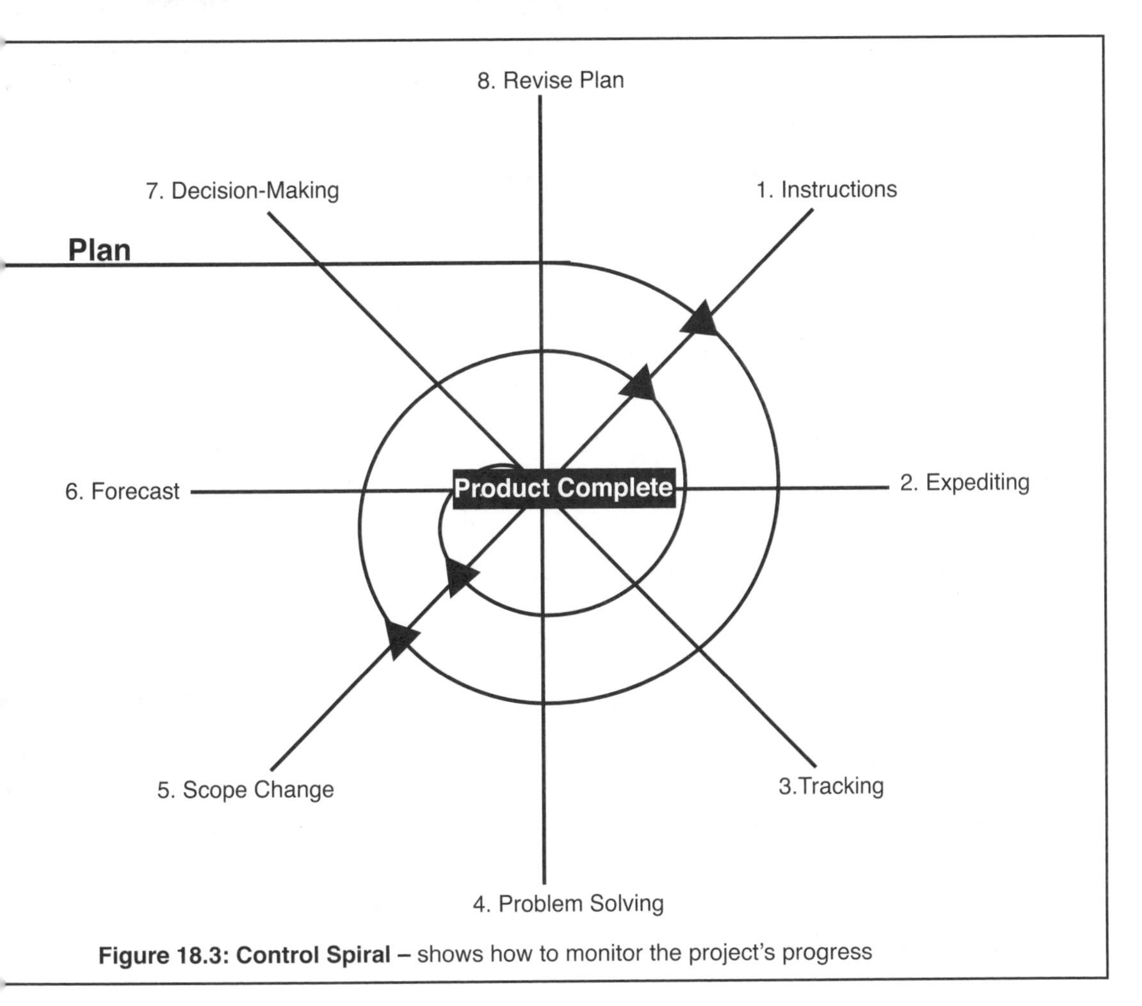

Figure 18.3: Control Spiral – shows how to monitor the project's progress

2. Project Control Cycle

The planning and control cycle can be subdivided into the planning cycle to develop the project management plan, and the control cycle which will be outlined in this chapter. The project control cycle is presented in figure 18.4 as a sequence of steps which need to be carried out each reporting period.

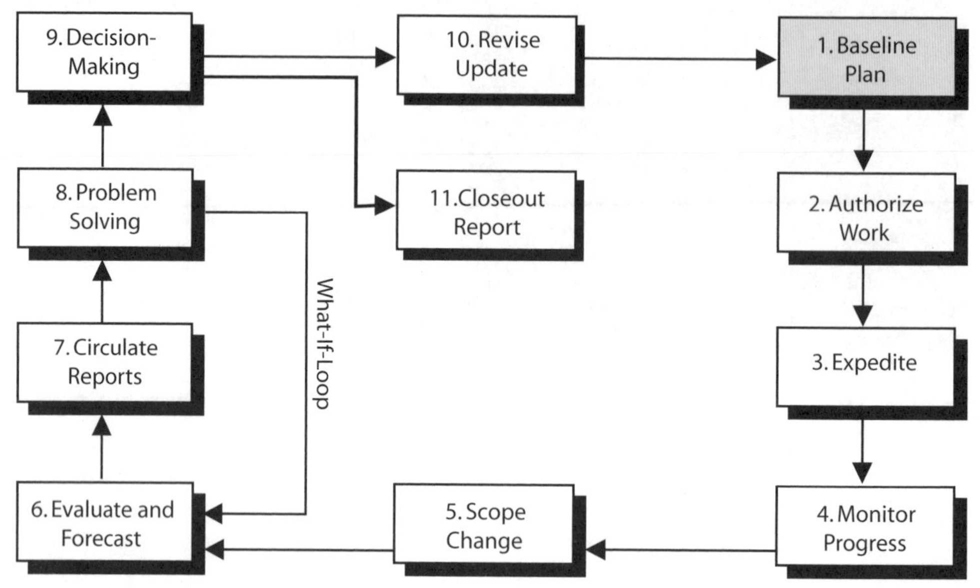

Figure 18.4: Project Control Cycle

Once the project starts it is inevitable that there will be some deviations; be it late deliveries, sickness, absenteeism, or scope changes. The project control cycle monitors the project's performance to date (time now) and compares it against the baseline plan. It also makes an assessment on future performance to determine that each parameter will be completed as planned.

The control cycle steps are explained as follows:

1. Baseline Plan	The baseline plan is the starting point for the project control cycle as it outlines the agreed plan for managing the project. The *Project Management Plan* chapter explains how to develop the baseline plan.
2. Work Authorisation	As the single point of responsibility the project manager is responsible for delegating and authorising the scope of work. The issuing of instructions to the appointed contractors and other responsible parties signals the start of the execution phase of the project. The methods for authorising work, reporting and applying control should be discussed and agreed at the handover meeting, so that all parties know how the project will be managed. A record of all decisions and instructions should be kept in the project office to provide an audit trail. Work authorization must ensure that the full scope of work is authorized to the responsible person(s); this is often achieved through a job card system.

3. Expedite	Once instructions, job cards, orders and contracts have been issued, project expediting takes a proactive approach to make the instructions happen. This involves the follow up function (usually by a team member) to confirm that; orders have been received by the contractor or supplier, materials have been procured, skilled labour is available, work has started as planned and the schedule completion dates will be achieved. Any variances should be reported through the data capture system to the responsible stakeholders. (see *Procurement Schedule* chapter.)
4. Monitoring Progress	The data capture system records the progress and, the current status of all the work packages and activities. The accuracy of the data capture has a direct bearing on the accuracy of all the subsequent reports (project status, trends and forecast).
5. Scope Change	The scope change control function ensures that all changes to the scope of work are captured and approved by the designated people before being incorporated into the baseline plan and communicated through the document control system. As the project is implemented, there will be changes to the scope of work which need to be considered, approved and authorized. These would tend to be authorized by issuing a revised document. Typical examples include: • Scope change, issue revised drawing • Planning change, issue revised schedule • Build-method change, issue revised build-method statement • Cost change, issue revised budget. Change control is also concerned with influencing the factors which create changes to ensure that any changes are beneficial (see *Scope Management* chapter).
6. Evaluation and Forecasting	The project's performance is analysed by comparing actual progress against planned progress within the CPM model, and extrapolating trends to forecast the project's position in the future (see *Earned Value* chapter).
7. Circulate Reports	Communicate information to all the nominated experts within the configuration management system.
8. Problem Solving	The problem solving function generates a number of feasible solutions, alternatives and opportunities for consideration.
9. Decision-Making	The decision-making function collates information and suggestions, and decides on an appropriate corrective course of action which commits the necessary resources. One of the main management functions is to make decisions which have the collective support of the team members and stakeholders. In fact, it may be argued that the sole purpose of generating information is to make decisions.
10. Revise Baseline Plan	If there are any changes within the project the baseline plan must be revised to reflect the current scope of work and incorporate any corrective action. By saving the old baseline plan an audit trail of changes can be archived. The control cycle is now complete and the next cycle will authorise the changes and corrective action.
11. Closeout Report	The lessons learnt must be documented in an agreed format and time frame, and communicated to the person responsible for the estimating data base and closeout report.

3. Baseline Plan

The baseline plan is the starting point for the project control cycle as it outlines all the project topics and their associated planning and control documents. This section will set out a list of the planning documents and the associated control documents.

Scope Management: The scope of work defines what the project is producing or delivering:

Planning Documents:	Statement of requirements, business case
	Project charter, project brief, project proposal
	Work breakdown structure (WBS)
	Activity list, bill of materials (BOM)
	Drawing register, specification register
	Contract
Control Documents:	Project communications
	Impact statements
	Variations and modifications
	Change requests
	Concessions
	Closeout report

Technical Support: The technical support from the design office extends from interpreting the client's statement of requirements to ensuring the project can be built (build-method), and ensuring the project will operate as per the client's needs (configuration management):

Planning Documents:	Client's brief
	Statutory regulations
	Specifications
	Design calculations
	Build-method
Control Documents:	Configuration control
	Impact statements
	Commissioning
	As-built drawings

Time Management: Outlines the sequence and timing of the scope of work:

Planning Documents:	Network diagram
	Scheduled Gantt chart
	Milestone schedule
	Rolling horizon Gantt chart
Control Documents:	Progress report (actual vs planned)
	Revised Gantt chart
	Earned value
	Trend documents

Procurement Management: The procurement function identifies all the bought-in items. These must be procured to meet the required specification, time schedule and budget:

Planning Documents:	BOM, parts list
	Procurement schedule
	Material requirement planning (MRP)
	Procurement budget
Control Documents:	Purchase order
	Expediting status report
	Revised procurement schedule and budget

Resource Management: Resource management integrates the resource estimate with time management to produce the resource forecast. This is usually related to manpower requirements:

Planning Documents:	Resource forecast
	Resource availability
	Resource levelled manpower histogram
Control Documents:	Time sheets
	Revised manpower histogram

Cost Management: Cost management allocates budgets and cashflows to the work packages:

Planning Documents:	Cost breakdown structure
	Activity budgets
	Department budgets
	Project cashflow statement
Control Documents:	Expenditure reports (actual vs planned)
	Committed costs and cost-to-complete
	Revised budgets
	Earned value

Change Control: As the project progresses the scope of work is revised and controlled through the following documents:

Change Control:	Project communications
	Impact statements
	Non conformance reports (NCR)
	Change requests and concessions
	Drawing revisions
	Modifications and variation orders (VO)
	Extras to contract
	Specification and configuration revisions.

Quality Management: Outlines how the company determines the required condition and how the company will ensure the product will achieve the required condition:

Planning Documents:	Project quality plan (ISO 9000)
	Quality control plan
	Parts lists, specifications and standards
Control Documents:	Inspection reports
	Non conformance reports (NCR's)
	Concessions
	Change requests
	As-built drawings
	Data books and operation manuals
	Testing and commissioning

Communication Management: The communication function is to disseminate information and instructions to the responsible parties:

Planning Documents:	Lines of communication
	List of controlled documents
	Distribution list
	Schedule of meetings and agendas
Control Documents:	Transmittals
	Minutes of meetings

Human Resource Management: This function sets the framework for the human factors:

Planning Documents:	Project organization structure
	Responsibility assignment matrix (RAM)
	Job descriptions
	Work procedures
Control Documents:	Time sheets
	Performance evaluations

Environmental Management: This function considers all the external issues that could impact on the project.

Planning Documents:	Laws and regulations
	Environmental issues
Control Documents:	Environmental report

Issues Management: This function considers all the issues that are outside the control of the project manager and the project team.

Planning Documents:	Issues log
Control Documents:	Issues progress

Baseline Plan (Project Management Plan): The baseline plan may be considered as a portfolio of plans and documents that outlines how to achieve the project's objectives. The level of detail and accuracy will depend on the project phases and complexity. The baseline plan should be a coherent document to guide the project through the execution and project control cycle.

4. Monitor Progress (Data Capture)

Monitoring through data capture is part of the progress reporting cycle where information is regularly reported back to the project manager on the project's progress and status. The data capture function may be assumed to be at the start of the information cycle and, therefore, the accuracy of the subsequent calculations is based directly on the accuracy of the data capture. For this reason, it is extremely important for the data capture to be at an appropriate level of accuracy. Consider the following points:

Responsibility	The project team member responsible for the accuracy of the data capture needs to be clearly identified by the project manager. One method of improving data capture is to make the department that uses the information responsible for updating it. This should encourage the users to ensure that the data input is accurate.
Critical Activities	A higher level of accuracy is required on critical activities, because any delays on these activities will extend the project's duration.
Simple Report Format	The design of the reports needs to be negotiated with the managers who will be using them. The reports need to be simple and easy to use; this will help to ensure accuracy and commitment.
Written Communication	The use of written communication should be encouraged because it addresses the human failing of misinterpretation and forgetfulness.
One Page Reporting	Managers should have a propensity for one page reporting, giving quality not quantity. People are more likely to read a one page document in preference to a 50 page report.
One Item Communication	Communications should be kept to one item or one question per email. If there are two items or two questions it would be best to consider sending two emails. This approach will help gain a separate reply for each question and also help to ensure that each question is filed under its topic area.
Timing	If information is received after a decision has been made, the value of the information is reduced.
Accuracy	The accuracy of data capture will directly impact on the accuracy of any reports generated. Data capture with an accuracy of +/- 20% will give subsequent reports an accuracy of +/- 20%.
Milestones	A project manager at NASA advised, '*Use plenty of milestones to report against - people do not tend to lie, however they may be economical with the truth.*'

If these points are used as a guideline for monitoring and data capture, the quality and accuracy of the information should match the appropriate level of control.

Data Capture Format: Consider the following data capture format (table 18.1); the WBS (left hand column), the headings are time, cost, quality, procurement etc., and below the headings each cell records the status, trend and forecast, where:

- *Status* is progress to date (timenow)
- *Trend* is an extrapolation of previous status reports (see *Earned Value* chapter)
- *Forecast* is what the project manager anticipates will happen in the future (see *Earned Value* chapter).

WBS	Time	Cost	Quality	Procurement	Resources	Others
100	Status Trend Forecast	Status Trend Forecast	Status Trend Forecast	Status Trend Forecast	Status Trend Forecast	Status Trend Forecast
200	Status Trend Forecast	Status Trend Forecast	Status Trend Forecast	Status Trend Forecast	Status Trend Forecast	Status Trend Forecast

Table 18.1: Data Capture Format

Although all the parameters have an impact on each other (see chapter on each topic) and need to be monitored, the three main deliverables are time (schedule), cost (budget) and quality (specification). This section will focus on progress with respect to time management and finishing the project as planned.

Time Status: Time status can be measured as:

- Milestone (start, intermediate point, finish)
- Percentage complete
- Remaining duration.

Revised Gantt Chart: The revised Gantt chart (see *Gantt Chart* chapter) showed how progress can be presented on a timeline as milestones, percentage complete and remaining duration. This section will show how progress can be presented as percentage complete at the WBS level and then rollup to the project level. Consider table 18.2:

1. WBS	2. Planned Hours	3. Progress (PC)	4. Earned Hours
100	100 hrs	100%	100 hrs
200	80 hrs	50%	40 hrs
300	50 hrs	10%	5 hrs
400	100 hrs	0%	0 hrs
Total	330 hrs	44%	145 hrs

Table 18.2: Data Capture Format

In table 18.2 the WBS (column 1) lists the scope of work. The planned hours (column 2) assigns the budgeted hours to each work package. The progress (column 3) measures the work done to date (timenow) as a percentage complete. And the last column, earned hours (column 4), is calculated as follows:

Earned Hours = Planned Hours x PC

Earned hrs (100) = 100 hrs x 100%
= 100 hrs

Earned hrs (200) = 80 hrs x 50%
= 40 hrs

Earned hrs (300) = 50 hrs x 10%
= 5 hrs

Columns 2 and 4 can be summed to give the total planned hours (330 hrs) and the total earned hours (145 hrs). The total progress can be calculated as:

Total Progress = Earned Hours / Planned Hours

Total Progress = 145 hrs / 330 hrs
= 44%

The total progress of 44% is fairly meaningless on its own, but when compared to planned progress to date, then it gives a measure of the schedule variance and an ability to forecast if the project will be completed on time (see *Earned Value* chapter).

The weak link in this method is the measurement of percentage complete. The *Earned Value* chapter discusses a number of options to improve the accuracy, one of which is to subdivide the work package's work into pre-agreed percentages. This will be explained in the following web design office example.

'Hello, please give me your percentage complete.'

Data Capture Example: Consider a website design office with a new IT project to produce 50 websites (table 18.3). By setting up a suitable data capture template the project manager will be able to quantify the performance to date to get a feel for the project's performance.

Data Capture Template (table 18.3)**:** The data capture template lends itself to spreadsheet calculations. The website design numbers are listed in the left hand column. The scope of work is subdivided into three headings (design, construct, and test site) and weighted (20% for design, 60% for construction and 20% to test the site).

As each website could require different hours to complete, this variable is accommodated in the planned hours column where, for example, the planned hours for website 10 is 80 hrs.

The progress is reported as a percentage complete of each section, therefore, if on website number 10 the design is 100% complete, this means the project has earned 100% of 20% of the work, which equals 20% of the allocated hours (80 hrs). If the construction of the website 10 is 50% complete, this means they have earned 50% of a 60% weighting, which equals 30% of 80 hrs. By adding the earned percentages (20% + 30%) website 10 is 50% complete of 80 hrs = 40 hrs.

Website Number	Design 20%	Construct 60%	Test Site 20%	Percentage complete	Planned hours	Earned hours	Actual hours
10	100% **20%**	50% **30%**	0%	(progress) **50%** (actual)	80	**40**	40
20	100% **20%**	90% **54%**	0%	(progress) **74%** (actual)	40	**30**	**40**
30	100% **20%**	50% **30%**	0%	(progress) **50%** (actual)	60	**30**	**35**
40	40% **8%**	0%	0%	(progress) **8%** (actual)	50	**4**	**5**
50	0%	0%	0%	(progressed) (actual work done)	60	**0**	**0**
Total				36%	290 hrs	104 hrs	120 hrs

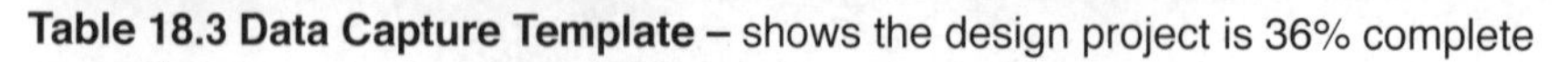

Table 18.3 Data Capture Template – shows the design project is 36% complete

Analysis: Overall, this project is 36% complete; this should be compared with the planned progress to see if the rate of work is sufficient to complete the project on time (for this example the data is not given – see *Earned Value* chapter).

To measure productivity the earned hours are compared with the actual hours. Overall the actual hours (120) are more than the earned hours (104). This means more hours are required to complete the work than was originally estimated. Therefore, action is required to either increase productivity or update the estimating data base.

Data Capture Format: Table 18.4 shows a typical data capture template for a maintenance type project. The job number refers to the items of work and the percentages refer to a typical maintenance routine. Where possible the planner should develop data capture templates tailored to the work.

WBS	Removal 10%	Inspection 5%	Repair 70%	Test 10%	Install 5%	Percentage Complete
Job 1						
Job 2						
Job 3						

Table 18.4: Typical Data Capture Template

Size of Activity: As the size of the work packages, activity and subsections are reduced so the accuracy of the data capture will increase, but so will the effort to capture the data – the project manager will need to strike a balance. As a guide, the subsections should be related to the reporting period, so that a subsection is completed within each reporting period. Therefore, for weekly reporting the subsection should not be greater than 30 to 40 hours per activity.

Gantt Chart Data Capture: The schedule Gantt chart itself can be used to capture progress data. The foreman or supervisor can mark up their progress on the schedule Gantt chart. The Gantt chart will then contain the planned work from the previous week, progress to date and planned work for the next week (this is similar to the rolling horizon Gantt chart in the *Gantt Chart* chapter). This information should be accurate and readily available as the foreman is at the operational end of the project.

'Hey Rachel, how do you like my website for the Neanderthal Weekly?!'

5. How to Apply Project Control

As this chapter has shown, there are many ways of applying project control. The table below offers a number of pointers:

Awareness	An effective way to achieve commitment is to make the contractor aware of the cost of any delay to the project.
Discuss	Any changes to the plan should be discussed with the foreman/supervisor first: • To ensure whether the changes are possible • To gather the foreman's input for the planning • To gain the foreman's commitment.
Too Busy	An excuse often used for not feeding progress back to the planner is, '*We don't have the time,*' or, '*We are too busy doing the work*'. This might be true, but the project manager needs to know that everyone is working in the right direction, and if more resources are needed.
Co-ordinate	Failure to coordinate and communicate information between departments might lead to a dissipation of company resources and duplication of effort. Effective coordination will help to cross check the work; this is a useful method for identifying discrepancies and future problems.
Priorities	It is the project manager's responsibility to establish priorities and differentiate between what is urgent and what is important. If the project manager allows the workforce to set their own priorities they might leave low paying jobs and jobs they dislike until last. This could adversely impact on the scheduling of the project.
Respond Early	It is the project manager's responsibility to respond quickly to any variation, to prevent small problems becoming disasters.
Exceptions	The project manager should encourage the team members to report on deviations and exceptions (management-by-exception) as a method of highlighting problem areas.
Estimate	The project manager should appreciate that as the schedule is only an estimate, the activities are unlikely to be exactly as per the schedule and, therefore, introduce a degree of flexibility.
Baseline Plan	Although plans should be revised to reflect the current progress, it is important not to forget the original baseline plan to guide the project to completion.

6. Reporting Frequency

The **frequency** of the reporting cycle should reflect the needs of the project; short reporting periods when there is a high level of change and uncertainty in the project, or long reporting periods when there is little or no change. For example, during the project start up and during the commissioning phase, the reporting cycle might need to be reduced to daily or even hourly while, under normal conditions, the reporting cycle is usually weekly or monthly.

As a rule of thumb, the reporting cycle should leave sufficient time to implement corrective action to bring any project deviation back on course without delaying any of the critical activities.

Exercises:

1. Develop a baseline plan for your project. Show how you name and number your documents.

2. Develop a control plan for your project. Show how you name and number your documents.

3. How do you determine the frequency of the reporting cycle?

Exercise 4: Given table 18.5, complete the earned value and percentage complete calculations (see Appendix 1 for the solution).

WBS	Planned Hours	Percentage Complete	Earned Hours
100	100	100%	
200	120	80%	
300	80	50%	
400	50	50%	
500	60	25%	
Totals			

Table 18.5: Data Capture Table - exercise 4

Exercise 5: Given table 18.6, complete the earned value and percentage complete calculations (see Appendix 1 for the solution).

Drawing Numbers	Design 20%	Drawing 70%	Checking 10%	Percentage Complete	Planned Hours	Earned Hours	Actual Hours
100	100%	0%	0%	(Measured PC)	100 hrs		20 hrs
				(Earned PC)			
200	100%	100%	0%	(Measured PC)	200 hrs		160 hrs
				(Earned PC)			
300	100%	100%	100%	(Measured PC)	80 hrs		100 hrs
				(Earned PC)			
400	50%	50%	0%	(Measured PC)	120 hrs		60 hrs
				(Earned PC)			
500	100%	50%	10%	(Measured PC)	100 hrs		40 hrs
				(Earned PC)			
Totals					600 hrs		380 hrs

Table 18.6: Data Capture Table - exercise 5

19

Earned Value

Earned value is a powerful planning and control technique that integrates costs and time, or man hours and time to give a true measurement of progress in comparable units.

Although the earned value technique was initially set up to track the progress of cost and time, in practice it is usually more appropriate to track progress measured as earned man hours and time. In fact, any variable that flows through the project can be used; tonnes of steel, cubic meters of concrete, metres of pipe, or pages of a document.

Earned value measurements compare actual work expended against work performed (earned), to give a measure of efficiency. It also compares work performed (earned) against the planned work, to give a measure of schedule performance.

1. Earned Value Terminology

Earned value, more than any other planning and control technique covered in this book, is shrouded in esoteric terminology. The key to mastering earned value is to understand the terms. It may be argued that to enter the field of project management, it is essential to speak the language of project management.

Budget at Completion	BAC
Planned Value	PV
Percentage Complete	PC %
Earned Value	EV = PC * BAC
Actual Value	AV
Estimate at Completion	EAC = (AV / EV) * BAC
Scheduled Variance	SV = EV - PV
Cost Variance	CV = EV - AV

Table 19.1: Table of Earned Value Terms and Abbreviations

Budget at Completion (BAC): Is the original cost estimate or quotation, indicating the funds required to complete the work. For example in figure 19.1 the BAC is $1,000. At the project management level the BAC does not include profit. The reason for this will become clear later when the actual costs are compared with the planned costs. The BAC becomes a generic term when man hours or another parameter are used.

Planned Value (PV): Is the integration of cost and time or, more commonly, man hours and time to give the characteristic S curve which forms the baseline plan (see the *Project Cashflow* chapter which explains how to draw an S curve).

Percentage Complete (PC): The PC is a measure of the activities performance and progress up to timenow and is required for the earned value calculation. For this example the PC is 40% at timenow. Timenow is the date up to which the progress is measured.

Earned Value (EV): Is a measure of achievement or value of the work done to timenow. The EV is calculated by the equation:

EV = PC (earned progress at timenow) x BAC
= 40% x $1,000
= $400

Actual Value (AV): Is the amount payable for the work done to timenow. It is the real cost incurred executing the work to achieve the reported progress. For this example consider at timenow EV is $600.

Estimate at Completion (EAC): The EAC is a revised budget for the activity, work package or project, based on current productivity. The EAC is calculated by extrapolating the performance trend from timenow to the end of the project. This value assumes that the productivity to-date will continue at the same rate to the end of the project. The productivity is defined by the ratio of actual (AV) to earned value (EV). If the actual (AV) are less than the earned value (EV), then the EAC will be less than the BAC and vice versa.

$$EAC = \frac{AV}{EV} \times BAC$$

But $$EV = PC \times BAC$$

Therefore $$EAC = \frac{AV}{PC \times \cancel{BAC}} \times \cancel{BAC}$$

$$EAC = \frac{AV}{PC}$$

For this example at timenow (see figure 19.1)

$$EAC = \frac{\$600}{40\%} \times 100$$

$$= \$1500$$

The budget variance is therefore BAC - EAC, \$1,000 - \$1,500 = -\$500 or (\$500). The project is forecast to be \$500 over budget.

Schedule Variance (SV): The schedule variance calculation is a measure of the time deviation between the planned value (PV) and the earned value (EV). The interesting feature about this time variance is that it is measured in money units.

$$SV = EV - PV$$
$$= \$400 - \$500$$
$$= -\$100$$

The sign of the variance will indicate if the project is ahead or behind the planned progress.

Negative variance: The project is behind the planned progress.

Positive variance: The project is ahead of the planned progress.

Cost Variance (CV): The cost variance is a measure of the deviation between the earned value (EV) and the actual value of doing the work (AV) (see figure 19.1).

$$CV = EV - AV$$
$$= \$400 - \$600$$
$$= -\$200$$

The sign of the variance will indicate if the costs are under or over the estimate.

Negative variance: The cost is higher than the original estimate (BAC).

Positive variance: The cost is lower than the original estimate (BAC).

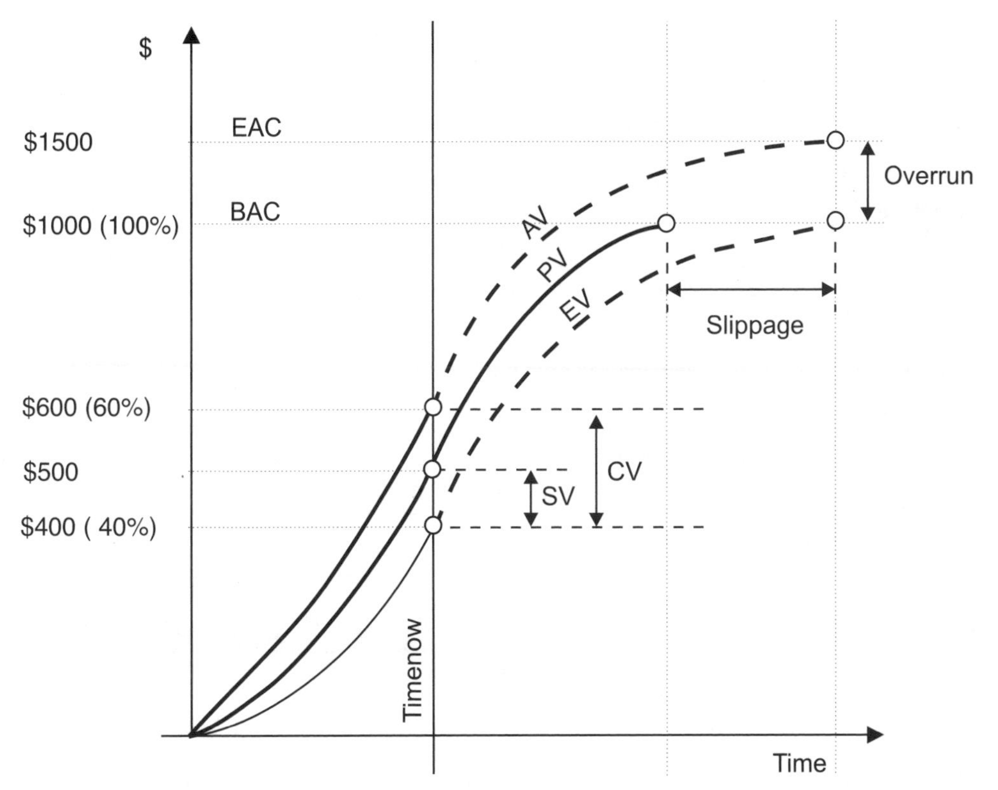

Figure 19.1: Earned Value Graph – shows the planned value, actual value and earned value curves, where the X axis represents the project's schedule and the Y axis represents the level of work or value

2. Earned Value Graph

The earned value graph (see figure 19.1) is produced using the following steps:

Step 1:	Draw the planned value (PV) curve (see how to draw an S curve in the *Project Cashflow* chapter). This shows the planned budget and completion date.
Step 2:	Draw the EV curve to timenow and extrapolate until the line intersects with BAC. This intersection will give a forecast completion date. This completion date, however, should not be looked at in isolation because it does not consider the network logic, critical path and timing of the activities.
Step 3:	Draw the actual value (AV) curve to timenow and extrapolate to the new end date of the project and EAC. Where EAC = (AV / EV) x BAC. This equation assumes progress to timenow will continue at the same rate to the end of the project.
Step 4:	Draw the SV and CV variances.
Step 5:	Determine how far the project is ahead or behind.

Table 19.2: Earned Value Graph - shows steps to draw the earned value graph

3. Earned Value Table

The earned value data can be presented in both a tabular format (table 19.4) and/or a line graph (figure 19.1). Consider the following steps to set up the earned value table 19.3:

Step 1:	Set up an earned value table (table 19.4) using the following abbreviated field headings (see table 19.1).
Step 2:	List the full scope of work in the WBS column.
Step 3:	Input BAC values for all work packages.
Step 4:	Calculate PV to timenow.
Step 5:	From the data capture sheet transfer the values for PC and AV.
Step 6:	Calculate EV = BAC x PC
Step 7:	Calculate SV, CV and EAC
Step 8:	Sum the following columns: BAC, PV, EV, AV and EAC.
Step 9:	Calculate the total PC, SV and CV.

Table 19.3: Earned Value Table – shows the steps to compile the earned value table

WBS	BAC	PV	PC	EV	AV	SV	CV	EAC
1000	$1000	$500	40%	$400	$600	-$100	-$200	$1500
2000								
Total								

Table 19.4: Earned Value Table - shows the values to timenow from figure 19.1

4. Project Control

Determining Percentage Complete: The weakest link in the earned value calculation is determining percentage complete (PC). If the activity has not started it is zero and if it is complete it is 100%, but all points in between are somewhat of a guess even if a more structured approach is used. A quick way to estimate PC is to use the 50/50 rule, if the job has started it is given 50% and when it is finished it is given 100%. This rule can be distorted to 40/60, 30/70, 20/80 and 10/90. If the work packages or activities are kept small, or short (less than 50 hours), then this method will work well. Consider the following:

- The earned value analysis should not be used in isolation. An activity with a large schedule variance may have plenty of float and not be a problem, while an activity with a small schedule variance may be on the critical path and need prompt action to prevent project over-run.
- Estimate-at-completion (EAC) is based on the ratio of past performance, but if the original estimate is fundamentally flawed, or performance is significantly different to planned, then the rest of the project should be re-estimated.
- NASA manager: '*Earned value is only useful if the difference between planned and actual is 10% to 15%, if greater then use other methods.*' [Perhaps use a rolling horizon barchart]

If the progress indicates that both SV and CV are negative and significant, then ordering a small contractor to increase their resources may actually put them into liquidation quicker! There is obviously a problem which needs to be investigated - is the contractor's estimate over optimistic or is the contractor's productivity under performing?

During the project the activities are usually at various stages of completion; some on target, some ahead of plan, some behind plan, some on budget, some overspent and some under spent. In this situation it is extremely difficult to quantify the project's overall status visually, and it may be argued that a subjective assessment on a complex project is bound to be inaccurate. This problem can be addressed by using the earned value model to roll-up all the activity data and report a bottom line for the project giving an overall position.

5. Client's View of Earned Value

So far the earned value technique has only been looked at from the sub-contractors' point of view. This section will consider the client's position:

Fixed Price	If the sub-contractors are working to a fixed price contract then EV and AC will always be the same.
Progress	The client can effectively use earned value to track the progress of the project in terms of man hours or costs.
Over Claim	The client must check that sub-contractors do not over claim. If a sub-contractor has claimed 80% of the contract by value, but only completed 50% of the work, there is little financial pressure that can be exerted.

This section clearly shows that earned value can provide the client with a powerful planning and control tool.

6. Earned Value Reporting

The earned value output lends itself to effective reporting for the following reasons:

Overall Status	The overall status of the project can be seen at a glance on a graph and the tabular reports present more detailed information at the work package and activity levels.
Responsibility	When reporting to functional management the report should clearly indicate the activities that fall under their responsibility. This information can be separately reported if a responsibility field has been included in the data base.
MBE	The reports can use a management-by-exception (MBE) technique to identify problem areas. The MBE thresholds can be set using any of the following: a) Threshold variance SV and CV. b) Activity float = 0 days, identifies the critical path, or sets activity float < 5 days, to identify activities which could go critical in the next week.
Trends	Plot trends wherever possible to indicate the direction of the project.
Extrapolate	Extrapolating trends will give an indication of future events and a quick feedback on recent actions. Even if the variances are negative, but reducing, this will show a positive trend indicating that the project is coming back on course.

Earned Value Example

The earned value technique is best learned by walking through an example. Consider a simple project to decorate the four walls and ceiling of an office. The contractor estimates that the preparation and painting will take 2 man days (16 hrs) per wall plus 2 man days for the ceiling. The contractor plans to complete the ceiling in one day, and complete one wall per day, starting work on Monday and finishing on Friday. The materials are available and charged separately.

Activity	BAC	Mon	Tues	Wed	Thurs	Fri
100 – Ceiling	16 hrs	16 hrs				
200 - Wall 1	16 hrs		16 hrs			
300 - Wall 2	16 hrs			16 hrs		
400 - Wall 3	16 hrs				16 hrs	
500 - Wall 4	16 hrs					16 hrs
Daily Total	80 hrs	16 hrs	16 hrs	16 hrs	16 hrs	16 hrs
Running Total PV		16 hrs	32 hrs	48 hrs	64 hrs	80 hrs

Table 19.5: Painting Example Plan - shows the scope of work and schedule

Progress Report: Timenow 1 (end of Monday)

WBS Activity	Percentage Complete	AV Actual Hours
100 – Ceiling	100%	16 hours
200 – Wall 1		
300 – Wall 2		
400 – Wall 3		
500 – Wall 4		

Earned Value Table at timenow 1 (end of Monday)

WBS Activity	BAC Budget	PV Plan	PC	EV Earned	AV Actual	SV	CV	EAC
100	16 hrs	16 hrs	100%	16 hrs	16 hrs	0	0	16 hrs
200	16 hrs	0						16 hrs
300	16 hrs	0						16 hrs
400	16 hrs	0						16 hrs
500	16 hrs	0						16 hrs
Totals	80 hrs	16 hrs 20 %	20%	16 hrs	16 hrs	0 hrs	0 hrs	80 hrs

Table 19.6: Earned Value Table at Timenow 1 - shows the project is on time and within budget

Exercise 1:

Continue with this painting example and produce the earned value table similar to table 19.6 for timenow 2, and timenow 3 (see Appendix 1 for solutions).

Progress Report: Timenow 2 (end of Tuesday)

WBS Activity	Percentage Complete	AV Actual
100 – Ceiling	100%	16 hrs
200 - Wall 1	50%	20 hrs
300 - Wall 2		
400 - Wall 3		
500 - Wall 4		

Progress Report: Timenow 3 (end of Wednesday)

WBS Activity	Percentage Complete	AV Actual
100 – Ceiling	100%	16 hrs
200 - Wall 1	100%	24 hrs
300 - Wall 2	100%	8 hrs
400 - Wall 3	50%	4 hrs
500 - Wall 4		

20

Project Risk Management

Learning Outcomes

After reading this chapter, you should be able to:

- Understand the risk management technique
- Identify project risks and opportunities
- Respond to project risks
- Produce a risk management plan
- Control project risks

The *Project Risk Management* knowledge area includes the risk planning and control techniques required to identify, analyse, respond to and control project risk. The risk management objective is to maximize the impact of positive events and minimize the impact of negative events.

Project management and project risk management both strive to achieve the project's objectives, but in different ways. Where project management strives to maximize the chances of success, project risk management strives to minimize the chances of failure.

Company success is achieved by pursuing business opportunities to gain competitive advantage. Projects and new ventures are typically set up to take advantage of these opportunities – to make something new, to enhance an existing facility, or to respond to market needs. Consequently, risk is an intrinsic part of project management. With increasing market competition, increasing technology and an increasing rate of change, risk management is gaining more significance and importance.

Project Management Plan Flowchart

The project management plan flowchart (figure 20.1) shows the relative position of risk management with respect to the other topics. One could argue that risk management should follow estimating (time and costs), as an estimate is a prediction of the future and all predictions, by definition, have inherent risks and uncertainties but, in practice, risks are embedded in every plan. For ease of presentation, risk management is presented here after all the planning and control techniques have been explained, and grouped together with the quality management, communication management, and human resource management chapters.

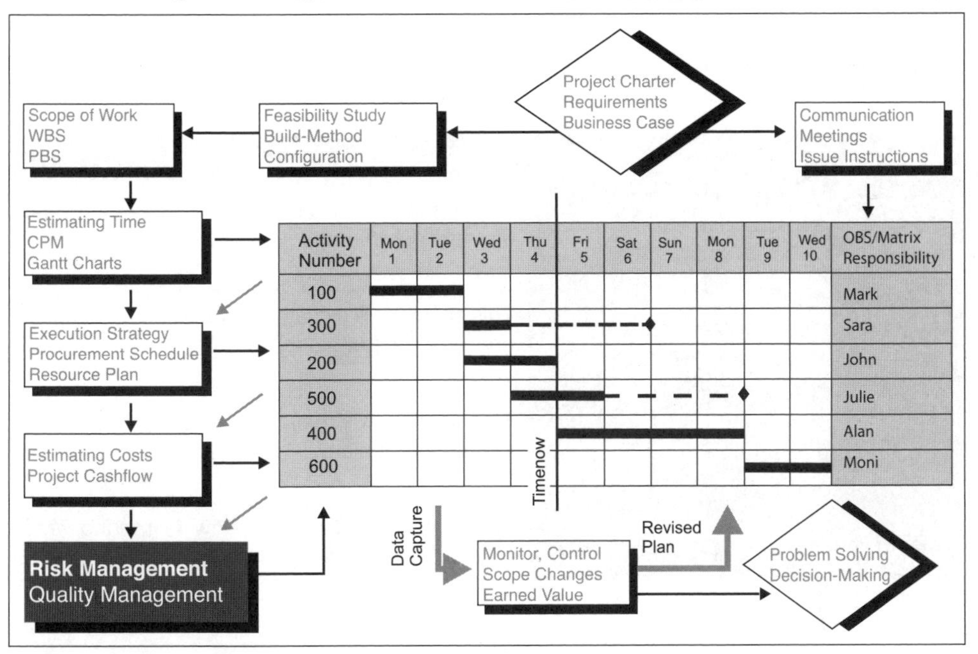

Figure 20.1: Project Management Plan Flowchart – shows the relative position of risk management with respect to the other knowledge areas

Body of Knowledge Mapping

PMBOK 4ed: The *Risk Management* knowledge area includes the following:

PMBOK 4ed	Mapping
11.1: Plan risk management	This chapter will explain how to develop a risk management plan.
11.2: Identify risks	This chapter will explain how to identify project risks.
11.3: Perform qualitative analysis	This chapter will explain how to analyse the probability and impact of a risk occurring.
11.4: Perform quantitative analysis	See PMBOK 4ed for details on quantitative analysis.
11.5: Plan risk responses	This chapter will explain how to develop a response to the identified risks.
11.6: Monitor and control risk responses	This chapter will explain how to control the risk management process.

APM BoK 5ed: The *Risk Management* knowledge area includes the following definition:

The APM BoK 5ed defines **Project Risk Management** as, *a structured process that allows individual risk events and overall project risk to be understood and managed proactively, optimising project success by minimising threats and maximising opportunities.*

Unit Standard 50080 (Level 4): The unit standard for the *Risk Management* knowledge includes the following:

Unit 120374: *Contribute to the management of project risk within own field of expertise*

Specific Outcome	Mapping
120374/SO1: Identify and recognize potential risks that could affect project performance.	This chapter will explain how to identify and quantify risk events.
120374/SO2: Contribute to the assessment of the impact and likelihood of identified risks.	This chapter will explain how to assess the impact and likelihood of identified risks.
120374/SO3: Contribute to the development of risk management statements and plans.	This chapter will explain how to develop a risk management plan.
120374/SO4: Monitor and control the project risks.	This chapter will explain how to monitor and control the risk management plan

1. Risk Management Model

The generally accepted risk management model subdivides the risk management process into the following topics as shown in figure 20.2.

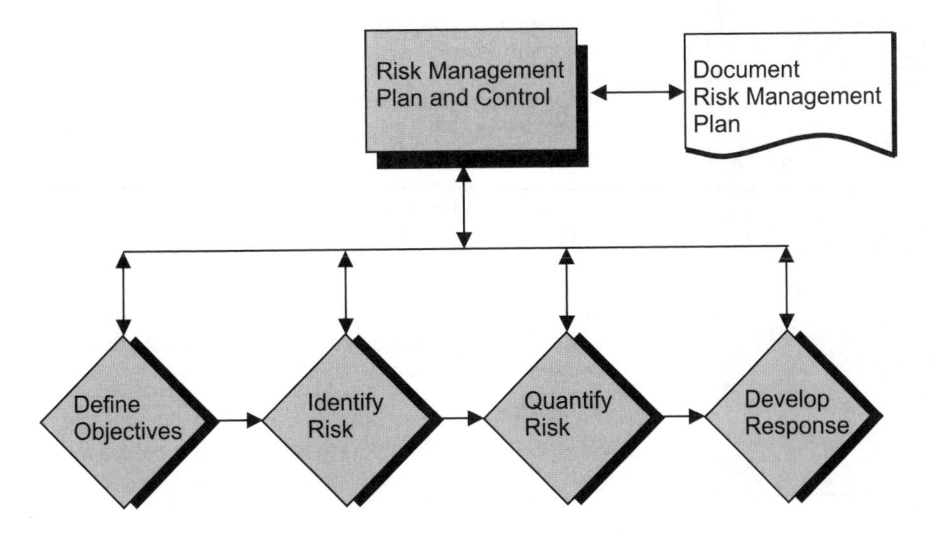

Figure 20.2: Risk Management Model – shows the main components of risk management

Define Objectives	The project's objectives define the deliverables the project sets out to achieve. A WBS can be used to define objectives at the work package level.
Identify Risks	Risk identification identifies areas of risk, uncertainty and constraints, which might impact on the project and limit or prevent the project achieving its objectives.
Quantify Risks	Quantifying risk evaluates the risks, prioritizes the level of risk and uncertainty, and quantifies the frequency of the risks occurring and the level of impact on the objectives.
Develop Response	Response development defines how the project will respond to the identified risks; this could be a combination of elimination, mitigation, deflection or acceptance.
Risk Management Plan	The risk management plan documents how the project team propose to manage project risk by identifying, quantifying, responding to and controlling project risk.
Risk Control	The risk control function implements the risk management plan. This might involve in-house training and communication to the other stakeholders. As the risks and the work environment are continually changing, it is essential to constantly monitor and review the level of risk and the project team's ability to effectively respond to the risks.

Risk Management Plan Example: This example develops a risk management plan for a fashion project to design, manufacture, deliver and sell a range of garments for the winter season. The risk management plan could be structured as follows:

Define Objectives	Design, manufacture, deliver and sell a new range of garments for the winter season that reflects the latest trends and is competitively priced for the high street market.
Identify Risks	Late delivery – if the garments are delivered late and miss the winter season selling period, this will reduce the sales figures and devalue the garments to sale price.
Quantify Risks	Late delivery of garments can be caused by the late deliveries of fabrics and trims to the manufacturer. The frequency of this occurring is low, but the impact would be high.
Develop Response	Order fabrics and trims early in the production cycle and expedite the order to ensure the delivery dates are met. Develop a contingency plan to use alternative available fabrics and trims.
Risk Control	Monitor the risk management plan to quickly identify and respond to any unexpected problems. Closeout the project and make recommendations for next season's manufacturing project.

'As I was saying, I've got a great book on risk management.'

2. Define Objectives

A risk may be defined as any risk event or situation that prevents the project achieving its defined goals and objectives (deliverable), conversely, an opportunity is an event that enables the project to exceed its original goals and objectives. It is, therefore, necessary at the outset to define as accurately as possible the project's goals and objectives. This is best achieved through a work breakdown structure matrix, where the WBS column of work packages lists the full scope of work, and the other columns list the associated activities and the objectives including any vendor specifications and statutory requirements. This can be expanded to include the following objectives:

WBS	Description	Objectives
1.1	Paint structure	Paint to manufacturer's requirements (usually shown on side of paint tin).
1.2	Fashion Show	Be ready for the event on the planned date and time (schedule requirement).
1.3	House extension	Build house extension to the assigned budget.
1.4	House extension	Build house extension to comply with building regulations and other statutory requirements.
1.5	House extension	Complete house extension on time.

Table 20.1: Objectives by WBS – shows types of objectives at the work package level

3. Risk Identification

Having defined the project's objectives, the next step is to identify the risks and uncertainties that could:

- Prevent the project achieving the stated objectives and deliverables
- Prevent the project making the most of the opportunities that could enable the project to exceed the stated objectives.

Experience: Project managers' risk identification skills grow with experience from the problems and situations they have encountered or observed. Greater awareness and appreciation is followed by greater knowledge and judgment.

However, project managers new to project management will have limited work experience and, therefore, need to develop risk identification techniques to identify potential risks, perceived risks, risk triggers and risk causes and effects. It is essential risks are identified early on otherwise they will be excluded from further analysis and will only be identified once they have occurred and impacted on the planned objectives.

The process of risk identification should not be a one-time event, but rather a continuous process, its frequency depending on the project's level of risk and the schedule of the progress meetings.

Using the list of objectives developed in table 20.1 as a starting point, two more columns are added for risk and opportunities.

WBS	Description	Objectives	Risk	Opportunity
1.1	Paint structure	To meet manufacturers requirements	Paint system fails	Referrals for more projects
1.2	Fashion show	Present on time	Presentation late, miss slot	Increase sales to buyers present
1.3	House extension	Build within budget	Price increase	Better discounts
1.4	House extension	Comply with regulations	Work rejected by council	Referrals for more projects
1.5	House extension	Finish on time	Late finish penalty	Early finish bonus

Table 20.2: Risk Identification

Risk identification should be a systematic process to ensure nothing significant is overlooked. Techniques for identifying risk include:

- Analysing historical records (closeout reports)
- Structured questionnaires
- Structured interviews
- Brainstorming
- Structured checklists (see the *WBS* chapter)
- Flow charts (build-method, walk-through)
- Judgment based on knowledge and experience
- Scenario analysis (what-if?).

Structured questionnaires, structured interviews and brainstorming are all methods used to generate ideas and receive feedback from colleagues, stakeholders, clients, engineers, suppliers, and governing agencies. Checklists, breakdown structures and flow charts (CPM) are used to group and subdivide information for collation and presentation.

Potential, perceived and actual risk events should be identified, documented and communicated in consultation with the appropriate stakeholders. Risk triggers, causes and effects, and risk owners should be identified in consultation with the appropriate stakeholders and recorded in accordance with project procedures.

The success of these techniques depends on how the risk management team has been selected and brought together. A balanced team that incorporates experience, knowledge, judgment, internal members and external consultants, will stand the best chance of success.

Why Projects Fail: Historically one of the main reasons for developing project management techniques was to address the high rate of project failure; usually identified as late delivery, over budget, under performance, poor quality and the deliverables failing to solve the client's problem or need. Listed below are some of the problem areas:

Topic	Problem
Client	The client was dealing with too many managers, causing confusion and duplication of work.
Feasibility Study	The project was built before a prototype was tested, resulting in unrealistic expectations and expensive retro-fitting.
Build-Method	The build-method was not thought through before the start of the project, leading to expensive changes in the way work was carried out.
Scope	The scope of work and objectives were unclear, causing confusion.
WBS	The WBS was not developed into work packages, leading to the project being managed at a high level (less detail).
Cost	The budgets were not defined, creating uncontrollable expenditure.
Time	The time estimates were inaccurate and the schedules were not defined, leading to insufficient time allowed to complete the work.
Procurement	The long lead items were not identified at the outset, leading to late ordering and, therefore, late delivery delaying the associated activities.
Resources	The resource loadings were not anticipated, leading to insufficient resources that delayed the associated activities' completion.
Quality	The quality requirements were not identified, leading to unsatisfactory work, disputes and rework.
Leadership	There was no single project leadership, resulting in a number of functional managers working independently.
OBS	The work was passed from one functional department to another with no-one integrating and co-ordinating the project.
Team Work	There was no team leader, resulting in dysfunctional teams and interpersonal conflict.
Communication	The lines of communication were unclear and documents were not controlled, resulting in workers using old revisions.
Instructions	The instructions were given verbally and not backed up in writing, leading to confusion and misunderstandings.
Progress Measurement	The monitoring and tracking of progress was poor which meant the project manager did not know where best to allocate the resources.
Stakeholders	There was disagreement amongst stakeholders regarding the needs and expectations for the project, causing dissatisfaction with the end result.
Operators	There was a lack of user involvement in the design which led to a product facility which was difficult to operate.

Table 20.3: Why Projects Fail – shows risks identified from previous closeout reports

4. Risk Quantification

Having identified a range of project risks, the next step is to quantify the probability of the risk occurring and the likely impact or consequence to the project, or the amount at stake. Likelihood of risk causes and impact are assessed in consultation with the appropriate stakeholders. Risk quantification is primarily concerned with determining what areas of risk warrant a response. Where resources are limited, a risk priority will identify the areas of risk that should be addressed first.

Quantifying risks sounds great on paper, but it is often very difficult in practice to put an exact number to a risk. To overcome this problem, a risk can be ranked as a high, medium, or low risk - this should highlight areas of high risk which warrant further investigation. With finite resources, it is essential to establish which risks should be addressed first so that the project team members can prioritize and focus their efforts on the high risk areas.

5. Risk Response

Having identified, quantified and prioritized the risks, the next step is to develop a risk response plan which defines ways to address adverse risk and enhance opportunities before they occur. Consider the following possible responses:

- Eliminate risk
- Mitigate risk
- Deflect risk
- Accept risk (with a contingency)
- Create an opportunity.

These responses are not mutually exclusive - the response might be to use a combination of them all. A natural sequence would be to first try and eliminate the risk completely; - failing that, at least mitigate the risk. And for the remaining risks the options are to try and deflect and/or accept with a contingency. Responses cost money, so a cost-benefit analysis should be performed, as there might be situations where it is more cost effective to accept a risk rather than taking expensive steps to eliminate it.

Eliminating Risk: Eliminating risks looks into ways of containing or avoiding the risk completely by either removing the cause or taking an alternative course of action. This should be considered during the concept and design phases, where the level of influence is high and the cost to change is low (see Burke R., *Project Management Techniques*).

Mitigating Risk: To mitigate a risk means reducing the risk's probability and impact. Activities to reduce or prevent a risk threat arising should be identified and documented in an agreed format. Mitigating a risk can be achieved by using proven technology and standards to ensure the product will work. Developing prototypes, simulating and model testing are three methods that share the notion of using a representation to investigate selected aspects of requirements in order to be more certain of the outcome or suitability. A prototype is a working mock-up of areas under investigation in order to test its acceptance, while a model is a miniature representation of physical relationships (often used in car and ship design).

Deflecting Risk: Deflecting a risk transfers the risk in part or whole to another party which can be achieved through contracting and outsourcing.

Contracting: Project contracts are a means of deflecting risk, usually deflecting away from the client to the contractor or supplier. There are a number of different types of contracts used which include:

- Fixed price contract
- Cost plus contract
- Unit rates contract
- Turnkey contract
- BOOT contract.

Acceptance: Accepting the risk means the project manager or client accepts the consequence of a risk occurring, also referred to as self-insurance. If this is the case, the project manager might need to develop a contingency plan to protect the business from the risk event.

Contingency: A contingency plan defines the actions the project manager will take ahead of time - if *'A' happens, we will do 'B'*. Activities to recover from a specific risk threat event should be identified and documented in an agreed format. For example, when planning a delivery trip, if there is an accident on route 'A', the contingency would be to have an alternative route 'B'.

The APM BoK 5ed defines **Contingency** as, *the planned allotment of time and cost or other resources for unforeseeable risks within a project. Something held in reserve for the unknown.*

Opportunity: Taking advantage of opportunities to enhance project success brings out the project manager's entrepreneurial flair. Activities to enhance an opportunity should be identified and documented in the agreed format. For example, during the course of the project there might be an opportunity to change to a more cost effective build-method, or change to a different supplier who is offering better discounts for the building materials.

A summary table should be developed to compile all the identified risks and their associated responses (see table 20.4 below).

WBS	Description	Objectives	Risk	Response
100	Project documents	Safe keeping	Fire and floods	Back-up and store off site
200	Production	Meet schedule	Insufficient resources	Approach list of contractors
300	Sales	Meet sales target	Increase competition	Reduce prices

Table 20.4: Risk Response – shows the response to identified risks

Risk Management Plan: The risk management plan documents the output from the previous sections:

- State objectives
- Identify risks
- Quantify risks
- Respond to risks.

The risk management plan assigns responsibility for implementation of the plan. The next section on risk control implements the risk management plan to create a working document.

6. Risk Control

The risk control function implements the risk management plan and makes it happen – surprisingly this part of risk management is often neglected! The risk management plan needs to be communicated to all the project participants in an agreed format and, where necessary, followed up with appropriate training and practice runs. The training should not only ensure that the risk management plan is understood, but also develop a company wide risk management culture and attitude. Risk control might include the following:

- Potential project risk events should be monitored to enable anticipation or recognition of an occurrence.
- Risk issues arising should be responded to and monitored until resolved.
- Risk variances should be identified and reported to higher project authority.
- Agreed risk responses should be implemented in accordance with the risk management plan.
- Lessons learned should be documented and communicated to the relevant parties managing the phase reviews and closeout report.

The risk management plan should be monitored and updated on a regular basis as part of the progress meeting to incorporate changes in the project.

WBS	Objectives	Risk	Respond	Control
100				
200				
300				

Table 20.5: Risk Control – shows how the risks will be controlled

Exercises:

1. Discuss the need for risk management. Where possible relate to risk management issues on projects you are familiar with.
2. Explain how you identify and quantify project risks together with your appropriate response.
3. Identify opportunities you have taken advantage of to enhance the building or operation of a project.

Further Reading:

Burke, R., *Project Management Techniques*, will develop the following topics which relate to this knowledge area:

- Project risk lifecycle
- Risk responsibility organization structure
- Retention and performance bonds
- Methods of contracting out risk
- Disaster recovery.

Burke, R., **Barron**, S., *Project Management Leadership*

21

Quality Management

Learning Outcomes

After reading this chapter, you should be able to:

- Understand a range of quality definitions
- Determine the project's quality requirements
- Establish the project's quality capability
- Produce the project's quality control plan

The *Project Quality Management* knowledge area includes the techniques to determine the project's quality policies, objectives, and responsibilities so that the project will satisfy the needs for which it was undertaken.

Quality is a much misused and misunderstood word. In everyday language quality implies superior characteristics and a better product but, in the project management environment, quality means a management system that covers:

- Determination of the project's quality requirements
- Setting up a process to develop and establish the capability to achieve the quality requirements
- Setting up a control process to confirm the project has achieved the quality requirements.

Project Management Plan Flowchart

The project management plan flowchart (figure 21.1) shows the relative position of the quality management knowledge area with respect to the other topics. Although there is an element of quality management embedded in every plan, for ease of presentation, quality management is grouped together with the risk management, communication management and organization structure chapters.

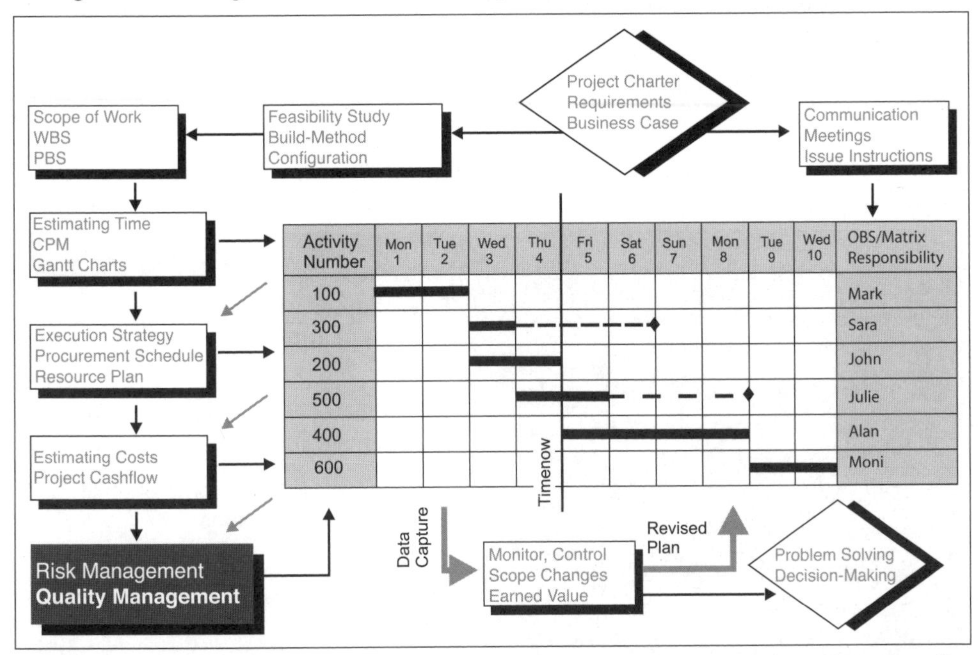

Figure 21.1: Project Management Plan Flowchart – shows the relative position of quality management with respect to the other knowledge areas

Body of Knowledge Mapping

PMBOK 4ed: The *Quality Management* knowledge area includes the following definition and sections:

The PMBOK 4ed defines the **Project Quality Management Process** as, *the processes and activities of the performing organization that determine quality policies, objectives, and responsibilities, so that the project will satisfy the needs for which it was undertaken.*

PMBOK 4ed	Mapping
8.1: Plan quality	This chapter will explain how to determine the quality requirements.
8.2: Perform quality assurance	This chapter will explain how to set up a quality assurance system to ensure the project has appropriate quality standards.
8.3: Perform quality control	This chapter will explain how to develop a quality control plan.

APM BoK 5ed: The *Quality Management* knowledge area includes the following definition:

> The APM BoK 5ed describes **Project Quality Management** as, *the discipline that is applied to ensure that both the output of the project and the processes by which the outputs are delivered meet the required needs of stakeholders.*

Unit Standard 50080 (Level 4): The unit standard for the *Quality Management* knowledge area includes the following:

Unit 120383: *Provide assistance in implementing and assuring project work meets quality requirements*

Unit 120377: *Identify, suggest and implement corrective actions to improve the quality of project work*

Specific Outcomes	Mapping
120383/SO1: Describe and explain the need for consistent processes and standards to achieve quality.	This chapter will explain the need for consistent processes to achieve the required level of quality.
120377/SO1: Describe and explain how quality management impacts on a project.	This chapter will explain how quality management impacts on the project.
120377/SO2: Identify and record corrective actions for improvement to project work.	This chapter will explain how to identify and record corrective actions.
120377/SO3: Disseminate corrective actions to appropriate stakeholders.	This chapter will explain how to disseminate corrective actions.
120377/SO4: Implement corrective actions to improve quality of project work.	This chapter will explain how to implement corrective actions.

'Royce, old chap, please make sure the car is of the highest quality.'

1. Quality Definitions

It is important not to confuse quality with degree of excellence or grade, where grade is a category of rank given to products which have the same function but different quality requirements. For example, Rolls Royce and Mini cars are often quoted as being at opposite ends of the quality continuum - the Rolls Royce being built to a much higher quality than the Mini. However, if a person wishes to buy an economical small car that will do 50 + miles to the gallon and is easy to park, then the Mini is the quality car that '*conforms to the client's requirements*'.

The PMBOK 4ed defines **Quality** as, *the degree to which a set of inherent characteristics fulfil requirements.*

To explain the definition, if a Rolls Royce is found to have a number of defects it would be said to have failed to '*fulfil the requirement*' and, therefore, be considered a low quality product. Meanwhile, if a Mini were found to be defect free it would be said to have '*fulfilled the requirements*' and, therefore, be a high quality product.

Quality Planning: Is the process of identifying the quality standards the project needs to comply with to achieve the required condition and satisfy the terms of the contract.

Quality Assurance: Is a systematic process of defining, planning, implementing and reviewing the management processes within a company in order to provide adequate confidence that the company is able to consistently manufacture the product to the required condition.

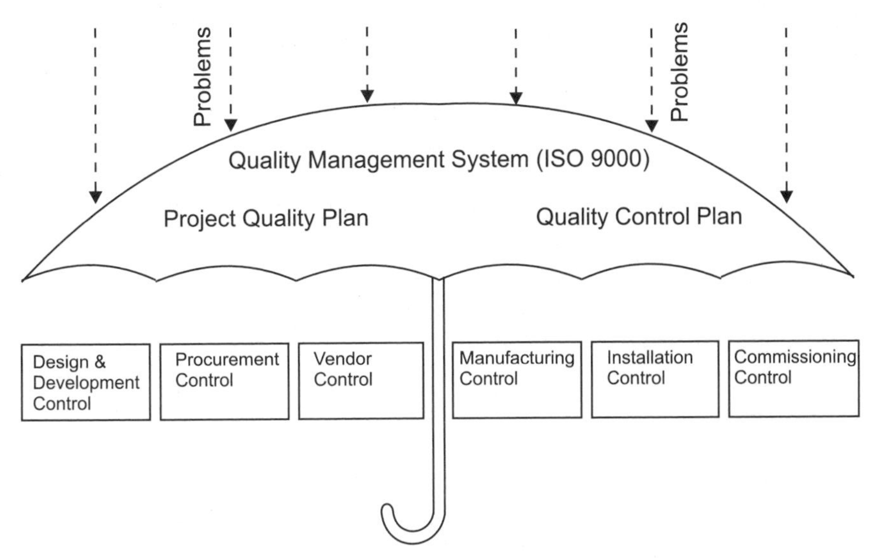

Figure 21.2: Quality Assurance Umbrella - shows how the quality management system protects the project from an array of problems

Quality Control: Is the process companies go through to confirm that the product has reached the required condition as determined by the specifications and the contract.

Quality Control Plan: Table 21.2 shows how the quality control plan integrates the three variables of:

- The project's scope of work
- The project's schedule
- The level of inspection.

Quality Circles: Quality circles are a management concept the Americans set up for Toyota cars in Japan after the Second World War, to continuously improve their manufacturing process by bringing all the people in a production line together to identify and solve problems.

Quality Audit: An audit in the project management context should be seen as a search for more information to assist problem solving and decision-making. This might be motivated by a non-conformance report (NCR).

The APM BoK 5ed defines an **Audit** as, *the systematic retrospective examination of the whole, or part, of a project or function to measure conformance with predetermined standards.*

Quality Training: Quality training is a company wide issue, from the CEO to the receptionist. All employees should undertake quality training so that they understand and can contribute to the quality of the management system and the quality of the project.

Project Quality Plan: The project quality plan is a detailed document explaining how the company will ensure that the project will be made to the client's requirements. The sub-headings from the ISO 9000 quality management system can be used to structure the document.

Total Quality Management (TQM): TQM considers the wider aspects of quality by integrating all of the quality management features. Total quality has a people and outcome focus. It first identifies what the client really wants and how it can best be achieved. TQM keeps an emphasis on continuous improvement, but also aims to keep the customer satisfied. For quality to be effective it needs to be introduced to all members and to all aspects of the company's operation. TQM advances the rationale that just as each organization has unique products, so each company needs a unique quality management system.

2. Quality Planning

Quality planning is the process of identifying the quality standards to be achieved and the testing required to confirm the project has achieved the required condition to satisfy the terms of the contract. This will require input from the stakeholders to determine their needs and expectations.

Satisfying the stakeholders' requirements is a combination of:

- Conformance to the requirements - this might be quantified in terms of the deliverables, scope, time, cost and quality standards.
- Testing and commissioning to confirm the project has reached the required quality.
- Fitness for purpose, or fitness for use - this might be qualified in terms of the benefits the company receives from the project – does the project solve the problem, and/or can the client make a return on his investment?

The PMBOK 4ed defines **Quality Planning** as, *the process of identifying quality requirements and/or standards for the project and product, and documenting how the project will demonstrate compliance.*

Table 21.1 shows a quality planning summary sheet where:

- Column 1: Lists the scope of work
- Column 2: Identifies the stakeholders needs and expectations
- Column 3: Determines the required condition
- Column 4: Develops a capability to make the project
- Column 5: Confirms the project has achieved the requirements.

1. WBS	2. Stakeholders	3. Objectives	4. Quality Assurance	5. Quality Control
100	Event	Ready on time	Develop accurate time planning capability	Monitor and control progress
200	Fixed price contract	Fixed budget	Develop project accounting capability	Monitor and control accounts
300	Military project	Quality requirements	Develop quality management capability	Monitor and control quality

Table 21.1: Quality Planning Summary Sheet

Table 21.1 shows how the quality requirements for time, cost and quality can be planned. This template can be used for the other knowledge areas; scope, procurement, resources, communication and risk.

Benchmarking: Compares actual and planned project practices to those used on similar projects to identify best practice, generate ideas for improvement, and provide a basis for measuring performance.

Quality Management Plan: The quality management plan outlines how the project team will implement the quality management system. This will include determining quality requirements, quality assurance, quality control and continuous improvement.

3. Quality Assurance

Quality assurance is the systematic process of defining, planning, implementing and reviewing the management processes within a company in order to provide adequate confidence that the product will be consistently manufactured to the required quality or condition.

The APM BoK 5ed defines **Quality Assurance** (QA) as, *the process of evaluating overall project performance on a regular basis to provide confidence that the project will satisfy the relevant quality standards.*

Quality assurance is a very sound but misunderstood process. It asks the basic question, '*What does the project team need to do to complete the project successfully to ensure the team is performing the job to the required level of quality, and to achieve this level first time?*'

To answer the above question, the project's quality assurance methodology is a dynamic iterative approach which monitors and integrates:

- Quality requirements
- Quality control
- Quality audits
- Quality trends
- Missing processes
- Continuous improvement.

Quality Requirements: The latest quality requirements are determined by the client and stakeholders in the scope of work, specification and contract. See previous section on quality planning.

Quality Control and Quality Audits: Quality control inspections and quality audits provide feedback on the project's present capability and performance.

Quality Trends: By plotting and extrapolating quality trends the project manager is able to see an indication of the general direction the project is heading. This can provide more information than the snap-shot given by a project status report.

Missing Processes: Quality control feedback on performance might highlight the need for certain missing or inefficient processes, procedures and documentation. The rational is that by fixing these deficiencies this should reduce production costs (less re-work), and this pro-active approach should give the client and stakeholders more confidence in the project team's capabilities.

Continuous Improvements: Striving for continuous improvements is an iterative approach to enhance the project's competitive advantage while cutting out the deadwood. Pareto's Principle, the 80-20 Rule, is a useful technique for identifying the best and the worst aspects of the project. Continuous improvement enforces what went right and introduces new techniques to continually improve future performance, while reducing waste and eliminating activities that do not add value.

4. Quality Control

Quality control is the process companies go through to confirm the product has reached the required quality as determined by the scope of work, the specifications, the contract, and in accordance with organizational standards and practices or recognised industry practice. The quality control system defines the method of inspection and testing which could be:

- Pre-inspection and in-process inspection
- Hold points and witness points
- Final inspection, testing and commissioning.

The APM BoK 5ed defines **Quality Control** (QC) as, *the process of monitoring specific project results to determine if they comply with relevant quality standards and to identify ways to eliminate causes of unsatisfactory results.*

The required condition (scope acceptance criteria and verification) should be laid down in the scope of work, specifications, and contract which roll up to form the project quality plan. The method of testing should be agreed with the stakeholders and outlined in the project quality plan. This could involve; checklists, inspections, testing, reviews, verification and validation against standards and requirements. Analysis might include statistical analysis, and comparison of results against a baseline or specification.

Non-Conformance Report (NCR): When a non-conformance, deviation or exception is identified, the resulting non-conformance report (NCR) is recorded in the required format and communicated to the nominated quality stakeholders in accordance with agreed procedures. This might initiate a quality audit to gather more information before corrective action is authorised to eliminate the causes of unsatisfactory performance of the product. In some cases, the corrective action might call for quality awareness training.

In the past, quality management used to focus on the inspection of the product after it was built. There was little involvement with the manufacturer's quality systems. The emphasis was on *'catching'* defects before they were released – the management tended to be inward looking. Now there is a general acceptance that **you cannot inspect quality into a product** if the product was not made properly in the first place. The emphasis has shifted to the workers at the *'coalface'* ensuring they have the support, equipment and training to do the job right the first time.

It is important to appreciate that quality comes from the manufacturing process and not from the inspection process. It should be recognised that it is far cheaper to perform the work right first time, rather than incur the following additional costs:

- Replanning and rework (labour, materials and equipment) and reinspection
- Knock-on disruption and delays
- Resource retraining and certification
- Reduced productivity.

5. Quality Control Plan (QCP)

The purpose of the quality management system is to ensure the project delivers the product the client ordered or required. This should be quantified in the contract as the required quality or required condition. The quality control plan will then outline how the required condition can be achieved, planned, inspected and documented.

The quality control plan links the quality requirements to the build-method and scheduled Gantt chart. The quality control plan offers the project team members the facility to impose the predetermined work sequence that they want, rather than what the production department or contractor might determine as resource efficient. This can be imposed with a quality control plan which lists the sequence of work and the level of inspection and testing. The sequence of work is determined by the build-method and network diagram. The level of inspection is determined by the level of risk and the level of control required, and can be regulated by: ***surveillance, inspection, witness, testing*** or ***hold points.*** Consider the following painting project:

WBS	Description	Required Condition (Specification)	Level of Inspection	NCR	Sign Off (Accepted by Client)
100	Permission to work	Health and safety	Issue gas free certificate [1]		
200	Surface preparation	SA 2.5 standard	Hold point Inspection [2]		
300	Paint undercoat	Check ambient temperature and humidity. Check thickness of paint	Hold point inspection [3] and [4]		
400	Paint topcoat	Check ambient temperature and humidity. Check thickness of paint	Hold point inspection [5] and [6]		

Table 21.2: Quality Control Plan – shows the sequence of work, required condition and level of inspection

Table 21.2 and figure 21.3 show the painting project has four activities. Activity 100 (hold point [1]) is to give the painters permission to work which means the work place complies with health and safety requirements. Activity 200 is to prepare the surface by grit blasting to SA 2.5. When activity 200 is complete, the surface is inspected with a gauge before it can be painted (hold point [2]).

Activity 300 is to paint the undercoat. Before painting the temperature and humidity are checked (hold point [3]). On completion the surface thickness is measured (hold point [4]). This hold point means the top coat cannot be painted before the undercoat has been approved. Activity 400 is to paint the topcoat, and is basically a repeat of activity 300 with hold points [5] and [6].

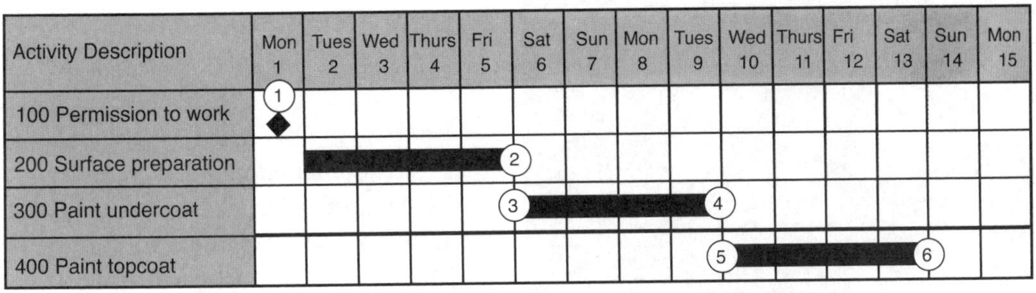

Activity Description	Mon 1	Tues 2	Wed 3	Thurs 4	Fri 5	Sat 6	Sun 7	Mon 8	Tues 9	Wed 10	Thurs 11	Fri 12	Sat 13	Sun 14	Mon 15
100 Permission to work	1 ◆														
200 Surface preparation		▬	▬	▬	▬ 2										
300 Paint undercoat						3 ▬	▬	▬	▬ 4						
400 Paint topcoat										5 ▬	▬	▬	▬	6	

Figure 21.3: Quality Control Plan - shown as a Gantt chart with hold points 1 to 5

In table 21.2 the non-conformance report (NCR) column enables the quality control to highlight a non-conformance. For example, if the surface preparation failed the inspection, the quality inspector would raise an NCR and the grit blasting would have to be performed again.

The QCP is an excellent document to plan and control the work, clearly stating the sequence of work and the required condition.

Exercise:

1. Discuss how you establish the required quality or condition for your project.
2. Discuss how you monitor and test the quality of the work on your project.
3. Discuss how you organise for corrective action to be taken when you find an item that does not conform to your project's plan.

Case Study Exercise: Your company does not have a presence on the Internet. Your job is to produce a quality control plan for the design and hosting of a website.

Further Reading:

Burke, R., *Project Management Techniques*, will develop the following topics which relate to this knowledge area:

- How to quantify the cost of quality
- Quality circles
- Quality audits
- Quality organization structure
- QTM.

22

Project Communications Management

Learning Outcomes

After reading this chapter, you should be able to:

- Understand communication theory
- Develop the project communication plan
- Develop a project's lines of communication
- Develop a filing system for the project
- Identify, file and retrieve documents

The *Project Communications Management* knowledge area includes all the techniques required to plan, execute and control the project's communication needs. This typically involves identifying the communication needs; the generation, the sending, the distribution, the receipt, the storage, the retrieval, and ultimately the disposal of project information. This chapter will explain how the project manager and the project team members use the project's communication system to manage the project's planning and control system.

Communication is one of those subjects that is hard to separate from what we do naturally everyday, so why does it warrant being a knowledge area? For a project to succeed there is a continuous need for effective communication to issue instructions, solve problems, make decisions, resolve conflicts, and keep all the stakeholders involved with the project supplied with the latest information.

The project manager and project team members working through the project office are in the key position to develop and maintain all the project's lines of communication, both inside the company with the functional managers, and outside the company with the client, contractors, suppliers and other stakeholders. The project office is like the '*front door*' to the project. It is estimated that project managers and project team members spend about 90% of their working time engaged in some form of communication, be it; meetings, writing emails, reading reports, or talking with project stakeholders.

The ability to communicate well, both verbally and in writing, is the foundation of effective team work. Through communication, team members share information, exchange ideas and influence attitudes, behaviours and understanding. Communication enables the project manager to develop interpersonal relationships, inspire team members, handle conflict, negotiate with stakeholders, chair meetings, and make presentations.

It, therefore, makes sense to rank communication management along with the other knowledge areas because, without effective communication, project success will be self-limiting.

Cost of Information: It is often stated that, *'Information costs money'* but, conversely, lack of information could be even more expensive. For example, the cost of communication failure could be due to the following:

- Poor problem solving and poor decision-making (both based on incomplete information)
- Rework due to the shop-floor using old drawings and instructions
- Downtime due to managers not being advised of the late delivery of materials
- Managers turning up for meetings that have been cancelled.

A trade-off needs to be established between the cost of mistakes and the cost of supplying useful information.

Projects are particularly prone to communication difficulties because of their unique nature and the matrix organization structure through which they are generally managed. There could be overlapping responsibilities, decentralized decision-making and complex interfaces all placing a strain on the communication system. However, if the communication system is well managed it could be the single most important factor determining product quality, efficiency, productivity and customer satisfaction.

The Internet and mobile phones have enhanced the communication mediums - a silent revolution has taken place as we move away from post (snail mail) and faxes. The mobile office and virtual office are becoming more popular for the project team. Consider some of the communication facilities available:

- Email (one-to-one, or one-to-many)
- Web sites
- B2B procurement
- Real time progress reports
- Video conferencing
- Mobile email and internet connections
- Mobile communication nationally and internationally.

Body of Knowledge Mapping

PMBOK 4ed: The *Project Communications Management* knowledge area includes the following requirements:

PMBOK 4ed	Mapping
10.1: Identify stakeholders	See the *Feasibility Study* chapter which will explain how to identify project stakeholders and assess their communication needs and expectations.
10.2: Plan communications	This chapter will explain how to plan the project's communications.
10.3: Distribute information	This chapter will explain how to distribute information.
10.4: Manage stakeholders' expectations	See the *Feasibility Study* chapter.
10.5: Report performance	This chapter will explain how to report project performance.

APM BoK 5ed: The *Communications Management* knowledge area includes the following definition:

APM BoK 5ed	Mapping
The APM BoK 5ed defines Communication as, the giving, receiving, processing and interpretation of information. Information can be conveyed verbally, non-verbally, actively, passively, formally, informally, consciously or unconsciously.	This chapter explains the communication theory.

Unit Standard 50080 (Level 4): The unit standard for the *Project Communications Management* knowledge area includes the following:

Unit 120376: *Conduct project documentation management to support project processes*

Specific Outcomes	Mapping
120376/SO1: Use a paper based and/or electronic filing system for a project.	This chapter will explain how to set up a project filing system.
120376/SO2: Use standardised processes for identifying, securing and finding documents.	This chapter will explain how to standardise the filing system.
120376/SO3: Provide project templates to team members.	This chapter will explain how to develop reporting templates.
120376/SO4: Assist in preparing project documents for handover at the end of a project or a project phase/stage.	This chapter will explain how to prepare handover documents.
120376/SO5: Describe and explain project documentation management processes.	This chapter will explain how to manage a document control system.

1. Communication Theory

The purpose of a project communication system is to transfer information from one person to another, or one stakeholder to another. Communication is essentially the interpersonal process of sending and receiving messages and information. The key components of the communication process are shown in figure 22.1. They include the sender who encodes and sends (transmits) the message, and the receiver who receives, decodes and interprets the message. The receiver then feeds back a response to the sender and closes the communication loop. The communication model focuses on each element of the process to identify what should happen and prevent misunderstanding.

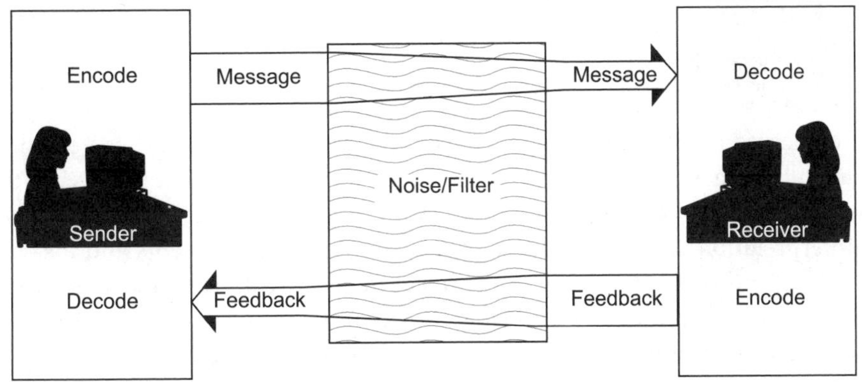

Figure 22.1: Communication Process - shows the communication process as a closed loop

Sender: The sender is the originator of the message and the starting point of the communication process or cycle. The sender will have a purpose for communicating. This might be to:

- Request information, send/request information
- Ask a question to clarify a point and gain knowledge or to solve a problem
- Communicate needs and expectations
- Issue an instruction or, more subtly, use persuasion to influence action
- Negotiate to encourage a mutual agreement between stakeholders
- Resolve conflict and limit negative impact on the project
- Encourage team building
- Network or make a courtesy call.

The sender is responsible for making the communication clear and complete and ensuring it is received by the intended person(s) and understood.

Encoding: Encoding is the process of converting thoughts, feelings and ideas into '*code or cipher*'. Although this may sound more appropriate for *James Bond* or *Spooks* (BBC drama series), in its broader sense, code and cipher are the words, numbers, phrases and jargon that are used in communication, which can sometimes sound like a foreign language.

Medium: The medium is the vehicle or channel used to convey the message. Project communications can be transmitted in many forms; formal or informal, written or verbal, planned or ad-hoc. Consider the following:

Formal Written	Letters, faxes, emails, memos, minutes, drawings, specifications and reports
Formal Verbal	Telephone, voice mail, meetings, video-conferencing *'It's for you!'*
Informal Verbal	Casual discussion between colleagues and stakeholders
Non-Verbal	Body language between parties.

The choice of medium will influence the impact of the message; for example, an email will not have the same impact as a face-to-face discussion. That said, the use of written communication should be encouraged on the project because it addresses misinterpretation and forgetfulness. All the important agreements and instructions should be confirmed in writing. The project manager will be thankful for keeping a written trail of agreements, if and when problems develop later in the project.

Written communications are acceptable for simple messages that are easy to convey and for those messages that are widely disseminated to all the stakeholders. However, verbal channels work best for complex messages that are difficult to convey, might need explanation, or where immediate feedback to the sender is valuable. Verbal communications are also more personal, this helps to create a supportive and inspirational climate.

Receiver: The receiver is the person who receives the transmitted message. The ability to receive will depend on the listener's hearing and listening skills, selective listening, eyesight and reading skills, tactile sensitivity and extra sensory perception.

Listening skills are a key part of project communication because both the speaker and the listener play a part in the communication process. In a conversation there has to be a speaker and a listener. Instead of being a passive recipient, the listener has a considerable influence on shaping the conversation.

Decoding: Decoding is the process of converting the message back into a readable format.

'Do I make myself clear?'

Noise, Filters and Perceptions: These are the factors that interfere with the effectiveness of the communication process. Distortions occur during encoding and decoding; communication channels can be blocked by too many messages, and filters and perceptions could influence the interpretations. Physical distractions can interfere with the communication, such as:

- Telephone interruptions
- Drop-in-visitors
- Even lack of privacy in an open plan office.

In these circumstances it is important to have a less disruptive space or office to be able to concentrate on the work, the issues and solve the problems.

Our backgrounds, culture, education and personality introduce communication filters and perceptions. Consider the following; language (lost in translation), social background, semantics, innuendos, intelligence, technical expertise, knowledge base, religion, politics, personal values, ethics, reputation, environment background and organizational position.

Other factors which will influence responses are preconceived ideas, frames of reference, needs, interests, attitudes, emotional status, self-interests, assumptions about the sender, existing relationships with the sender and lack of responsive feedback from previous communications with the sender.

Feedback: It is good manners for receivers to acknowledge receipt of communication, and give senders a time frame for a reply to any questions. It is important to feed back to the senders so that they can gauge how effectively the message was understood, and also for the receivers to confirm they have interpreted the message correctly. No effective communication will have occurred until there is a common understanding.

2. Communication Plan

The project manager and project office are at the heart of the project's information and control system. It is the project manager's responsibility to not only develop the project organization structure, but also to develop the project's communication plan and lines of communication. The communication plan should outline the following:

Who	Who are the project stakeholders? Establish the lines of communication (see stakeholder analysis in the *Feasibility Study* chapter).
What	What information do the project stakeholders need? This includes the scope of communication, content and format.
How	How do the stakeholders want the information transmitted? This could be by email, document, telephone, meeting, or presentation.
When	When do the stakeholders need the information? This could be linked to the schedule of meetings.
Document Control	What documents need to be controlled and who should they be sent to?
Storage	Where and how should the information be filed for storage, backup, and disaster recovery.

The above table outlines the structure of the communication plan. The following sections will explain how to manage each of the topics.

3. Communication Stakeholders (Who)

Identifying the project stakeholders is the first step in designing the communication plan, because communication is all about transferring information between the stakeholders so that they can carry out their project responsibilities. The *Feasibility Study* chapter includes a section on identifying the project stakeholders.

Lines of Communication (Who): A line of communication may be defined as a formal or informal link between two or more people, departments, companies, suppliers, contractors or stakeholders. The lines of communication tend to follow the organization chart which not only outlines the position of the project manager, the project team members and the project office, but also implies responsibility, authority and who reports to whom. Further, the stakeholder analysis will identify all the other interested parties that are both internal and external to the company (see figure 22.2).

Every effort should be made to include all the key people in the project's lines of communication. To leave out a key person will not only limit his knowledgeable contribution to the project but might also result in him adopting a hostile and negative attitude to the project. If senior people are included in the circulation list this will add weight to the document's perceived importance.

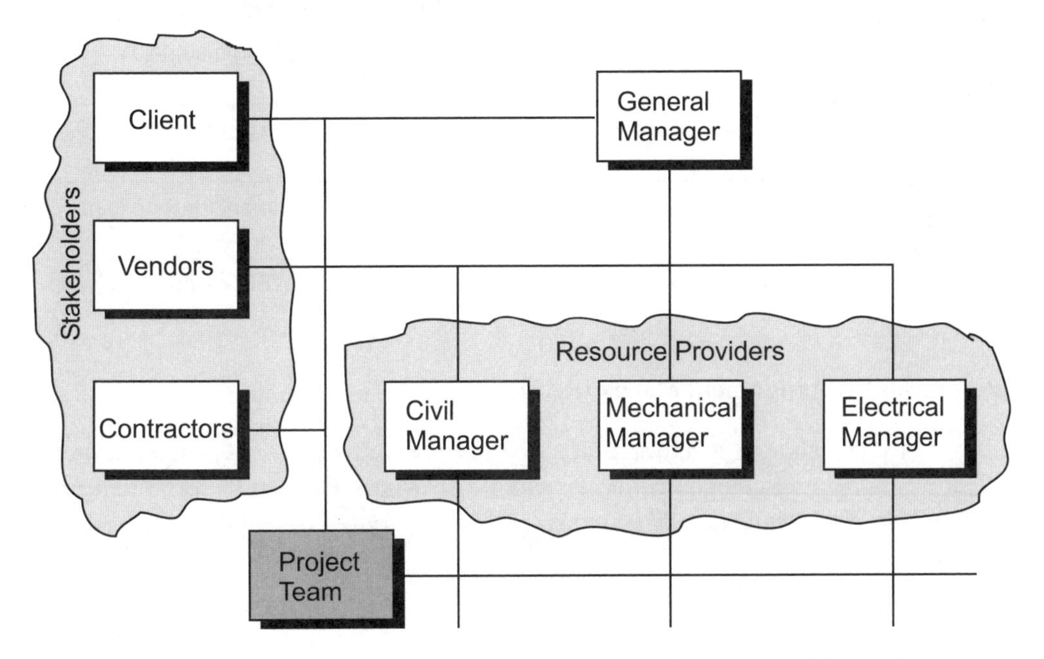

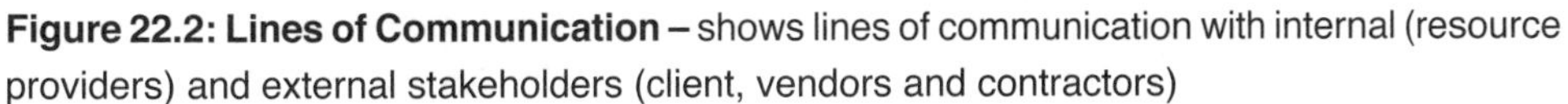

Figure 22.2: Lines of Communication – shows lines of communication with internal (resource providers) and external stakeholders (client, vendors and contractors)

4. Communication Content (What)

The communication content determines the scope of communication - what should be communicated. This is a tricky issue - if the project manager and the project team members filter the information they might be accused of being manipulative. However, if the project office gives everyone all the information they will be overloaded and are unlikely to read it. The objective should be to communicate sufficient information for the recipient to solve problems, make good decisions and feel involved and part of the project. Certain information should be controlled by the project manager; contract, specifications, drawings, instructions and scope changes. The controlled list (people and content) should be developed by agreement. The art of good communication is to strike a balance with the value of information supplied against the cost and time it takes to collect, process and disseminate the information.

The scope of communication can be an additional field on the Responsibility Gantt Chart (see table 22.1) where responsibility is linked to the scope of work. If communication can be linked to a WBS work package, then it should be easier to identify the interested parties.

WBS	Description (SOW)	Responsibility	Stakeholders to be Informed
1001	Production information	Production Manager	Vendor supplying materials
1002	Quality information	Quality Manager	Client's quality representative
1003	Transport information	Transport Manager	Traffic Department for wide loads

Table 22.1: Scope of Communication – shows the link between the WBS, responsibility and the stakeholders

Responsibility Assignment Matrix (RAM):

The PMBOK defines a **Responsibility Assignment Matrix** (RAM) as, *a structure that relates the project organization breakdown structure (OBS) to the WBS to help insure that each component of the project's scope of work is assigned to a responsible person.*

RACI Chart: The RAM is also referred to as the RACI chart that takes its name from the following:

- **Responsible** (for performing the work)
- **Accountable** (for the completion of the work)
- **Consult** (opinions sought, two-way communication)
- **Inform** (keep up to date).

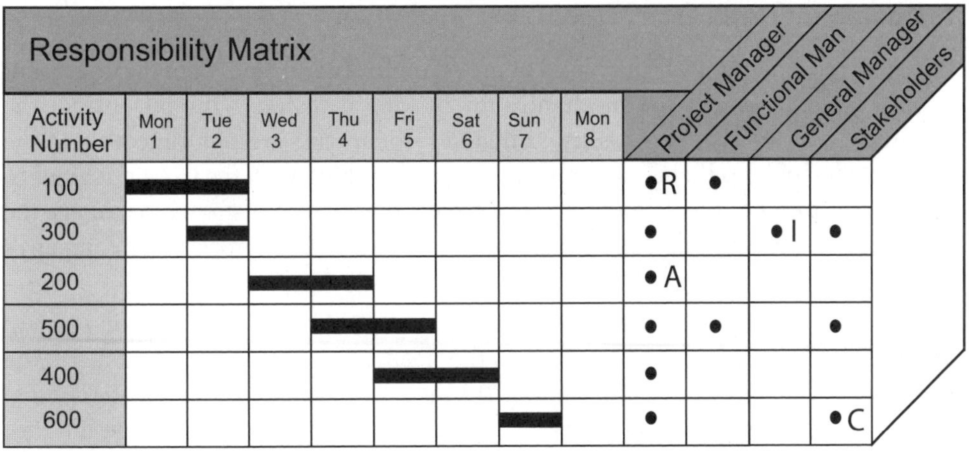

Responsibility Matrix

Activity Number	Mon 1	Tue 2	Wed 3	Thu 4	Fri 5	Sat 6	Sun 7	Mon 8	Project Manager	Functional Man	General Manager	Stakeholders
100									•R	•		
300									•		•I	•
200									•A			
500									•	•		•
400									•			
600									•			•C

Figure 22.3: RACI Chart – shows the link between the WBS and the OBS (**Burke**, R., **Barron**, S., *Project Management Leadership*)

5. Communication Method (How)

The method of communication determines how the information and the reports are transmitted. Where possible the client should be encouraged to accept the contractors' standard forms (templates) which will have been developed over previous projects. The information presented should be in a comprehensive format so that the recipient can quickly assimilate the situation and take appropriate action if required. Table 22.2, Communication Matrix shows what information/documents are communicated to whom and by what method (how).

Communication Matrix:

What/Who	Client/ Sponsor	Project Manager	Designer	Company Lawyer
WBS 1000	Contract	Contract		Contract
WBS 2000	Drawings	Drawings	Drawings	
WBS 3000	Minutes	Minutes		Minutes

Table 22.2: Communication Matrix – shows what information/documents are communicated to whom and by what method (how)

6. Communication Timing (When)

The frequency of reports and turnaround time for responses should be discussed and agreed at the handover meeting. An information sequence should be established; for example, the project's progress could be captured on a Friday, processed on a Monday and reported on a Tuesday at the project progress meeting. Reporting frequency is discussed in the *Project Execution, Monitoring and Control* chapter.

7. Document Control

If project information is the lifeblood of the communication plan, then the document control system outlines the mechanics for making it happen. The purpose of the document control system is to ensure that key documents are sent timeously to the nominated people so that all the stakeholders are working with the latest documents, and there is proof of the delivery. It is the project office's responsibility to manage the document control system. This would be presented to the client as part of the company's document control policy.

Project Filing System: It is the project manager's responsibility to set up a filing system exclusively for the project which reflects the needs of the project, and manages the movement of all the project information. The purpose of a filing system can be defined as the collecting, sorting, cataloguing (naming and numbering), filing (storage, library), retrieval, issuing, and tracking of project related documents. It is essential that all the stakeholders are working with the latest documents and all the old documents are removed or cancelled. For the project office to control the flow of information effectively a hub and spoke arrangement should be considered so that all the information is channelled through the project office (in and out). This means the project office can track and monitor all the communications, and control scope changes and issue new versions (revisions) (See **Burke**, R., *Project Management Techniques*).

Collecting: The *Project Execution, Monitoring and Control* chapter discusses how project data can be collected, processed and transmitted to and from the project office.

Sorting and Cataloguing: The project's filing system should be structured and indexed (named and numbered) to reflect the needs of the project, to enable ease of filing and retrieval, and to ensure they are current and up-to-date. For example, the filing system should be linked to the WBS structure.

An inventory of the project documentation, including multiple versions should be prepared, maintained, correctly filed and annotated in accordance with the agreed filing system.

The actual layout of the filing system might be set out by the client in the contract document and confirmed at the handover meeting. This is one of the ways that the client ensures the contractors are managing their filing system correctly. This layout also gives a clear required condition for a document control audit.

Confidential Documents: The document control system must accommodate sensitive and confidential documents ensuring their safe transmittal and safe custody. For example, businesses operating in a competitive market need to be able to develop new products away from the prying eyes of their competitors. And similarly, military procurement needs to be able to develop new weapon systems in secret, away from their potential opponents.

Circulation: At the outset of the project the circulation list of controlled documents should be agreed with the client and other stakeholders. The circulation list would typically be identified by the project organization structure's lines of communication (for the client, the contractors and the suppliers) and the responsibility matrix.

What Documents are Controlled?: The documents to be controlled would be agreed with the client, but would typically include; contract, specifications, drawings, schedules, reports, certain correspondence and materials (traceability).

Controlled Copies: An audit trail of transmittals should demonstrate that the controlled documents (usually marked **controlled copy**) have been sent to the right people and there is a signature confirming the date of receipt. It is important that the documentation system confirms that everyone is working to the latest revisions and that old revisions are removed from the system or stamped 'cancelled'.

On large projects, document control would be the responsibility of the configuration department or a document controller. However, on small projects, this would typically be the responsibility of the project manager and the project team.

Transmittal Note: A transmittal note or delivery note is sent with every controlled document. The addressee must sign the transmittal note and return a copy to the project office to confirm the document has been received – this is similar to the process used for couriered documents (see figure 22.4). This process gives an audit trail and traceability for all movements of project documents.

TRANSMITTAL NOTE	
NUMBER:	DATE ISSUED:
FROM:	TO: RECEIVED BY: DATE:
CONTENT:	

Figure 22.4: Transmittal Note (to confirm receipt of document)

It is the project office's responsibility to manage the flow of the controlled documents. The project office must ensure that the controlled documents reach their destination timeously. A control sheet (of the controlled documents) is essential (see table 22.3). Each week the document controller should confirm receipt by telephone or email and ensure the signed transmittals are returned. A list of non-returned transmittals can be tabled at the next progress meeting to encourage compliance.

Transmittal Number	Date	Document Type	Document Number	To/ Destination	Date Transmittal Returned

Table 22.3: Document Control Sheet – shows a transmittal summary

Information Back-Ups: The document control system should include an information back-up procedure. If information is channelled through a central computer system then the whole system can be saved daily (or even hourly) and a safe copy held off site. This would be central to the project's and company's disaster recovery system.

Handover Documents: One of the areas that is often conveniently forgotten by the contractors is the handover of project documents when the project is complete. At the handover meeting the client, contractors and suppliers should agree on what the handover documents include; their content, format, timing and location.

The savvy project manager should identify the handover documents as a work activity early on in the project (handover meeting) so that they can be planned, monitored and controlled. These documents should be identified in the contract, but would typically include; as-built drawings, quality control inspections, equipment manuals, operator manuals and service manuals.

8. Project Reporting

Project performance monitoring can be subdivided into data capturing and reporting. In this context, project reporting communicates the project's performance and scope changes to the stakeholders so that the project team can make decisions and apply corrective control, if necessary, to keep the project on track.

Project performance data can be collected, processed and reported in many ways. This section will outline a few of the commonly used formats and templates. The format (structure), frequency and circulation of reports needs to be established during the start up phase of the project. The reports should be designed to assist problem solving and decision-making by the various levels of management so that they can ensure the project will meet its stated goals and objectives.

Status Reports: Status reports simply quantify the current position of the project. This data capture function is the first link in the information and control system - all subsequent evaluations are based on this data. Status reports may be specific and focus on the key areas of the project, such as, time, cost and quality, or they may be general and include a much wider scope (see table 22.4).

Activity	Description	Status
100	Website	Concept approved
200	Marketing	Brochures approved
300	Procurement	Shipment arrived

Table 22.4: Status Report - shows the current position of the project

Variance Reports: Variance reports quantify the difference between actual and planned; for example, a revised budget being compared with the original budget. The variance is simply the difference between the two values (see table 22.5).

Activity	Original Budget	Revised Budget	Variance	Variance %
100	$20,000	$22,000	($2,000)	(10%)
200	$25,000	$20,000	$5,000	20%
300				

Table 22.5: Budget Variance Report

When a variance is reported as the difference between two values, it does not take the size of the parameter into consideration. This problem can be addressed by converting the variance into a percentage of the planned value. Now the variance is expressed as a percentage of the original base.

Trend Reports: The status report gives the manager a snapshot of the project, but does not give an indication of the project's direction. To address this, the trend report uses historical data and extrapolates this forward to give the project manager a feel for the direction of the project. Figure 22.5 shows the earned value graph where both PV (planned value) and AV (actual value) are extrapolated to show their current trends.

Earned Value: The earned value report integrates the variable parameters of cost with time, or man hours and time. The integration of data enables the planner to model the various parameters more realistically (see figure 22.5 and the *Earned Value* chapter).

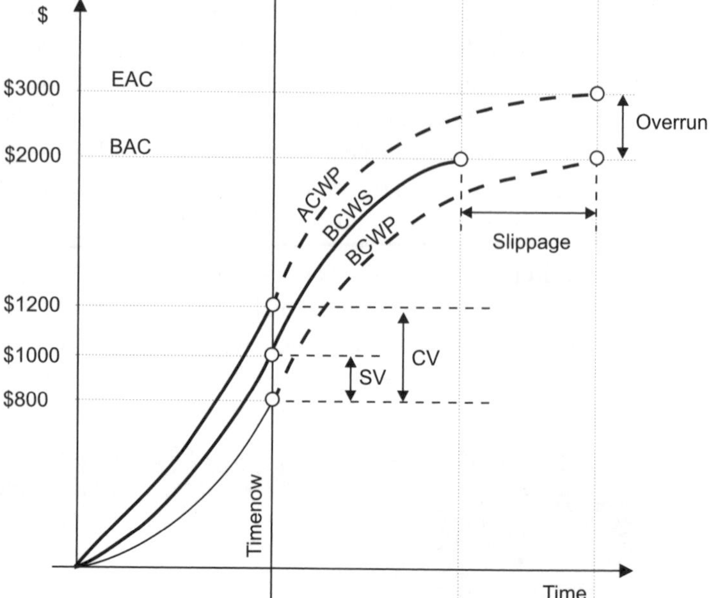

Figure 22.5: Earned Value - shows how trends can be extrapolated

Exception Reports: Exception reports are designed to flag an occurrence or event, which falls outside predetermined control limits. This threshold can be set by the project manager as a guideline for the project team members to follow and highlight the important information. For example, the project team member could be requested to report:

- All activities that have a float less than 5 days. This would highlight all the activities that could go critical in the next week.
- All the deliveries that are due in the next week. This would focus on deliveries that could disrupt the work and the resource planning.
- All non-conformance reports (NCR) where the project has not attained the required condition as outlined by the specifications or contract. This would focus on the workforce that needs training, or equipment that needs upgrading.

Minutes: Minutes of progress meetings are one of the main documents for linking the project together (see *Project Meetings* chapter).

Monthly Reports: On a long term project, monthly reports give the project manager an excellent forum to present what is happening on the project and report it to senior management and key stakeholders. The monthly report should roll-up the weekly progress meetings and any other special meetings to give an overall picture of the project. The report should highlight any significant trends or variance, particularly those where the CEO's action is required. It should also identify all major events happening in the next month so that the client, CEO and other stakeholders can plan ahead.

Reporting Period: The agreed timing and frequency of the reports should link-in with a schedule of meetings and report roll-ups. This would generally be weekly, but should also include key milestones and be adjusted to accommodate risk and the level of control required.

It is important to use structured reports as much as possible, as ad-hoc reports might not include the information the project manager needs - people naturally tend to report the good news while they are economical with the bad news. However, if managers are asked to report against a structured format, they will generally answer accurately and honestly.

Reports should be quick, and easy to prepare and read. The managers should not need to spend too much time reporting - about an hour per reporting period would be a reasonable amount. If it takes longer then managers should rightly question the reporting requirements.

Exercises:

1. Using the communication theory figure 22.1 apply this template to one aspect of your project environment.
2. Discuss how you determine the official lines of communication on your projects.
3. Progress reports keep the project office and the project manager informed as to what is happening on the project. Discuss the report templates you use to monitor and control work on your projects.
4. Document control ensures the right people receive the right documents at the right time and there is proof of transmittal. Discuss how you achieve this on your project.

Case Study:

For this case study you have been appointed project manager to design and host a website. Your presentation should outline how you will approach the following:

1. Communication cycle - data capture, processing, dissemination and storage
2. Lines of communication
3. Handover meeting
4. Progress meetings
5. Document control.

Further Reading:

Burke, R., **Barron**, S., *Project Management Leadership*, will develop the following topics which relate to this knowledge area:

- Non-verbal communication
- Communication channels.

Burke, R., *Project Management Techniques,* will develop the following topic which relates to this knowledge area:

- Hub and spoke arrangement.

23

Project Meetings

Learning Outcomes
After reading this chapter, you should be able to:
Organize a project meeting
Organize a project handover meeting
Organize a project progress meeting
Organize a project workshop and brainstorming session

This chapter will explain how to administer *Project Meetings* which are a key technique within the *Project Communication Management* knowledge area. Communication management uses project meetings and project workshops to collect and disseminate project information, generate creative ideas, solve problems and make decisions.

Project meetings are a key part of the project communication process. Some managers might prefer to hold many small meetings, while other managers might prefer to hold the occasional big meeting with all the stakeholders in attendance, others might prefer many informal ad hoc meetings or more formal structured meetings. Whatever the project manager's preferences, there are five basic reasons for holding a project meeting:

1. Information sharing - exchange of data
2. Problem solving - brainstorming, generating ideas, options and alternatives.
3. Decision-making - selecting a course of action, gaining support and commitment from the team members.
4. Planning and execution - issue instructions – what, who, when, how, where and why.
5. Control - evaluation, monitoring, measuring, feedback, reviewing and forecasting.

As project managers are responsible for establishing the project communication plan, they are also responsible for setting up the schedule of project meetings. Project team members typically attend more meetings than their functional colleagues. This is because projects tend to be multi-disciplined with a high degree of uncertainty, and therefore, require a greater amount of communication to keep everyone informed.

Although a high percentage of the communications and meetings will be informal and ad hoc, it is also essential to have a formal structure to confirm all the agreements and instructions are being addressed, and to keep all the stakeholders informed.

For project meetings to be effective they require good planning to ensure genuine participation from the entire team. Advance notice of the purpose and time of the meeting must be given to those who will be attending so that they have sufficient time to prepare.

Body of Knowledge Mapping

PMBOK 4ed: *Project Meetings* fall under the *Project Communication Management* knowledge area which does not outline any specific requirements.

APM BoK 5ed: *Project Meetings* fall under the *Communication* knowledge area which does not outline any specific requirements.

Unit Standard 50080 (Level 4): The unit standard for the *Project Meetings* knowledge area includes the following:

Unit 120382: *Plan, organize and support project meetings and workshops*

Specific Outcomes	Mapping
120382/SO1: Explain the purpose, objective and scope of project meetings and/or workshops.	This chapter will explain the purpose of project meetings.
120382/SO2: Plan for a project meeting and/or workshop.	This chapter will explain how to plan for a project meeting.
120382/SO3: Arrange and support a project meeting and/or workshop.	This chapter will explain how to support a project meeting.

1. How to Prepare and Run a Meeting (Generic)

Whatever the purpose of the meeting, all meetings benefit from advance preparation and structure. This section will outline how to prepare and run a typical project meeting:

Call Meeting	As the single point of responsibility the onus falls on the project manager to call project related meetings. The project manager needs to be clear about the precise objectives of the meeting. If the project manager only wants to convey a limited amount of information then sending an email might suffice.
Agenda	The agenda is a list of the topics to be discussed at the meeting. The agenda makes sure everyone knows exactly what is going to be discussed. The agenda should be prepared as a logical sequence of items, with time allocated to each item on the basis of its importance – this gives the chairperson a means of controlling the meeting to limit the discussion digressing to non-relevant topics and becoming a talking shop. If time runs out, then at least the important issues will have been covered.
Attendance	The purpose of the meeting will determine who should attend. There needs to be a balance between the size and the purpose of the meeting. As meetings grow in size they tend to become increasingly less productive. If everyone is expected to participate the attendance needs to be less than ten people. The size of the meeting will influence the input from certain people. Some individuals do not like to speak in large groups, preferring to contribute in smaller groups, or to the project manager directly. They may well approach the project manager after the meeting with some valuable input.
Book Venue	The venue needs to be selected and booked. The facilities of the venue need to be organized – table, chairs, white board, projector, DVD player, refreshments, etc.
Circulate Background Information	Information and material relevant to the discussion needs to be circulated in advance to everyone attending the meeting. People should know what is expected of them so they can come prepared.
Time Limit	A formal meeting should have a time limit, as meetings can be a great time waster. Sufficient time needs to be allocated to each item on the agenda to set the speed of the meeting, and to ensure the meeting ends on time (attendees might have other meetings to attend or deadlines to meet).
Start on Time	Starting on time sounds obvious, but if meetings start late then attendees will tend to arrive late for the future/subsequent meetings. Starting on time is also symbolic that the meeting is important. It is a good idea to schedule meetings 10 minutes after the hour to allow people time to arrive from their previous meetings.
Chairperson	The chairperson needs to identify himself at the beginning of the meeting so that all attending know who is running the meeting.

Purpose	The chairperson should start the meeting by stating the explicit purpose of the meeting and clarify what should be accomplished by the time the meeting is over.
Rules	The chairperson should set out the rules, which might include; mobile phones off, and speaking through the chair.
Minutes	The chairperson should assign someone to take the minutes.
Encourage Participation	Good meetings have lots of interactive discussion and a cross-flow of ideas. The chairperson's challenge is to encourage interactive discussion and encourage full participation. This is a balance between drawing out the silent and controlling the talkative - redirecting the discussion so that everyone has their say is usually more effective than trying to limit the talkative. A few subtle techniques go a long way toward increasing participation. A good meeting is not a series of dialogues but should be a cross-flow of discussion and constructive debate. The chair should guide, mediate, stimulate and summarize. The chair should refrain from preaching, or one-on-one dialogue - the point is to listen and facilitate discussion. Because junior team members are usually reluctant to disagree with senior people, it is best to encourage the junior people to get their ideas on the table first. Credit must be given where credit is due - the people who suggest ideas should get the credit. Giving due credit encourages continued participation.
Summarize	Before moving on to the next item on the agenda, the points that have been discussed need to be summarized, together with the actions that need to be taken. It also needs to be confirmed who is responsible for what, and what deadlines must be achieved. These points should be included in the minutes.

Table 23.1: Preparing a Meeting – shows a checklist for preparing and running a meeting

Minutes: Minutes of a meeting are a permanent certified record of what was said and agreed by the team members. Minutes should be taken for all meetings and produced as soon as possible after the meeting, preferably the next day, but certainly within the agreed time frame. The minutes should be communicated to the key people as per the communication plan and document control. The minutes should document discussions and agreements taken during the meeting, together with actions to be carried out before the next meeting. The minutes of one meeting usually form the agenda of the subsequent meetings.

If problems arise, such as disputes and legal action, the minutes can be used as evidence of decisions made up to that point in the project. It is, therefore, important that the minutes are an accurate and a true reflection of the meeting; this is why the first item of an agenda should be the approval of the previous meeting's minutes. The certified copy should then be filed in the project office's library as an historical reference and information for the closeout report.

Minutes		
Type of Meeting:	Progress Meeting	
Meeting Number:		Date:
Next Meeting:		
Compiled By:		
Circulation:		
Attendance:		
Agenda	Approve minutes	Action
	Discussion	

Table 23.2: Minutes of Meeting Template – shows the main headings to structure the minutes

'It's your fault we're late!'

'No, it's your mistake!'

2. Handover Meeting

The purpose of the handover meeting is to formally commence the project, the project phase or subcontract. The meeting would normally include the client, the project sponsor, the senior management, the project team members and other concerned stakeholders contractors and suppliers. The purpose of the handover meeting is to set the scene for the project and outline the project's objectives, how they will be achieved and how the project will be managed. It is the client's prerogative to chair the handover meeting.

A certain amount of pre-planning will help get the meeting off to a good start. Project managers need to establish their leadership and management style. This will serve as a model for future meetings and set the tone of professionalism. A typical agenda for the handover meeting might include:

Purpose of the Meeting	Confirm the purpose of the meeting and agree on the agenda, assign a person to take the minutes.
Project Charter	Confirm the identity of the project (name and number). Outline the purpose of the project and the scope of work.
Contract	Discuss contractual requirements, scope of work, specification, retention, bonds, penalties and warranties.
Scope of Work	Outline the subdivision of work (PBS, WBS, OBS, responsibility matrix), who is **responsible** for what, and who has the **authority** to issue instructions. Outline how the **scope of work** will be documented - this might include; a drawing register, parts list, specification, brief, proposal, feasibility study etc.
Project OBS	Identify and introduce all the relevant project participants and stakeholders. Discuss the scope of work and contracts - who does what, people's responsibilities and authority.
Project Team	Discuss the importance of the **project team** and the need for team members and stakeholders to work together to achieve the project goals. Encourage the team members to participate - introduce themselves, identify their areas of expertise and explain what they feel they can contribute to the project.
Communication Plan	Discuss the lines of communication and the document control system. Discuss who needs what, when and how information should be transmitted. Discuss what documents will be controlled, who they will be sent to, and how they will be transmitted.
Build-Method	Discuss how the product will be made, walk through the sequence of events. Discuss the required condition and hold points.
Closeout Reports	Review relevant closeout reports from previous projects and discuss what can be learnt from previous achievements and problems.
Baseline Plan	Discuss the baseline plan, particularly the project's schedule and milestones, the procurement schedule, the resource loadings, the budgets and cashflows.

Execution and Instructions	Explain the procedure for issuing instructions, the format and who has authorization (client and contractor). List the documents that will be used for issuing instructions; these could be drawings, schedules, minutes, memos, letters, faxes, emails and job cards.
Reporting	Discuss the project's monitoring and reporting requirements; the content, format, frequency and circulation.
Project Meetings	Discuss the schedule of project meetings, attendance, venue, agenda and minutes. At the start of the project there will be a juggling of time tables, meeting schedules and venues to suit the availability of the project's participants, and link in with the contractors' reporting periods. The frequency of the progress meetings should reflect the needs of the project.
Configuration Management	Discuss how changes will be incorporated and communicated. Outline procedures for scope changes and identify the people with approved signing power.
Payments	Discuss how the progress will be measured and how payments will be made.
Client Supply	List of client supplied items.
Inclusions and Exclusions	List of inclusions and exclusions.
Commissioning	Discuss how the product will be run-up, tested, verified, accepted or rejected. Discuss how the product will be handed over and the need for operator manuals and training.
Issue the Minutes	Issue the minutes timeously (preferably the following day) to those who were present and any other key stakeholders who did not attend the meeting. It is important to make a formal record of all decisions made, actions discussed, tasks assigned and comments made during the meeting. Although the minutes should be brief and clearly indicate actions required, consider including some history for complicated items.
Follow-up	Follow-up to ensure the minutes will be actioned timeously and not turn into a rush before the next meeting.

Table 23.3: Handover Meeting – shows a checklist for running a handover meeting

It is important to get the handover meetings right, to set the framework and tone for the project, and then follow-up with progress meetings to keep the momentum going and to keep the project on track.

Although the handover or start-up meeting is outlined here as one discrete meeting, in practice, this may be conducted over a series of meetings.

3. Project Progress Meetings

Progress meetings are generally held every week to monitor progress and guide the project to a successful completion. Progress meetings provide an effective forum for the project manager to co-ordinate, integrate and manage the project's participants. Meetings provide a dynamic environment where interaction and innovation enhance the crossflow of ideas and help solve problems. The meetings should also provide the venue for consensus and decision-making. A typical progress meeting would include the following:

Agenda	Circulate the agenda before the meeting to the list of participants who are responsible for the topics outlined in the agenda. The agenda should list action points so the participants can prepare for the meeting.
Minutes	Approve the minutes of previous meeting. Confirm who will take the minutes for this meeting.
Actions	Report on actions from the previous meeting. An action item is a problem that is logged and assigned to a person to be resolved - and tracked until closeout. Confirm the activities to be performed to solve the problem, identify the owner of each activity. List dependencies of each activity and their owner. Estimate the duration of each activity, outline how the activities will be tracked - even track on a daily basis if appropriate, but never lose focus on completing the project.
Progress	Report progress by work packages or deliverables - prioritize if necessary.
Configuration Management	Discuss scope changes and concessions - the implications and approval.
Document Control	List controlled documents transmitted and police signing of transmittals.
Claims	Discuss any claims since the last meeting.
Quality	Discuss any NCRs and quality issues.
Payments	Approve invoices for payment.
Minutes	Issue the minutes of the meeting as soon as possible, preferably the next day.

Table 23.4: Project Progress Meeting – shows a checklist for running project progress meetings

4. Brainstorming Workshops

They say '*two heads are better than one*' – so five or six heads interacting together in a brainstorming workshop should be better still to generate a stream of innovative ideas.

'*Brainstorming was born on Madison Avenue in the 1950s; brainstorming was long considered the preserve of those wild and crazy folk in advertising. In more recent years, however, it has spread into the mainstream and is now used by businesses of all kinds, not to mention everyone from civil servants to scientists and engineers, or, indeed, anyone with a problem to solve.*' Sunday Times 9 July 2000.

Brainstorming workshops are a great technique for generating creative ideas. Generally performed in groups, it is a fun way to get lots of fresh ideas on the table and get everyone thinking and pulling together. The participants should be relatively at ease with one another, so that they are comfortable shouting out off-the-cuff zany ideas - this often generates the best results. A typical brainstorming session would be run as follows:

Venue	Meet in a room that is conducive to the success of the workshop, use white boards and flip charts to capture the ideas for all to read.
Team Size	Restrict the workshop to about five to ten people; any less and there might be a low volume of ideas and cross-fertilization, any more than ten and people cannot get a word in edgeways.
Purpose	The chairperson should state the purpose of the session and define the problem and, where necessary, give the background to the problem and recent history of decisions.
Rules	The chairperson should outline how the brainstorming session will be run.
Suspend Criticism	All ideas, no matter how crazy they might seem, should be encouraged and recorded on a flip chart without comment or criticism from the group. The general goal of brainstorming is to collect as many ideas as possible, making quantity much more important than quality at this initial stage. If anyone's ideas are criticized that person might clam up and refrain from generating any more ideas. The ideas can be evaluated later.
Freewheel	Team members need to free up their minds of any inhibiting constraints. They should be encouraged to think laterally to generate extreme views - the crazier the ideas the better. Use word association to generate ideas.
Cross Fertilize	Encourage the team members to listen to other people's ideas, and try to piggyback and build on the ideas to generate more ideas (synergy). Cross-fertilization enables them to combine and improve on ideas that can sometimes result in surprising twists and turns.
Reverse	Reverse brainstorm - *'In how many ways can this project fail'*.
Quantity	The more ideas the better – run the session at a fast pace - aim for 50 ideas in 15 minutes. Work on the rationale that no idea is a bad idea.

Note Ideas	Use flipcharts and white boards to note the ideas, maybe grouping the ideas if possible. Leave the ideas up for incubation. When the session is finished, copy the ideas on to paper and circulate.
Time	Restrict the session to about 30 minutes, any longer and the ideas drop off and people get bored.
Follow-up	After the session circulate the ideas to the team and ask them to consider the problem and send any further suggestions to 'Mr Minute'.

Table 23.5: Brainstorming Workshop – shows a checklist for running a brainstorming workshop

Brainstorming Limitations: Brainstorming is one of the most popular team idea generating techniques that has achieved general acceptance. It should be recognised that one of the limitations of brainstorming is that it is rarely preceded by in-depth, prolonged and detailed study that underpins most creative endeavours. Preparation is a vital part of the creative process, such as, looking at closeout reports and re-defining the problem.

Exercises:

1. Discuss the purpose and reason for holding project meetings.
2. Discuss how you would prepare and run a handover meeting.
3. Discuss how you would prepare and run a project progress meeting.
4. Discuss how you would prepare and run a workshop to generate creative ideas.

Further Reading:

Burke, R., **Barron**, S., *Project Management Leadership*, will develop the following topics which relate to this knowledge area:

- Networking
- Teamwork.

Burke, R., ***Entrepreneurs Toolkit,*** will develop the following topics which relate to this knowledge area:

- Entrepreneurial spiral
- Innovation process
- Catching the wave.

24

Project Organization Structures

Learning Outcomes

After reading this chapter, you should be able to:

- Understand the advantages and disadvantages of using a functional organizational structure to manage multi-disciplined projects
- Understand the advantages and disadvantages of using a matrix organizational structure to manage multi-disciplined projects
- Understand the responsibility-authority gap

The *Organization Structure* knowledge area describes how the organization structure manages the way projects are staffed, managed and executed.

Projects are performed by people and managed through people, so it is essential to develop an organization structure, which reflects the needs of the project tasks, the needs of the project team and, just as importantly, the needs of the individual. To achieve this, projects are managed through three types of organization structures:

- Functional organization structures
- Matrix organization structures
- Team structures (next chapter).

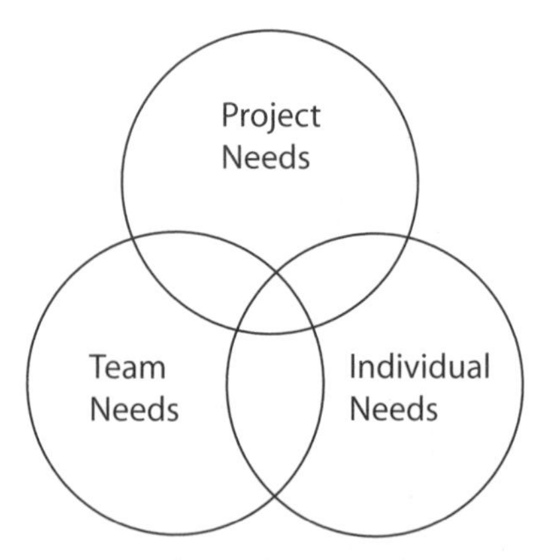

Figure 24.1: Intersecting Needs - shows the overlap between the different project needs, the team needs, and the individuals' needs

Projects by definition are temporary and unique. This begs the question, are project organization structures used to manage the project also temporary and unique?

Body of Knowledge Mapping

PMBOK 4ed: The *Organization Structure* knowledge area includes the following:

PMBOK 4ed	Mapping
2.4.2: Organization structure is an enterprise environmental factor that can have an effect on the availability of resources and influence how projects are conducted.	This chapter will explain how project organization structures manage company resources.

APM BoK 5ed: The *Organization Structure* knowledge area includes the following requirements:

APM BoK 5ed	Mapping
6.7: The organization structure defines the reporting and decision-making hierarchy of an organization and how project management operates within it.	This chapter will explain how to manage the functional OBS and the matrix OBS.

Unit Standard 50080 (Level 4): The unit standard for the *Organizational Structure* knowledge area includes the following:

Unit 120378: *Support the project environment and activities to deliver project objectives*

Specific Outcomes	Mapping
120378/SO2: Suggest appropriate structures, methods and processes to projects.	This chapter will explain how to manage projects through a functional organization structure and a matrix organization structure.

1. Functional Organization Structure

The functional organization breakdown structure (OBS), which has been handed down from the military and the church, is the most common type of company organization structure. The functional OBS at a high level subdivides the company into divisions and departments which typically align with the company's locations and specializations. These departments are then further subdivided into sub-departments and sections.

This **traditional** organization structure is based on the subdivision of product lines or disciplines into separate departments, together with a vertical hierarchy. The figure 24.2 outlines a typical structure with a number of functional departments reporting to the general manager.

Figure 24.2: Functional Organizational Structure

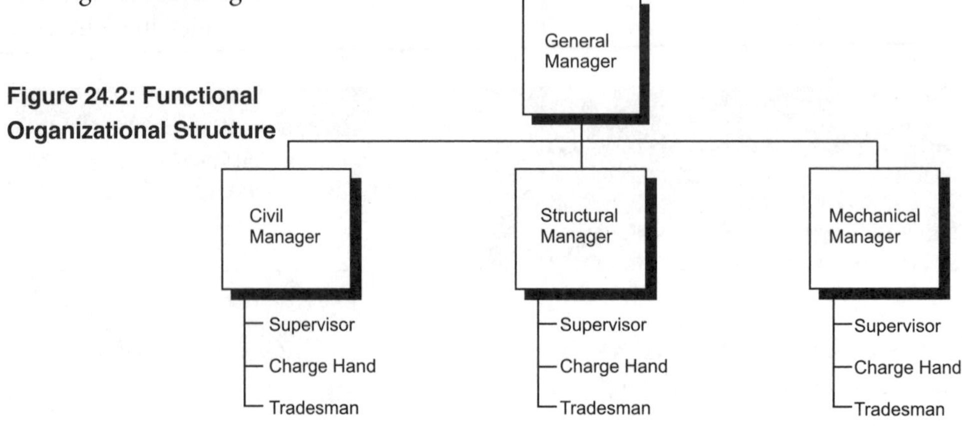

The **advantages** of the functional organization structure (particularly for managing projects within their own department) are:

Simple	Managing single discipline projects tend to be simpler.
Flexible	The functional organization structure can achieve a high degree of flexibility, because people in the department can be assigned to a project, then immediately re-assigned to other work. Switching back and forth between projects is easily achieved.
Home	The functional department provides a home for technical expertise which offers technical support and continuing development.
Support	The technical expertise provide excellent support as the work is usually carried out within the department.
Career Path	The functional department provides the normal career path for advancement and promotion.
Estimate	The functional department's work is simpler to estimate and manage, as the scope of work is usually restricted to its own field. The functional database should contain information from previous projects (closeout reports).
Communication	The lines of communication within the department are short and well established.
Reaction Time	There is quick reaction time to problems within the department.
Consistent	Some employees work best in a consistent work routine, rather than the challenge of diverse projects.
Responsibility	The employees are given clearly defined responsibility and authority for work within the department.

The **disadvantages** of the functional organization structure (particularly when being used on multi-disciplined projects) are:

Responsibility	On multi-disciplined projects there is no single point of responsibility as the project's scope moves from one department to another - this can lead to co-ordinating chaos.
Communication	On multi-disciplined projects there are no formal lines of communication between the people within the different departments. Generally, the only formal line of communication is through the functional managers which will lengthen the lines of communication and slow down the response time. With these long communication cycles, problem solving and decision-making will be negatively impacted upon.
Conflict	Competition and conflict between the functional departments could limit effective communication of important project information.
Priority	Departmental work might take priority over project work. If there is a resource overload the project's schedule could be pushed out. This would negatively impact on the handover to the next department and delay the project's completion.
Client	For functional managers the project is not always the main focus of concern, particularly when the scope has moved to another department. The client could well feel like a football being passed from one department to another. Clients prefer to deal with one person - the project manager.
Stakeholders	The responsibility for external co-ordination with the client, suppliers and other stakeholders might become muddled because of departmental overlap, underlap and inadequately defined responsibilities.
Co-ordinating	Without a clear project manager, the client might end up co-ordinating the different functional departments themselves.
Myopic	The department might myopically focus on its own scope of work in preference to taking a holistic view of the project. A departmental solution to a problem might not be the best solution for the project as a whole.
Motivation	The motivation for people assigned to the project can be weak if the work is not perceived to be mainstream.
Multi-Discipline	Multi-disciplined projects call for horizontal forms of co-ordination; a characteristic that is foreign to vertically orientated functional hierarchy structures.

The functional organization structure does offer excellent facilities within its own department but, where a multi-disciplined scope of work calls for interaction with other departments, the system could be found lacking. To address this problem the matrix organization structure offers an interaction of both functional and project interests.

2. Matrix Organization Structures

The topology of the matrix organization structure has the same format as a mathematical matrix. In this case the vertical lines represent the functional department's responsibility and authority, while the horizontal lines represent the project's responsibility and authority, thus giving the matrix structure its unique appearance and name.

The matrix structure is a temporary structure created to respond to the needs of the project where people from the functional departments are assigned to the project on a full-time or part-time basis - thus the matrix structure is initially superimposed on the existing functional structure.

The matrix structure is considered by many practitioners to be the natural project organization structure as it formalises the informal links between departments. On multi-disciplined projects people need to communicate at the operational level to perform their tasks. Where the lines of responsibility intersect, this represents people to people contact, thus providing shorter, more formal lines of communication.

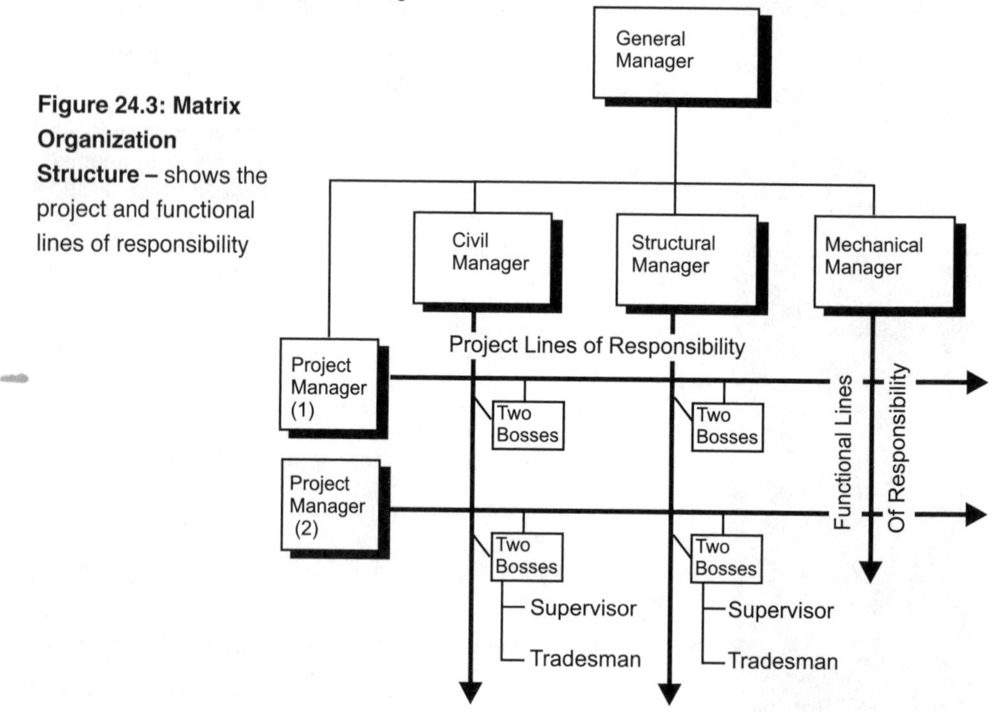

Figure 24.3: Matrix Organization Structure – shows the project and functional lines of responsibility

A characteristic of project management is that it relies on a number of functional departments and contractors to produce the project, each appearing to act autonomously and yet requiring strong communication bonds with each other. The project manager needs to cut across organizational lines to co-ordinate and integrate specific resources which are located in different departments. To achieve this, the project manager must have appropriate tools; particularly an information system that not only accommodates interdisciplinary tasks but also the cross-functional capability of retrieving data from the different departments.

Consider some of the **advantages** inherent in the typical matrix organization structure:

Responsibility	The project has a clear single point of responsibility – the project manager.
Resources	The project can draw on the entire resources of the company. When several projects are operating concurrently, the matrix structure allows a time-share of expertise, which should lead to a higher degree of resource utilisation.
Equipment	By sharing the use of equipment, the capital costs can be shared between projects and functional departments.
Seconded	With seconded resources, project termination is not necessarily a traumatic event. The resources do not need to worry about continuing employment as they can return to their original functional departments.
Client	There is a rapid response to client needs. The client communicates directly with the project manager.
Corporate Link	The corporate link will ensure consistency with company policies, strategies and procedures, yet give the flexibility to tailor these to the project's needs.
Job Descriptions	The matrix organization structure can be tailored to the needs of the project with respect to; job descriptions, procedures, work instructions and lines of communication.
Trade-off	The needs of the project and functional departments can be addressed simultaneously by negotiation and trade-off. The project is mainly concerned with **what** and **when** (scope and planning), while the functional department is concerned with **who** and **how** (resources and technical).
Problem-Solving	Problem solving can draw on a much wider base for ideas and options - brainstorming.
Experts	Teams of experts within the functional department are kept together even though the projects come and go. Therefore, technology, know-how, expertise and experience are not lost when the project is completed and the project team disbanded. Specialists like to work with other specialists in the same discipline thus increasing innovation, problem solving ability and synergy.
Multi-Disciplinary	The multi-disciplinary environment exposes people to a wider range of considerations and challenges.
Career Path	By retaining their functional home, specialists keep their career path. If they work well in a multi-disciplined environment then a new career in the project office might open up.
Training	The matrix organization structure is a good training ground for project managers working in multi-functional and cross-functional environments.
Integrates	The matrix organization structure integrates the PBS and WBS with the OBS.

The following **disadvantages** are inherent with the matrix organization structure:

Complex Structure	The organization structure is complex and more difficult for the participants to understand when compared to the simpler functional organization structure.
Dual Responsibilities	Dual responsibility and authority leads to confusion, divided loyalties, unclear responsibilities, and conflicts over priorities and allocation of resources.
Conflict	The two boss situation is a recipe for conflict, between both, managers and employees, and between managers over the allocation of employees.
Priorities	A company with a number of projects calling on the same resources faces real problems establishing priorities and resource allocation, as the project manager and functional manager will each claim their own work should have the highest priority. In this situation the priority should be made at a senior management level.
Cost	The cost of running a matrix organization structure is higher than a functional department because of the increased number of managers involved in the administration and decision-making process.
Integration	Project integration between departments is more involved and complex than when integrating people within one functional department. With functional projects it is clear who has the power to make decisions, however, with the matrix organization structure the power may be balanced between departments. If this causes doubt and confusion then the productivity of the project could suffer.
Sharing	The sharing of scarce resources could lead to inter-departmental conflict.
No Desk	After a secondment of a few years personnel may find they either do not have a functional department to return to, or their position has been re-appointed, '*Someone is sitting at my desk!!*'
Complex Situation	In the matrix organization structure the project manager controls the administration decisions, while the functional managers control the technical decisions; this division of power and responsibility could lead to an overly complex situation.
Personnel	The functional departments are unlikely to give up their best personnel to the project.

For the matrix organization structure to work successfully the functional departments will have to make major changes in the way they work. The matrix organizational structure introduces new management interfaces and increases the potential for conflict. New management skills are required for the functional managers to accommodate conflicting goals, priorities and resource demands.

3. Responsibility - Authority Gap

A characteristic of the matrix organization structure is that it relies on the functional departments for resources. Although the project office and functional departments may appear to work autonomously, for project success they need strong communication bonds with each other.

Project managers need to cut across organizational lines to co-ordinate and integrate specific resources located in the functional departments. To achieve this they must have both a fully integrated information and control system and the means of addressing the responsibility - authority gap.

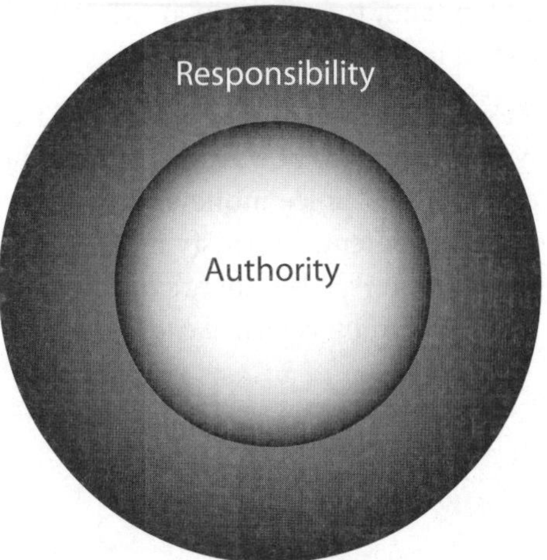

Figure 24.4: Responsibility – Authority Gap – shows the larger circle of assigned responsibility and a smaller circle of formal authority, the difference in the size of the circles being the responsibility-authority gap

Responsibility may be defined as feeling obliged to perform assigned work, while authority is the power to carry out the work. As shown in figure 24.4, the authority gap happens when project managers are assigned responsibility, but do not have sufficient authority to make the work happen. It is often stated that authority should be commensurate with responsibility, but feedback suggests that general managers are often reluctant to assign sufficient formal power to the project managers. In which case the project managers need to address the authority gap in other ways. Consider the following:

Formal Authority: Formal authority or position power is automatically conferred on project managers with their appointment to a project. While position power can be exerted over team members, this type of power is very limited in its acceptance and often does little to influence behaviour, particularly if the resources are owned by the functional managers and they do not report to the project manager.

Budget Authority: If project managers hold the purse strings this will confer some financial power over the functional managers, particularly if their departments are run as cost centres. Budget authority is certainly powerful when dealing with outside suppliers and contractors who depend on the project mangers' payments for their existence. Budget authority can lie with both the carrot and the stick - the promise of future business, an incentive bonus and threats of withholding payment for poor work.

Coercive Power: Coercive power uses the fear and the avoidance of punishment and threats to influence behaviour. This may be seen as power not to reward, or to threaten demotion, withhold overtime, limit salary increases, or transfer to another job. Use of coercive power is linked to organization position, tends to inhibit creativity and can have a negative impact on team morale.

Information Power: Expert knowledge and technical ability are effective if perceived to be valuable and are shared appropriately with functional managers and project participants. However, information power will erode trust and create resentment if hoarded. Project managers should be at the centre of the information and control system and, therefore, in an ideal position to capture, process, file and disseminate useful information.

Reward Power: Reward power is the ability to provide positive rewards for performance. To be effective, it must properly correspond to participants' values and expectations. Since money is not always available, project managers must consider a variety of potentially satisfying sanctions, especially those related to work challenge and recognition (see Herzberg's motivation and hygiene factors in **Burke** R., and **Barron** S., *Project Management Leadership*).

Cognitive Persuasion: The logical (cognitive) approach includes the use of reasoned argument, evidence and logical consistency. This way project managers can persuade the functional managers to contribute to their projects and part with their some of their best people for the projects' success, their department and their company. This approach works well on large capital projects.

Personal Power: Charismatic project managers can compel people to listen and follow their leadership. They are natural leaders who can persuade and encourage others to accept their vision or proposed course of action. They often have a sense of mission, a sense of purpose, a good sense of humour, are empathetic to staff needs, enthusiastic and self-confident.

Exercises:

1. The matrix structure is considered by many practitioners to be the logical project organization structure. Discuss how you can use this structure to manage your projects.
2. Clients prefer to deal with one person - the project manager. Discuss how this applies to your projects as a contractor, and as a client.
3. Acquiring power outside formal authority is the key to successful project management. Discuss what techniques you use to gain power.

Case Study: The designing and hosting of a website involves people from different departments and different outside companies. These would include (but not be limited to) the marketing manager, the Webmaster, the website hosting company and company IT trainers to train the staff to use the Internet.

Outline how you would use a matrix organization structure to manage a website project for your company.

Further Reading:

Burke, R., *Project Management Techniques*, will develop the following topics which relate to this knowledge area:

- Job description
- Project organization structure continuum
- Pure project organization structure.

Burke, R., **Barron**, S., *Project Management Leadership*, will develop the following topics which relate to this knowledge area:

- Leadership
- Motivation
- Pure project organization structure.

25

Human Resource Management (Project Teams)

Learning Outcomes After reading this chapter, you should be able to:
Understand the purpose of project teams
Write a team charter
Understand the team development phases
Understand the levels of team building

The *Project Human Resource Management* knowledge area includes the techniques to organize, manage, and lead the project team. This chapter will explain how to lead and manage the dynamics of project teams.

The implementation of the project management plan is through people, therefore, to effectively implement the project management system the project manager must gain support and commitment from the project team and other stakeholders. Many projects fail to reach their optimum level of performance, not because of any lack of equipment or project management systems, but purely because the human factors were not addressed.

The growth of new technology, increased complexity, and competition, has generated a need for multi-disciplined teams to work closely together. Teamwork should aim to bring individuals together in such a way that they increase their effectiveness without losing their individuality - an orchestra is a good example of a team working on this basis.

A project team may be defined as a number of people who work together to achieve shared common goals. Through interaction they strive to enhance their creativity, innovation, problem solving skills, decision-making skills, team morale and job performance. A team implies a number of people working together to achieve results, while a group of people implies a collection of individuals who, although they may be working on the same project, do not necessarily interact with each other. This is often the case when the project manager co-ordinates the project with the people individually. Under such conditions, unity of purpose is a myth (see figure 25.1).

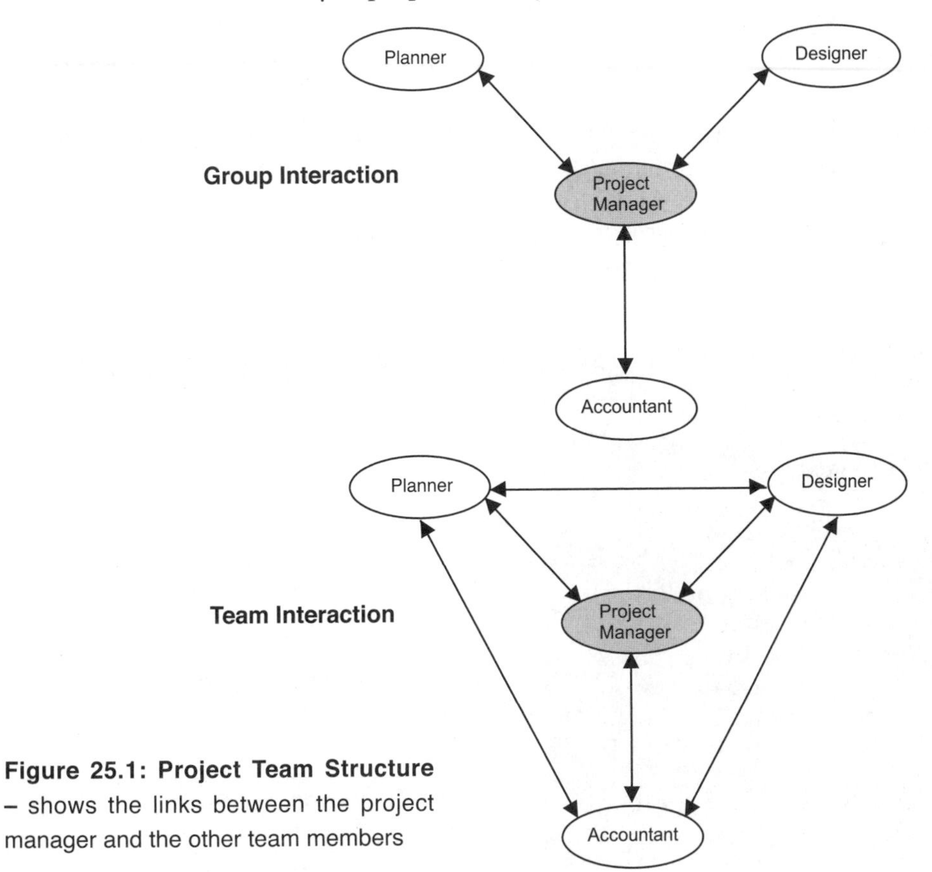

Figure 25.1: Project Team Structure – shows the links between the project manager and the other team members

Body of Knowledge Mapping

PMBOK 4ed: The *Human Resource Management* knowledge area includes the following:

The PMBOK 4ed defines **Human Resource Management** as, *the processes that organize, manage, and lead the project team.*

The PMBOK 3ed defines **Team Development** (managerial and technical) as, *both enhancing the ability of stakeholders to contribute as individuals as well as enhancing the ability of the team to function as a team.*

PMBOK 4ed	Mapping
9.1: Develop human resource plan.	This chapter will develop the team charter.
9.2: Acquire project team.	See **Burke,** R., **Barron,** S., *Project Management Leadership.*
9.3: Develop project team.	See **Burke,** R., **Barron,** S., *Project Management Leadership.*
9.4: Manage Project team.	See **Burke,** R., **Barron,** S., *Project Management Leadership.*

APM BoK 5ed: The *Human Resource Management* knowledge area includes the following definition:

The APM BoK 5ed defines **Teamwork** as, *when people work collaboratively towards a common goal as distinct from other ways that individuals can work within a group.*

Unit Standard 50080 (Level 4): The unit standard for the *Human Resource Management* knowledge area includes the following:

Unit 120379: *Work as a project team member*

Specific Outcomes	Mapping
120379/SO1: Demonstrate an understanding of working as a member of a team.	This chapter will explain the purpose of working as a team member.
120379/SO2: Collaborate with other team members to improve performance.	This chapter will outline the importance of team member collaboration.
120379/SO3: Participate in building relations between team members and other stakeholders.	See **Burke,** R., *Entrepreneurs Toolkit.*
120379/SO4: Respect personal, ethical, religious and cultural differences to enhance interaction between team members.	See **Burke,** R., **Barron,** S., *Project Management Leadership.*
120379/SO5: Use a variety of strategies to deal with potential or actual conflict in a project team.	See **Burke,** R., **Barron,** S., *Project Management Leadership.*

'Ladies and gentlemen may I introduce you to the Harmonious trio?!'

1. Purpose of Project Teams

Project teams are formed because they are an efficient and effective way of managing projects, where efficiency implies performing the work well, and effectiveness implies performing the right work. Consider the following points:

Volume of Work	To achieve the project schedule the volume of work must be distributed (shared) amongst a number of people. This will give an indication for team size.
Range of Skills	The scope of the project might require a range of skills which no one person is likely to have. Consider the orchestra - this is an excellent example of a set of complementary skills and talents (functional skills) which are required to produce the music.
Ideas	Brainstorming and discussions are a good example of interactive team work which, through synergy, generates creative ideas, solves problems and spots opportunities.
Decisions	Once a project team has made a collective decision by the agreed method, the team dynamics and cohesion commits all the team members to support their agreed course of action.
Risk	Project teams generally take riskier decisions than an individual would do on their own. This is because, in a team, there is a feeling of mutual support, and a greater confidence that they will succeed. But if they do not succeed then the team members know the risks will be shared amongst the team which will lighten the blow.
Motivation	Team dynamics enhances personal motivation to perform well - the individual team members do not want to let the side down.
Support	Project teams support other team members when they need help both technically and emotionally. This mutual support helps to build bonds and cohesiveness between team members.

Management tests show that teams repeatedly make better decisions than the team members would make individually with the same information. This has been attributed to the team's wider range of skills and experiences, and team synergy. The television programme '*Who Wants to be a Millionaire*?' frequently highlights this phenomena; when the contestant *'asks the audience'* a question the majority of the audience usually answer correctly and this, of course, is the basis of a democratic society.

2. Team Charter

The team charter outlines the purpose of the project team and lays the ground rules for the project team's effective operation. The team charter can be a simple document outlining, in a few words, what is required, or it can be a much larger document defining precisely what is required and how it should be carried out. This section will present a project charter template and give a brief description of all the headings.

The APM BoK 5ed defines the **Project Charter** as, *a document that sets out the working relationships and agreed behaviours within a project team.*

Ownership: Although the team charter is owned by the project manager, it is usually written in conjunction with the project team members. This involvement helps to ensure the project team members have a constructive input into the team charter's development which will help to encourage their buy-in, and also help to ensure they are assigned sufficient authority and power to use company resources to get the job done.

The team charter is a mechanism to:

- Initiate the project team selection
- Give the team an identity
- Outline how the project team will be led and managed
- Outline the project team's objectives
- Outline how the project team's objectives are to be achieved
- Outline team roles
- Outline the team building process.

The team charter should be established in the early stages of the formation of the project team. It should be sufficiently detailed to get the team started and should allow flexibility in operation while the team are settling into their roles.

During the project, the team charter serves as a contract, between the project team members and the stakeholders (client, sponsor, contractors, and suppliers), to identify the extent of the team's operation and authority. The team can tighten and enhance the team charter once its existence and purpose have been accepted.

Team Charter Template: A typical team charter template should include the following headings:

Heading	Purpose
1. Team Name	To give the project team an identity.
2. Team Motto	To give the project team a personality.
3. Team Objectives	To confirm what the project team has to achieve.
4. Team Leadership	To confirm who will lead the team.
5. Team Roles	To identify team roles and duties within the project team.
6. Responsibility and Authority	To delegate responsibility and authority.
7. Team Ethics	To establish accepted norms and standards of behaviour within the team.
8. Conflict Resolution	To establish a method of conflict resolution.
9. Stakeholders	To outline the stakeholders' needs and expectations.
10. Resources	To list available company resources.
11. Constraints	To identify constraints.
12. Problem Solving	To outline a method of generating ideas and solving problems.
13. Decision-Making	To outline a method to make collective decisions.

Table 25.1: Team Charter – shows the headings of a template which can be used to develop a team charter

These key points can be expanded as:

1. Team Name	Every thing in project management has a name and number! Giving the team a name helps to confirm the team's existence as a separate identity and link the team with the project. The individuals then become members of the project team.
2. Team Motto	Companies usually have a logo and a catch phase to help establish their brand. Project teams can also have a logo and a motto to establish their brand – this is usually associated with the project. This helps project teams to develop a personality and character of their own.
3. Team Objectives	The purpose of the project team should be outlined. Why have the individuals been bought together? What are the team's objectives and how will the success of the project team be verified and measured? The team's objectives are obviously related to the project's objectives - where the project manager's responsibility might be expressed as completing the project within time, cost and quality – the team members' responsibility might be expressed as completing the work packages within time, cost and quality. The team members might also be assigned administration duties which do not relate to any work package directly but are required for the functioning of the project work, these objectives might be harder to define. Where possible, it is important to define the objectives as they act as an aid for problem solving and decision-making.
4. Team Leadership	All teams have a leader, and for project teams this would usually be the project manager, but there could be projects with more than one team and, therefore, there would be a team leader for each team. It is the team leader's role and responsibility to lead, inspire and motivate their team to perform by giving the team members direction, vision and empowerment.
5. Team Roles	This section outlines the team members' roles and duties which can be subdivided into functional roles and team roles. Functional roles refer to a person's technical skills, product knowledge, work experience and practical ability. Team roles refer to the way team members behave and inter-relate with other team members.
6. Responsibility and Authority	This section explains how responsibility and authority will be assigned and delegated to clarify who reports to whom? It is essential the team members know what they are responsible for and what formal authority they have to use company resources.
7. Team Ethics	As a project team develops it naturally establishes its own norms and standards of expected behaviour. These need to be documented together with team members' ethics and governance.

8. Conflict Resolution	All project teams experience conflict from time to time. It is, therefore, essential for the team charter to outline a process to deal with the conflict and to resolve issues. Without some clear mechanism to deal with interpersonal conflict, small conflicts could soon escalate to a civil war and break up the team.
9. Stakeholders	The project team members need to know who are their stakeholders, both inside and outside the project, so that the stakeholders' needs and expectations can be established.
10. Resources	This section outlines what company resources and support the team members can use to help them to achieve their objectives.
11. Constraints	The team charter should identify the constraints (project, internal, and external) that could impact on the team members and limit their ability to achieve the team's objectives.
12. Problem Solving	One of the main reasons for working in a close knit project team is to solve problems and spot opportunities through collective interaction, innovation and brainstorming. This section should outline how the project team will solve problems.
13. Decision-Making	The project manager should discuss and agree how decisions will be made on the project. This section should outline where the team will sit on the autocratic to democratic continuum (see **Burke,** R., **Barron,** S., *Project Management Leadership*).

The team charter establishes a common vision that helps to keep the team focused on its purpose and goals as well as the way it will operate. The team charter should be seen as a subset of the project charter.

The project team should take ownership and tailor the team charter to address the individual needs of the team members and the team as a whole. Disagreement or misunderstanding about the charter should be quickly addressed and resolved to the mutual satisfaction of all team members. An effective team charter enables the project team to proceed in an environment of change and uncertainty.

3. Team Development Phases

Team development is a dynamic process where the relationships between the team members pass through a number of phases as they get to know each other. It is important for the project manager, as team leader, to be aware of these development phases so that the team members can be guided through the stages. This will increase the team's effectiveness and protect the team from inter-personal conflict which always threatens to implode a team.

Most of us have watched one of those reality television programmes where a number of '*ordinary people*' come together to compete in some way, usually in an alien environment. Whether it be; '*Survivor*', '*Big Brother*', '*Iron Age Man*' or the '*Apprentice*', the typical format is to bring a number of individuals together, who have never met before, and present them with a number of demanding tasks and, at the end of each programme, the 'weakest link' is voted off.

If you have been looking at the bigger picture you may have noticed a common thread running through all these programmes - as the team members get to know each other and interact they pass through a number of distinct team development phases. This section will discuss the classic model for team development; forming, storming, norming, performing model.

Forming	The team members come together to form a team and get to know each other. The team members will be mostly focusing on themselves as individuals, thinking about what is expected from them and how they fit into the team.	
Storming	As the team members begin to work together they start to express their opinions and perceptions about how the team should work together and how the job should be done. If these opinions are different it will certainly lead to a healthy debate, but could also lead to arguments and inter-personal conflict.	
Norming	There is consolidation within the team and an acceptance of differences and an agreement to work together as a team. The team establishes order and cohesion. The team leader helps to clarify team roles, norms and values.	

Performing	The team members are now working effectively together as a team. There is co-operation and role flexibility between the team members and effective problem solving. The team is now totally focused on the project.	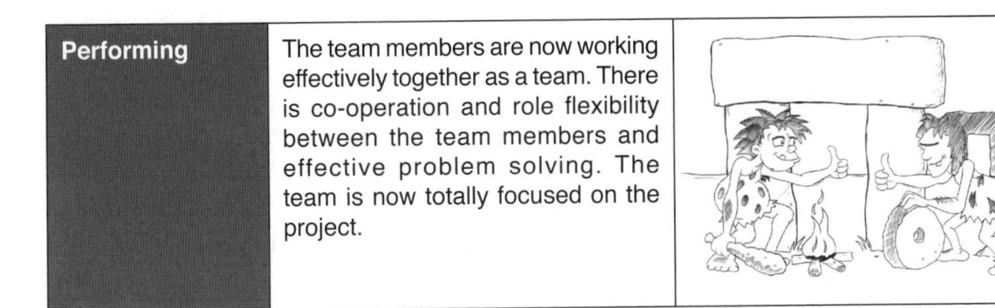

Team Performance: Teams are formed to carry out a project, to produce a product or a service. This section uses a line graph to show the relative performance of the team over the development phases.

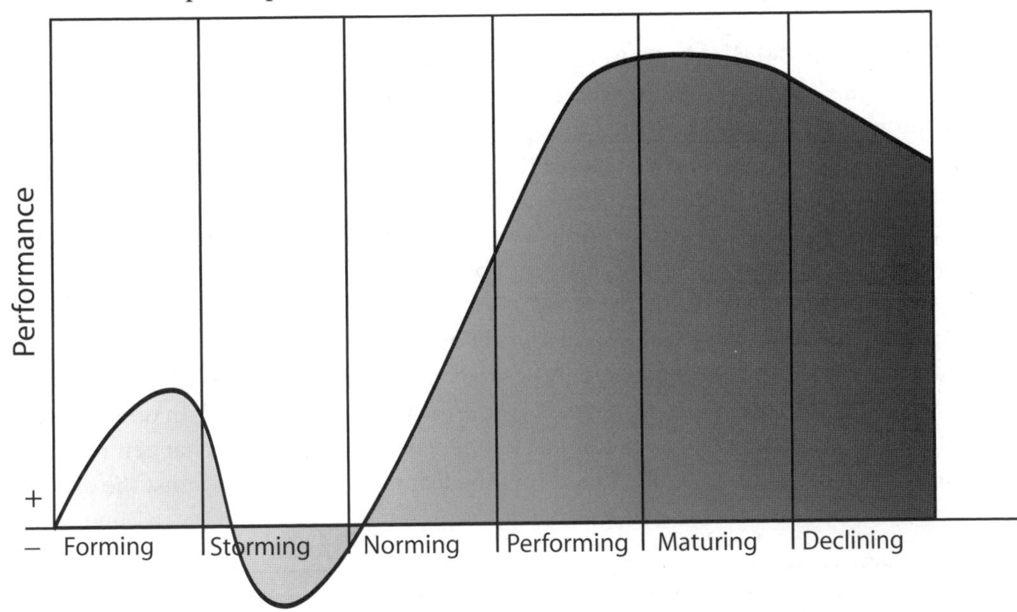

Figure 25.2: Team Performance – shows a line graph of the relative performance over the development phases

Forming: Initially the team may achieve moderate performance if there is a strong leader to give the team direction. But the team's performance is unlikely to be more than moderate because the team members do not know the other members' abilities and how they can work together as a team.

Storming: The performance falls off in the storming phase if the disagreements and arguments lead to inter-personal conflict, breakdown in communication, poor problem solving and little co-operation.

Norming: The performance starts to increase as rules and norms are established, and the team members want to work together.

Performing: The team is now working well together, totally task focused, and the performance continues to increase.

4. Team Building Techniques

Bringing a number of people together to perform a task does not necessarily mean they will work together effectively as a team even if they have all the necessary complementary skills, a balance of personalities and share a common goal. To be successful a team also needs effective leadership (to give the team direction), and also effective team building to enable the members to work together as a team.

The PMBOK defines **Team Building** as, *activities designed to improve inter-personal relationships and increase team cohesiveness. It is also important to encourage information communication and activities because of their role in building trust and establishing good working relationships.*

What is Team Building?: Team building enables a group of diverse people to work together effectively as a unit to achieve the project's goals. People generally tend to see the project in terms of their own discipline and background; if this causes the team members to go in different directions then the team's performance will be adversely impacted. The project manager's task is to encourage the individuals to see the big picture and to focus on achieving the overall project goals together. Obtaining this team spirit and commitment is what team building is all about.

Team building will occur naturally as people work together towards a common goal, but usually it is far **too slow** and too ad hoc a process to be of value for projects of short duration. The project manager needs to consider using team building techniques to accelerate the team building process as soon as the team is formed. In other words, the fully operational team does not just happen; it must be made to happen through effective team building. Figure 25.3 shows the teams performance against the project lifecycle.

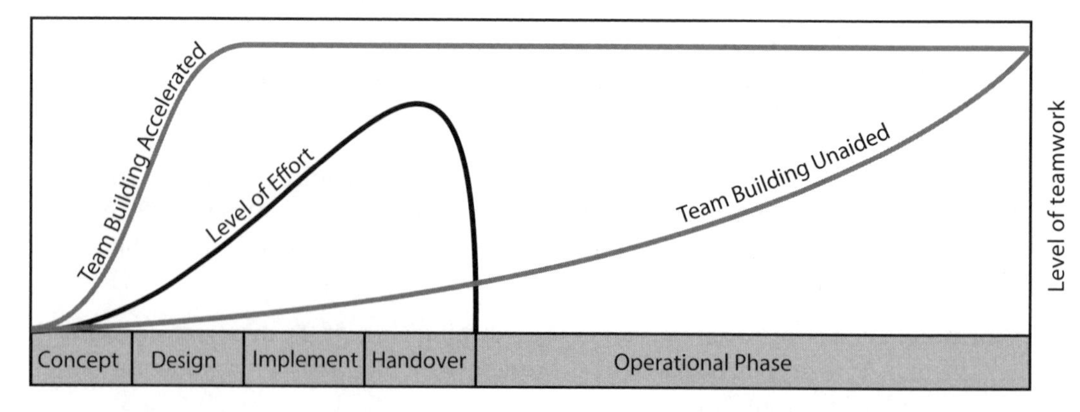

Figure 25.3: Team Building Lifecycle – shows the team building lifecycle superimposed on the project's lifecycle

Team building can be presented as a number of levels, consider the following four levels:

Level 1: Interpersonal Team Building	The first level of team building focuses on the team members simply getting to know each other. It is recognized that if team members have a better understanding of each other's personalities and behaviours they will be better able to communicate with each other, which in turn, means they will be able to work together more easily.
Level 2: Team Roles	The second level of team building focuses on the team roles each of the team members play within the team. Adventure team building exercises are designed to bring out the team member's leadership and team role skills. The main value of team role definitions lies in the way that it enables team members to see how they fit into the project team (like cogs in a wheel, or pieces in a jigsaw puzzle).
Level 3: Shared Vision	The third level of team building is to establish a shared and common vision. This is important because, although the team members may now know each other and know their individual team roles, they might have a different appreciation of the purpose of the project and, consequently, might be pulling in different directions.
Level 4: Task Focused	The fourth level of team building focuses on how the team members will carry out the project and explore ways to improve team performance (efficiency and productivity). The team building exercises focus on the team members' technical skills and competencies, and how they can apply them. The team building exercises are not adventure based any more, they are now totally focused on how to achieve the task. For example, a football analogy would be the team practice sessions where the players practice different tactics that involve a number of players working closely together.

Exercises:

1. Identify the project teams in your company and discuss why your company uses them.
2. Develop a team charter for your team.
3. Relate the team development phases to a team you are familiar with.
4. Discuss how your company uses team building techniques.

Further Reading:

Burke, R., **Barron**, S., *Project Management Leadership*

- Leadership
- Motivation
- Team building

26

Project Management Office

Learning Outcomes

After reading this chapter, you should be able to:

- Understand the principles behind the project management office
- Understand the benefits of a project management centre of excellence
- Set up a project office to offer a management-by-projects approach
- Understand the benefits of a mobile office in a project environment

The project management office (PMO) or project office (PO) enshrines the project management approach as a central point for managing a company's projects. Companies have design offices and IT departments as centres of excellence, so it also makes sense to have a project management office as a centre of project management excellence.

The PMBOK 4ed defines the **Project Management Office** (PMO) as, *an organizational body or entity that can be responsible for the centralised and co-ordinated management of those projects under its domain. The responsibilities of a PMO can range from project management support functions to actually being responsible for the direct management of a project.*

1. Site Office

To understand the purpose and workings of the project office it is interesting to look at how it evolved and developed. Historically the site office on a construction project was probably the first example of a project office. Out of necessity the site office was located on site at the work face close to the action. This enabled the site manager (project manager) to monitor the progress directly and make quick, on the spot decisions as and when required.

'Hey Roger, get me a chisel from the site office.'

2. Matrix Organization Structure

Multi-disciplined projects were a key motivator to create a project office. Multi-disciplined projects were originally passed from one functional department to another as the work progressed. The *Project Organization Structures* chapter has already discussed the problems associated with this arrangement. The outcome was to manage these multi-disciplined projects using a matrix organization structure, where the project manager would manage the project as it passed from one functional department to another functional department.

As the benefits of this type of management approach were realised, the project manager(s) needed a more permanent home – and this home became the project office. Once the home had been established other benefits followed, particularly those associated with the project management office as a centre of excellence, and ultimately channelling all the company's work through the project office using a management-by-projects approach.

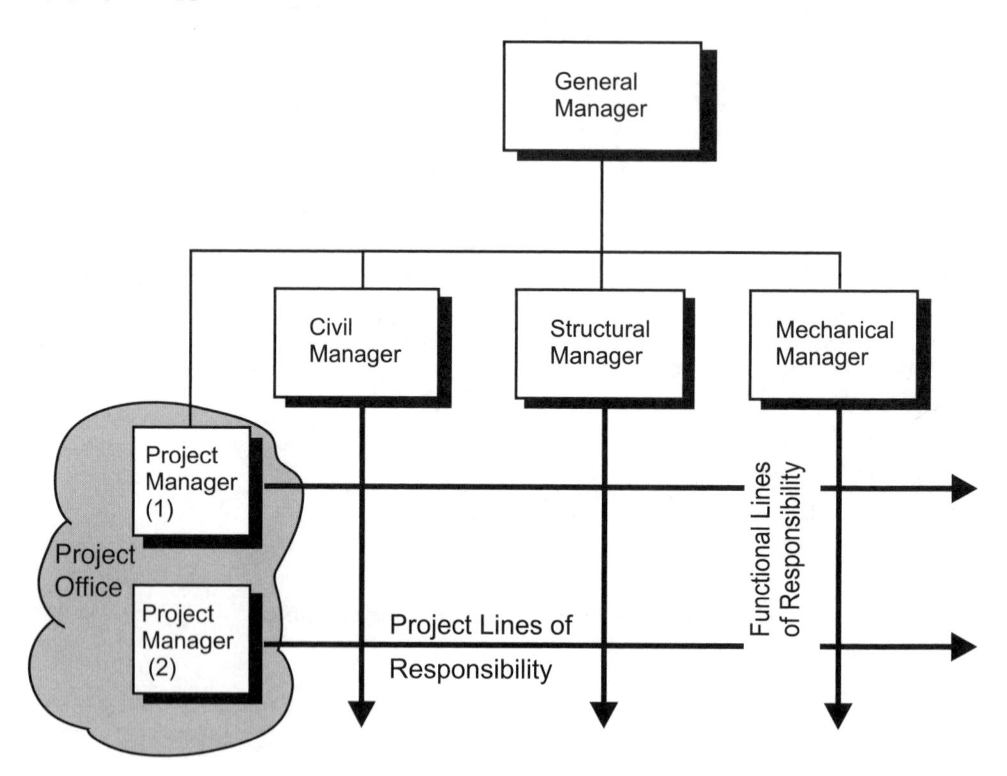

Figure 26.1: Matrix Organization Structure – shows the project office driving the project through the functional departments

3. Centre of Excellence

One of the benefits of having a project management office is to create a centre of excellence for all aspects of project management. Grouping the project managers and the project team members in the same office, creates a pool of project expertise, where they can draw from a large data base of project experience and information. Consider subdividing the project management office into the following subheadings:

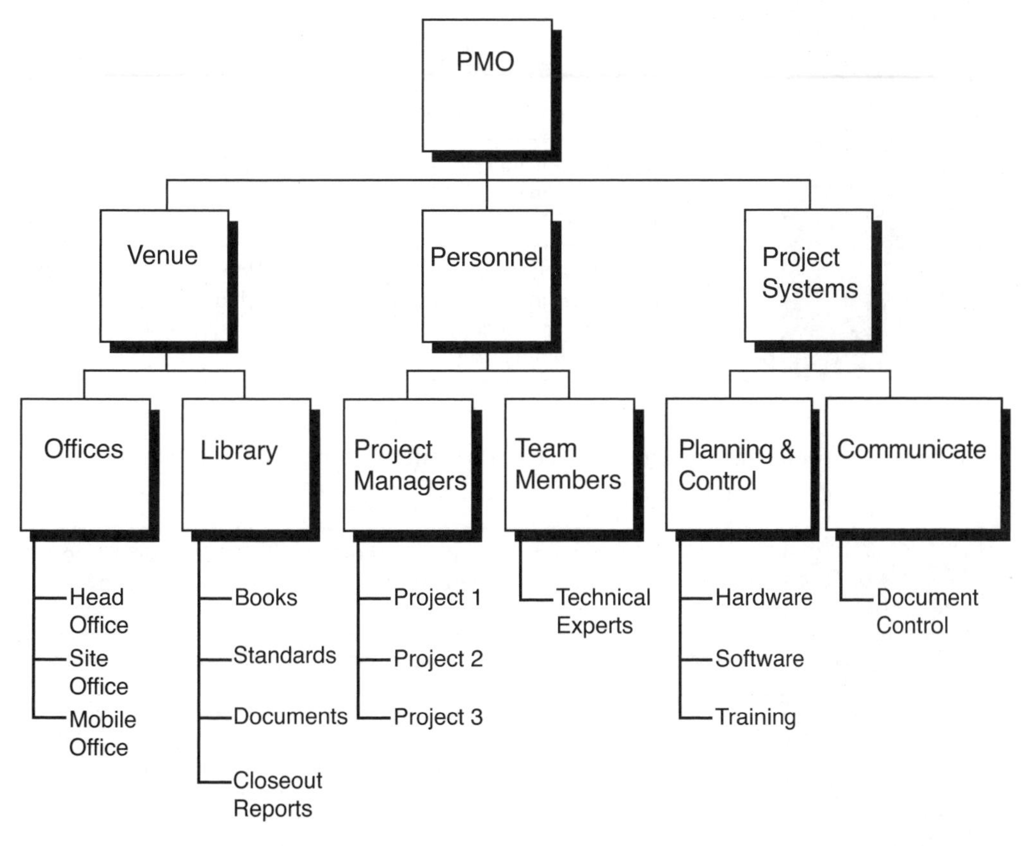

Figure 26.2: Project Management Office – shows subdivision into Venue, Personnel and Project Systems

Venue: The project management office (PMO) as a physical venue offers the following types of facilities:

Offices	The PMO gives the project managers and project team members office space and a home. For convenience, they could be grouped per project in the same area.
War Room	The PMO provides a central *'war room'* to manage projects where all the information, people and communication facilities are available.
Central Point	The PMO offers a central point for the company to network with stakeholders and potential clients.
Meetings	The PMO has meeting rooms with audio-visual/presentation facilities.
Communicate	The PMO provides a front door for the project and the company. If the client needs to communicate with someone on any project, the client will find them in the PMO.
Equipment	The PMO offers an economy of scale by grouping all the company's project activities. This means the company can warrant buying and installing the latest project management software and hardware, together with trained people to run and maintain them.

Library: The project management office library offers a central place to file, store and retrieve project information:

Central Point	The PMO offers a central point to collect, file, store and retrieve project information which can be accessed both in hard copy and electronically.
Standards	The PMO is the ideal place to keep the latest information on project management standards.
Books	The PMO library is a central place containing a range of project management literature and technical journals, together with a broad selection of relevant books, technical journals and media.
Document Control	The PMO is a central place to store controlled drawings, specifications and contract documents in a secure environment.
Closeout Reports	Once the project is over, many may wish to forget the experience. However, the library is the ideal place to house the closeout reports.

Personnel: The project management office offers a team of project experts. As projects are run by people the project team members are a key resource:

Resource Pool	The PMO offers a flexible resource pool of project managers and project team members.
Matrix OBS	The PMO offers the option to use a range of organization structures. A matrix structure is often used as a temporary organization structure which overlays the other functional departments and enables the PMO to share and co-ordinate resources across a number of projects.
Problem-Solving	The PMO offers a cross-section of experienced team members which enhances brainstorming workshops and problem solving.
Home	The PMO offers the project managers a home to go to between projects. This may be the ideal time to update their skills and prepare for the next project – perhaps becoming part of the bidding team.

Systems and Methodologies: The project management office offers a central place in the company to set up project management systems, methodologies and information formats:

Systems	The PMO offers a central point to design and develop project management systems appropriate for the type of projects the company runs.
Standards	The PMO has the expertise to identify the latest information on project management standards so they can comply with the latest standards.
Software	The PMO offers a central point to install project management software and associated hardware. This can be used for both projects managed by the project office and other projects managed by other departments within the company.
Project Quality Plan	The PMO often supplements the company's quality plan with a project quality plan which is tailored to the needs of the project.
Quality Control Plan	The PMO offers the expertise to develop the quality control plan (QCP). The quality control plan links the scope of work (WBS) with the project quality plan. The QCP identifies what needs to be inspected, the level of inspection and the hold points.
Audits	The PMO can carry out project management audits (internal and external) to confirm the project management systems are conforming to the project quality plan.
Administration	The PMO has admin staff to support the project managers. This enables the project managers to spend more time managing their projects and less time on processing data and admin chores. (Research indicates that project managers spend 50% of their time on admin).

Project Communication: Information is the life blood of projects. The PMO is in an ideal position to control all the lines of project communication and controlled documentation:

Lines of Communication	The PMO is responsible for identifying the project's lines of communication.
Directory	The PMO is the keeper of the project directories, of clients, stakeholders, suppliers, contractors and team members, etc.
Communication	The PMO can communicate project information to the project team, the client, contractor, suppliers and other stakeholders.
Document Control	The PMO provides a documentation control centre to manage the collection, storage and distribution of project documents. Controlling the flow of documents is often a weak link on small projects. The hub and spoke arrangement enables prompt expediting and follow up, with signed transmittals to confirm delivery and provide an audit trail.
Website	The PMO can set up a website to present the status on all its projects - there is a trend towards real time reporting.

Tendering: The project management office provides the facilities to tender for new work:

Marketing	The PMO is an ideal place to advertise (design brochures) and promote the company's capabilities and qualities.
Networking	The PMO should excel in networking skills to open the doors of business opportunities.
Quotation	The PMO quotations are based on an empirical data base of past performance built up over time from experience and documented in closeout reports. This ensures a respectable level of accuracy.
Legal	The PMO offers a central point to co-ordinate quotations. This would include legal support with the company's standard terms and conditions of contract.

Planning and Control: The project management office provides the facilities, project systems and personnel to plan and control the company's projects:

Baseline Plan	The PMO integrates all the components to produce a coherent baseline plan.
Instructions	The PMO issues work instructions and job cards to authorize and execute the scope of work.
Monitoring	The PMO measures progress through a structured data capture process.
Meetings	The PMO manages project progress meetings with the client, project team, contractors and suppliers.
Scope Changes	The PMO manages the scope change process by; evaluating the impact on the build-method and configuration, revising plans and issuing instructions.
Multi-Projects	The project office team is able to support several projects simultaneously from their pool of resources.

Training: The project management office offers a centre of project management expertise which can be used to train and mentor new project team members:

Project Training	The PMO is the ideal place to design, set up and run project focused training courses.
Training	The PMO offers project management training courses for the whole company which can be run in conjunction with the HR department and the training manager.
Fast Tracking	The PMO is able to run fast track training sessions in project management for newly formed project management teams.
Mentoring	The PMO has experienced project managers who can mentor, coach and support apprentices with on the job training. This will ensure the development of future leaders.

4. Management-by-Projects

The PMO is an ideal venue for controlling the management-by-projects approach. The benefit of the management-by-projects approach is that it is schedule focussed and not resource focussed. This means the workforce is scaled up or down to suit the workload schedule and not the schedule adjusted to suit the level of the workforce available. Clients can now plan ahead with some certainty knowing when their work will be completed.

5. Mobile Project Office

Mobile phones, portable notebook computers and wi-fi communication networks are making the mobile project office a feasible reality. These mobile facilities are not only enabling the project managers to stay in touch with the project office while on site trips, but actually enabling the project managers to take their office with them.

Mobile facilities/broadband enable communication with the head office, clients, suppliers and contractors while on the move. The mobile office could be a site office, an airport lounge (or any wi-fi hotspot), the project manager's car, or even his yacht. Consider the following points:

- The mobile office frees project managers from their corporate desk so they can actually spend more time on site visiting clients, contractors and suppliers.
- Mobile conferencing enables project managers to attend virtual meetings with their project team and clients. Regular virtual meetings should keep a scattered team linked and working together.

Project managers can even benefit from mobile office facilities while they are on an extended business trip or a holiday. They are able to relax knowing everything is okay back at the PMO and, if there is a problem, they have the opportunity to nip it in the bud. They can also keep on top of their emails while they are away to avoid being greeted by several hundred emails on their first day back at work. Holidays in the bush or on a tropical island, and flying time, no longer need to be vacuum periods without any business contact.

It is expensive to keep office space and a desk available for project managers who are working away from the office for long periods of time. A more cost effective option is to **hot desk!** This means returning project managers would use any desk that is available and log into the PMO's wireless network from their laptop. If the project manager has files, reference material and books which cannot be electronically stored, these can be kept safely in a locked cabinet.

Exercise:

1. Discuss the benefits for your company using a PMO approach.
2. Discuss the benefits of the mobile PMO and discuss how this facility can be used on your projects.
3. Discuss the problems associated with hot desking in the PMO.

'Tell the client I'm working on his project!'

27

Managing Small Projects

Learning Outcomes

After reading this chapter, you should be able to:

- Understand the definition of a small project
- Manage a small project
- Understand the different management skills required during the product lifecycle

Mention project management and most people will conjure up in their minds an image of large scale capital projects in the construction industry or defence procurement - but there is a wind of change happening – the prevalence and importance of small projects is being recognised. In the context of this book, new venture creation and the implementation of new products all have the characteristics of a small project, so it is appropriate, if not essential, to use project management techniques to implement them.

In recent years there has also been a trend towards proactive companies restructuring their work into a number of small projects to give them better planning and control. In this **management-by-projects** approach, project management techniques are being used to plan and control these new enterprises, new ventures and outsourcing.

Projects can range in size, scope, cost and time from mega international capital projects costing millions of dollars over many years (building a power station or establishing a new mobile phone network), to small business projects or new ventures with a low budget taking just a few weeks to complete (setting-up a small business or repairing a car). It is the techniques and problems associated with these small projects that this chapter will focus on.

Although large capital projects capture the imagination, in practice they only account for a few percent of project activity. At the other end of the scale, small projects account for a whopping 90% of project activity, making them by far the most pervasive type of projects. Therefore, it makes sense to focus on the tools and techniques that project managers and entrepreneurs can use to manage these small projects.

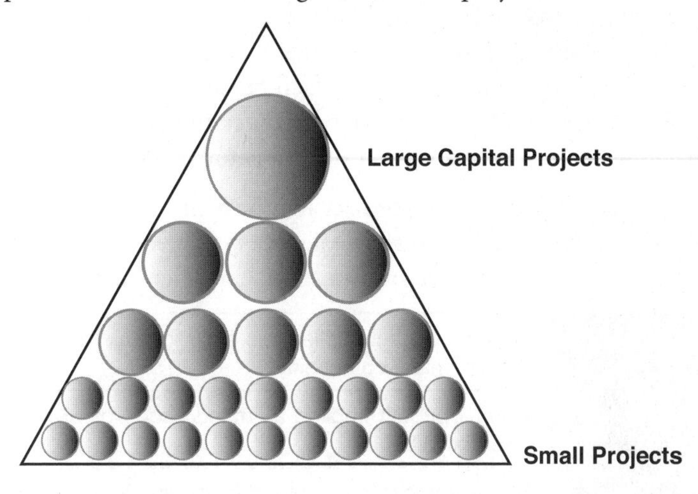

Figure 27.1: Pyramid of Projects – shows a few large projects at the top, but many small projects at the base of the pyramid

Some people might assume that project management techniques are only for managing large construction projects. This myth has developed because most of the project management methodologies and frameworks were originally designed for managing large capital projects. And using these frameworks and methodologies for small projects does not always work - in fact, they can be counter-productive. What most small projects need is a simple methodology to help guide them.

1. Define Small Project

It is common practice to quantify the size of a business by the number of employees and turnover. For example, SMEs (small and medium size enterprise) are defined as:

Micro firm	0 - 9 employees	Euro 2m turnover
Small firm	0 - 49 employees (includes micro)	Euro 10m turnover
Medium firm	50 - 249 employees	Euro 50m turnover
Large firm	over 250 employees	Euro <50m turnover

Using this structure and terminology SMPs (Small Medium size Projects) can be defined as:

Micro project	0 - 9 employees	Euro 2m turnover
Small project	0 - 49 employees (includes micro),	Euro 10m turnover
Medium project	50 - 249 employees	Euro 50m turnover
Large project	over 250 employees	Euro <50m turnover

It should be noted that project size is relevant to the industry sector. For example, a project of several million dollars would be considered small in the oil industry, but a project of only a few thousand dollars would be considered large in the fashion industry.

A project's perceived size may also relate to the type of work, type of industry, level of complexity and duration. Small projects would probably only have a few employees with a simple organization structure making communication rapid and efficient. And because there are only a few people, each person may be responsible for a number of positions. For example, the planner becomes the accountant, and the procurement manager becomes the quality control inspector.

Define Small Project: A 'small project' may be defined, for the purpose of this chapter, as a simple project with a limited scope of work managed by one person (the entrepreneur / project manager), or a small project team. Consider the following points:

Simple	A simple project implies it is easy to visualize all the links in the project - it has limited complexity.
Scope of Work	A small project implies a limited scope of work with a straightforward PBS and WBS.
Small Team	A small project team should mean the lines of communication are short and the team's reaction to the project's workload should be quick.
Short Duration	A small project implies a short duration, which means there may not be sufficient time to develop systems and learn from mistakes - the project has to *'hit the ground running'*.
Small Budget	A small project implies a small budget which should be tight and visible, with no excess funds for mistakes.
Skills	A small project implies the project manager or project team members are either a *'jack of all trades'* or specialises in a niche market which requires a limited range of skills.
Risk	A small, simple project implies limited risk - small companies do not have the capacity to accept risky projects.
Strategic	A small project implies limited strategic and political importance.
Decision-Making	A small project implies innovative problem solving and fast decisive decision-making.
Information	A small project implies a small volume of information and simple computer systems. The sheer volume of data is one reason that makes large projects complex.

Knowing the size of the project will enable the project manager to determine the appropriate project management system that best suits the project. However, deciding whether the project is small or large may be an over simplification, as the real question should be whether the project is simple or **complex**. As projects become more complex, the project management style needs to change from intuitive and informal to organized and structured.

2. Managing Small Projects

Managing a small project is not necessarily a scaled down version of a large project. Although small projects might appear to be simpler and more straight forward, they often have their own unique problems. Because of the informal way small projects are often set up they can be doomed to fail or become self-limiting. Consider the following points:

Scope of Work	The scope of work is inadequately and inaccurately defined - no drawings, no specifications and no contract.
Instructions	Instructions are given verbally - nothing in writing to confirm agreements. This could lead to misunderstandings.
Standards	Minimum quality standards not established or adequately defined at the outset - making it difficult to enforce quality verification and control to accept or reject the work.
Schedule	The schedule is not discussed - there is a perception that the work will be done straight away once the order is given.
Payment	The method of payment is not discussed, although the price of the project would usually be agreed at the outset.
Arbitration	There are no arbitration mechanisms - this makes it difficult to quickly and amicably sort out any disputes.
Exit	There are no exit strategies - this makes it difficult to terminate the contract.
Duration	The duration of the project is short - this does not give the project manager time to establish a management system and learn by mistakes.
One-off	Many small projects are one-offs; for example, building a house extension. Therefore, there is a lack of previous experience and knowledge to draw from.

The project manager has a multitude of challenges to face when implementing small projects and new ventures, so it is important to ensure innovative ideas and marketable opportunities are not handicapped at the outset by ineffective project management. Many creative ideas have floundered because of unrealistic expectations, poor estimating, communication breakdown within the team, poor co-ordination between the stakeholders and uncontrolled cashflow.

To overcome these shortcomings, project managers use a number of special project management techniques to plan and control the progress of their projects. These should form an important part of the project management entrepreneur's portfolio of project management skills.

3. **Product Lifecycle** (management skills)

The product lifecycle or new venture lifecycle is an excellent model for clearly showing the different phases or stages a product or company passes through from the cradle to the grave or, more realistically, from new venture opportunity to its eventual disposal or replacement.

The interaction between the phases has been developed and discussed in detail in the *Project Lifecycle* chapter. This chapter will outline how the different management skills are integrated and relate to each other on small projects.

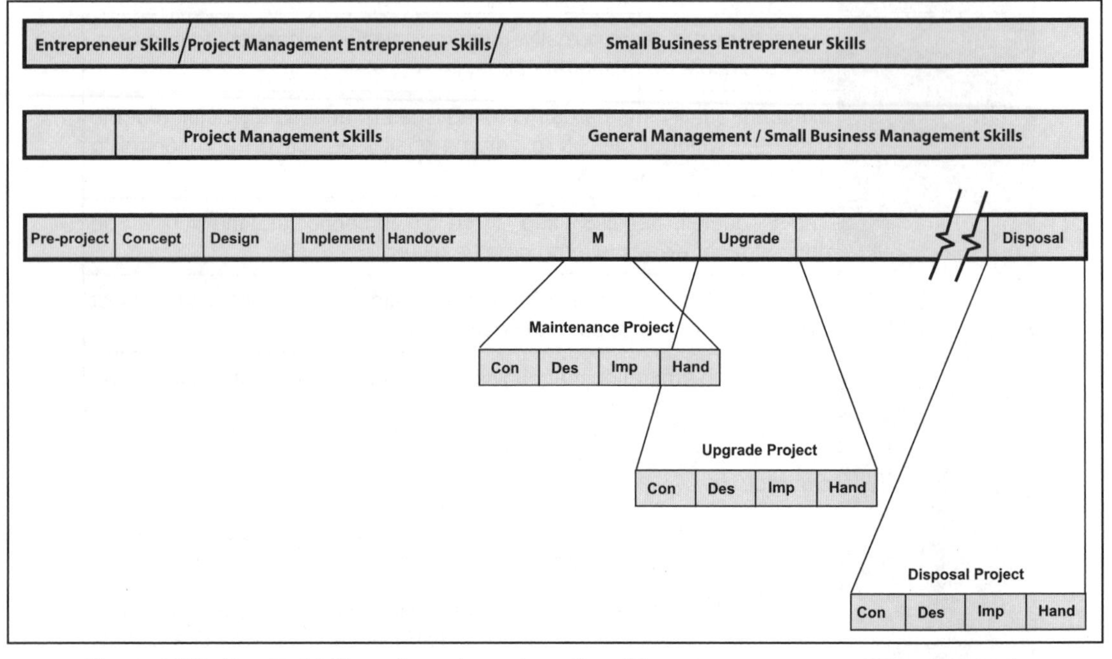

Figure 27.2: Product Lifecycle – shows how the different management skills relate to the product lifecycle phases

Entrepreneur Skills: It is the entrepreneur or the entrepreneurial skills which start the ball rolling. Entrepreneurs are experts at inventing new products, spotting innovative opportunities, solving challenging problems and making the decision to initiate new ventures or projects. Once the new venture has been identified, it is handed over to the project manager for implementation.

Project Management Skills: It is the project manager who uses project management skills to implement or set up a new venture. Project managers are experts at planning and controlling the scope of work to implement the new venture using a portfolio of special project management tools and techniques.

Project Management Entrepreneurial Skills: During the implementation of the project or new facility, the entrepreneurial project manager will be continually looking for better opportunities. This effectively delays the design freeze and keeps the door open to the latest technology, responding to competitors' products and responding to changing market trends. The entrepreneurial project manager will also be continually looking for ways to speed up production to beat the competition to market – this may involve shortening lead times and running activities in parallel. All these fast tracking skills rely on a quick response to better information as the project moves forward. However, it can also greatly increase risks. As the implementation or project phase is completed, the new venture or facility is handed over to the small business manager for the operational phase to make the product or service.

Small Business Management Skills: When the new venture is up and running, or created, the new facility is now complete and the venture moves into the next phase; the operational phase to manufacture the product or provide the service. This is when the product is made and distributed to the market and sold. This is the acid test of the marketing plan – will the customers or punters buy the product?

It is the small business management skills which run the new venture, facility or company on a day-to-day basis. Small business management skills are essentially inward looking, trying to continually improve efficiency and productivity.

Small Business Entrepreneur Skills: Where the small business manager is inward looking at efficiency and productivity, the small business entrepreneur is outward looking at the market and competition. It is a natural trait to continually monitor:

- The competitions' products
- The competitors' pricing strategy
- The latest market trends
- Ways to incorporate the latest cutting edge technology to maintain competitive advantage.

This is when a company needs small business entrepreneurial skills to be aware of changes in the market, incorporate new technology and respond to competition and, like links in a chain, all these skills rely on each other to ensure continuing success. This brief overview of the product lifecycle clearly outlines the different management skills that are required as the new venture moves through the growth phases.

4. Small Business Management

Entrepreneurs and small business managers are often thought of as being one and the same. But, in practice, entrepreneurship is the management of change, particularly when starting a new venture or introducing a new product or service. Whereas, small business management is the management of the company on a day-to-day basis, particularly with respect to repetitive jobs. Some of the key small business management functions include:

- Marketing the company and the products
- Accounts, budgets, book-keeping and cash flow
- Paying wages, invoices and debtors
- Complying with rules, regulations and taxes
- Buying or renting premises
- Buying, leasing or hiring plant and equipment
- Procurement of material and services
- Warehousing and stock control (JIT)
- Distribution
- Labour relations, recruitment and hiring
- Supervision and leadership
- Manufacturing the product (technical), and scheduling the workflow
- Quality control
- Customer service.

Entrepreneurship and small business management obviously go hand-in-hand, and may be seen as two sides of the same coin. Small businesses swing in and out of periods of entrepreneurial change as the small business responds to new technology, competitors and the latest market trends.

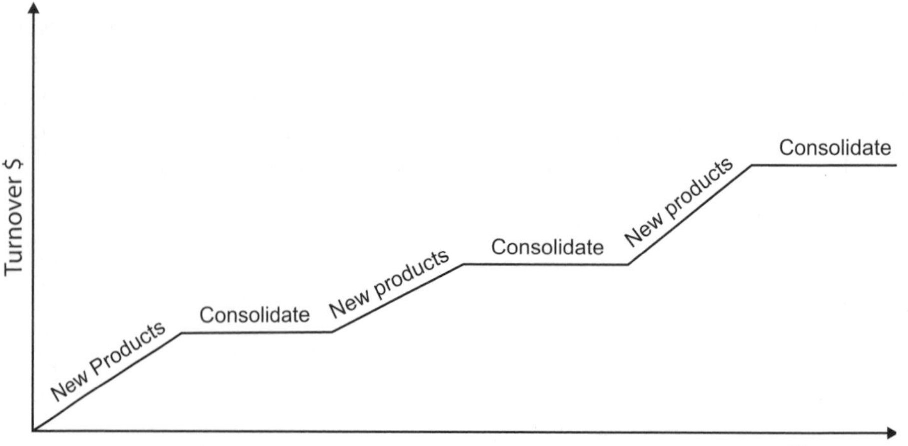

Figure 27.3: Entrepreneur / Small Business Manager Cycles - shows periods of entrepreneurship followed by periods of consolidation

Exercises:

1. Outline the difference between large and small projects.
2. Discuss why entrepreneurs need project management skills to set up new ventures.
3. Discuss the potential problems of managing small projects intuitively.

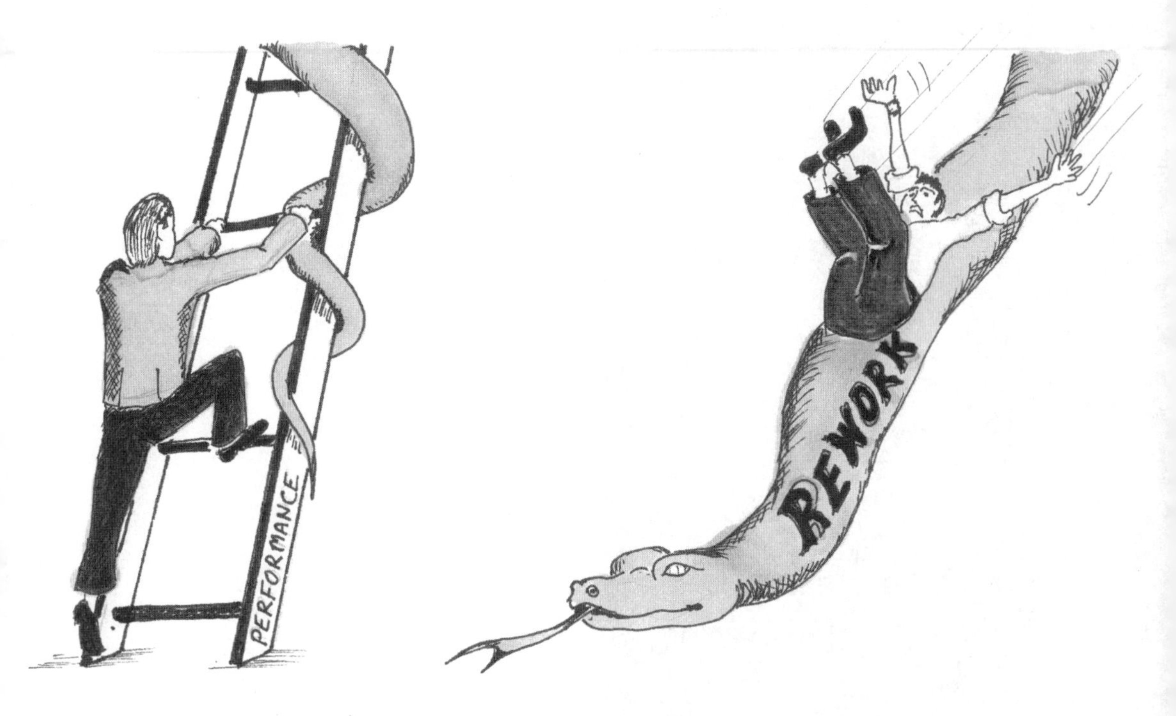

'I was doing really well until the quality inspector rejected my work!'

28

Event Management

Learning Outcomes
After reading this chapter, you should be able to:
Understand how event management can benefit from using project management techniques
Understand the different types of events
Understand that events are the ultimate in time-limited scheduling

Event management has all the characteristics of a project, which is why project management techniques are a core subject within the event management body of knowledge.

This chapter will outline how a range of special project management tools and techniques can be used to structure, plan and control an event.

Event management, also referred to as event co-ordination, must be the ultimate in time-limited scheduling – because if the project is only one day late, or even only one hour late it could have missed the whole event.

1. What is an Event?

An event has already been defined in the *Time Management* chapter as an activity with zero duration. At the activity level an event is only a minor part of a WBS work package. But, in the context of this chapter, an event is the whole purpose of the project – so it has a completely different meaning.

An event may be defined as a special occasion such as a wedding, a show, a business conference, or a sports match. Events range in size from a meeting of a few people to a mega event such as the Olympic Games, which involves thousands of people and costs billions of dollars even though the event only lasts a couple of weeks.

Events can be subdivided into a number of different types which have different purposes, different venues and different ways of interacting:

Exhibition	Exhibition events, such as a Boat Show or a Book Fair, are typical of events where a number of companies come together under one roof to present their products on stands/booths while the visitors walk around the venue.
Fashion Show	A fashion show is a typical example of a dynamic type of event where the buyers and clients sit and watch, and the models walk down the catwalk wearing the products.
Conference	A conference is a typical example of an event where the experts come to present their papers or opinions (product) while the audience sit and listen. These types of events are particularly popular with education and politics. And, most importantly, the key note speaker (who usually opens the conference) will have a precise start time which cannot be missed.
Sports Match	Sports matches are usually short intense competitions that are, generally, held in specially built stadiums or buildings (tennis court, ice rink, bowling alley).
Music Concert	Music concert events can attract huge crowds, where the artists come to perform a gig on stage and the audience sits, stands or even dances. Venues can range from theatres, parks and football stadiums, to old air bases.
Wine Tasting Event	Wine tasting events are a popular social and cultural event where vine growers and wine makers come to present their products and the guests sample their products.
Yacht Launch	Building a yacht would include a number of distinct events; laying the keel, stepping the mast, the launching and the sea trials.
Team Building Event	Team building events use a range of innovative locations to improve the interaction of management teams; locations and situations would include mountaineering, sailing and African safaris.
Holiday	A two week holiday in the sun is an annual event most people look forward to. These events are characterized by at least one precise deadline - the flight and a number of prerequisites (passport and visas).
Wedding	A wedding is probably the largest private event or function that many families have to organise; similar events would include a christening, a 21st, a graduation, a retirement leaving party and a funeral or a wake.

2. Project Management Techniques

The project lifecycle technique offers an overview of an event from the embryo of an idea, to the preparation, to the event and, finally, to the restoration and closeout. Events have a distinctive lifecycle profile which is characterized by a long preparation phase followed by a very short operational phase.

The event lifecycle can be represented by the product lifecycle where the project phase is the preparation of the event and the operation phase is the running of the actual event (see figure 28.1 and table 28.1).

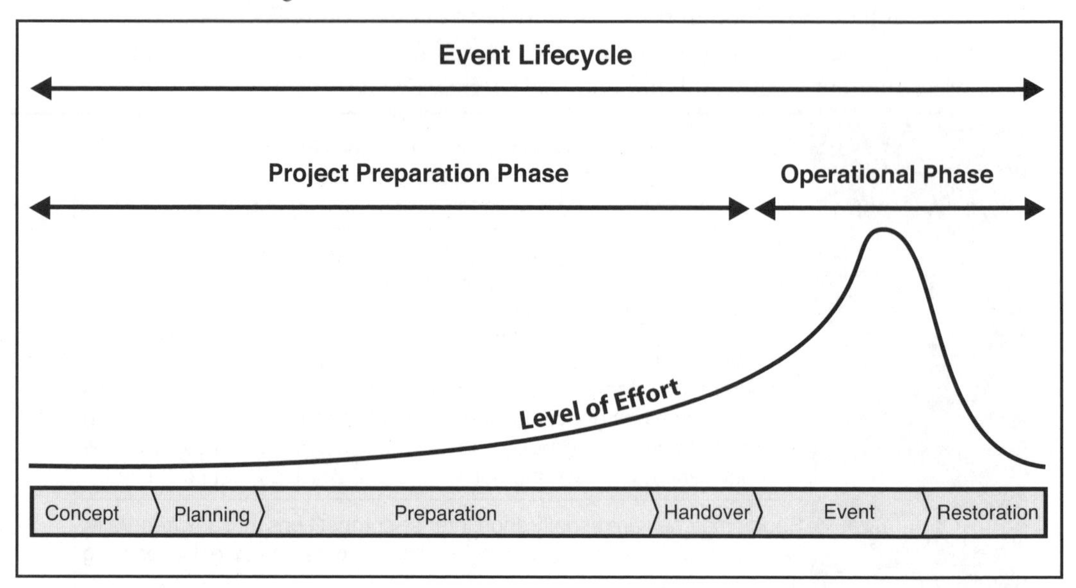

Figure 28.1: Lifecycle of an Event – shows the long preparation phase, followed by a very short operation phase

Concept Phase	Includes the feasibility study and the build-method.
Planning Phase	Includes detailed planning, booking the venue, organising the exhibitors, designing the layout of the exhibition, marketing and promotion.
Preparation Phase	Usually the longest phase which implements the plans to create the event facilities.
Commission Phase	Includes commissioning the event facilities and the handover to run and coordinate the actual event.
Operation Phase	Includes running and coordinating the actual event.
Disposal Phase	Includes removing the props, restoring the venue, and compiling the event closeout report.

Table 28.1: Event Lifecycle - shows the phases of the event lifecycle

An extreme case of an event would be the Olympic Games which has an eight year preparation phase, followed by a very short, two week competition phase. With such long lead times event managers need to beware they do not fall into the trap of delaying planning design decisions because the event is several years away, and then end up with panic decisions and a rushed building programme. This is probably what happened to the Greek Olympics which was touch and go whether they would finish on time.

The Football World Cup, which is the biggest sporting event in the world, has addressed this problem by staging the Confederations Cup one year before the World Cup. This is in effect a dress rehearsal for the main event, but it also ensures everything is up and running well in advance.

Strategy: The starting point of event management is to determine the purpose of the event – it may be in response to an opportunity for a new exhibition, or it may be an annual event such a Motor Show (shop window for the motor industry).

Feasibility Study: The feasibility study technique can be used by event coordinators to identify the stakeholders and determine their needs and expectations. This exercise encourages the event coordinators to cast the net wider than just over the people attending the event and the companies represented. The feasibility study outlines the build-method and confirms the event is commercially feasible and making the best use of the company's financial resources. The feasibility study should also consider the best time and venue to hold the event, considering availability of venue and the potential for clashing with other events.

'Come back!! I'm only a few minutes late!!'

Event WBS: The WBS technique can be used by event coordinators to subdivide the scope of work into manageable work packages and/or a checklist. The work packages and checklist of tasks become the backbone of the project to set up the event. Consider the WBS for a wedding event, which most people would experience in their lifetime.

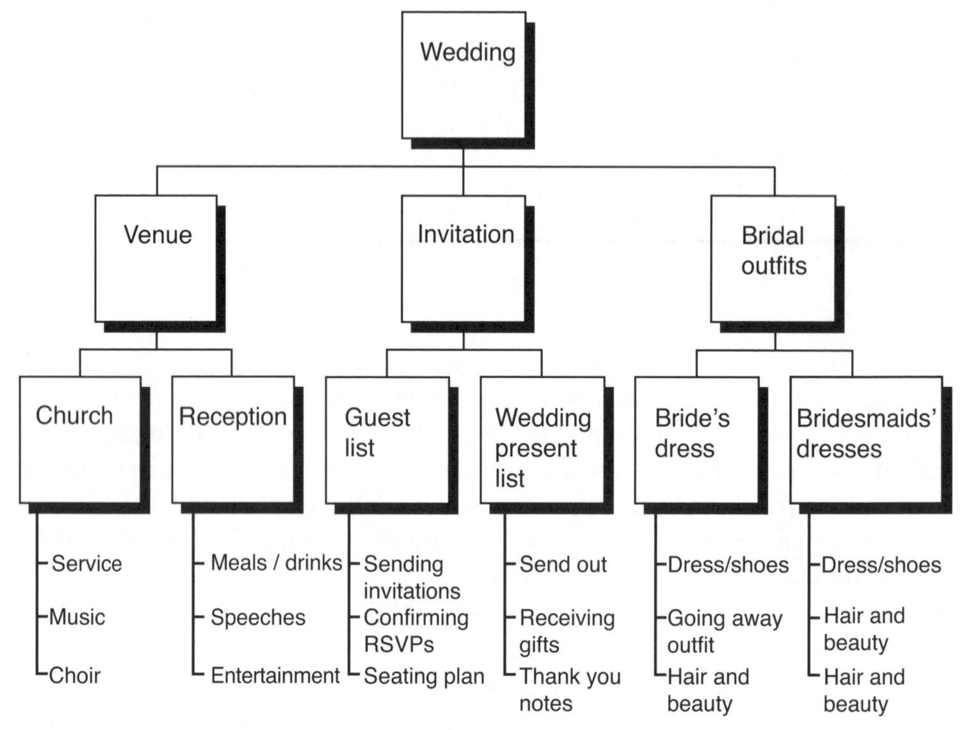

Figure 28.2: Event WBS – shows the subdivision of a wedding into a number of components

WBS Checklist	Description	Responsibility	Budget	Procure	Resource	Risk
100	Reception	Family	$60,000	Catering	Waiters	Rain
200	Bride's dress	Maid of Honour	$15,000	Dress	Bridesmaids	Dress Gets damaged
300	Groom	Best Man	$10,000	Ring	Hotel	Loses the ring

Table 28.2: WBS Spreadsheet – shows a wedding checklist which links the responsibility, budget, procurement, resources and risk to the WBS

CPM: The CPM techniques can be used by event organisers to establish the logical sequence of activities leading up to the event, together with the start and finish dates of all the activities and the critical path.

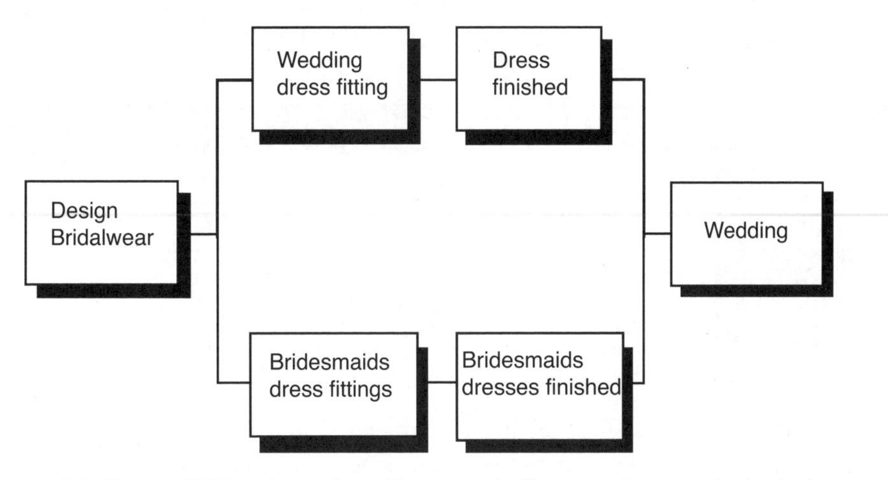

Figure 28.3: Event CPM – shows how the network diagram presents the logical sequence of the activities

Gantt Chart: The Gantt chart technique can be used by event coordinators to present the scheduling of the activities and the key dates. Events by definition are time-limited projects, so the management of the event has to focus on achieving a number of key dates.

For the wedding event the marriage proposal initiates the project and, once the couple have come to terms with the reality of their commitment, they need to start planning. One of the first decisions is the trade-off between the budget, number of guests, venue, and date, as each one influences the others.

Description	Jan	Feb	Mar	Apr	May	Jun	Jul	Aug	Sep
Marriage Proposal	◆								
Wedding Plan		▬	▬						
Book Venues			▬						
Guest List				▬	▬	▬			
Bridalwear						▬	▬	▬	
Wedding									◆

Figure 28.4: Event Gantt Chart – shows how the Gantt chart presents the key milestones and activities

Procurement Schedule: The procurement scheduling technique can be used by event coordinators to compile a procurement list and a procurement schedule. The procurement list is developed from the WBS work packages and checklists. The long lead items need to be identified early on so they can be actioned to prevent any delays. In the case of a wedding, the booking of the venues (the church and reception - probably in that order), determine the actual date of the wedding.

Procurement List	Date Required (from Gantt chart)	Lead Time	Order By Date
Bridal Dresses	Two weeks before wedding	3 months	
Flowers	Two hours before the wedding	1 month	
Car	One hour before the wedding ceremony	2 months	
Reception	Immediately after the wedding ceremony	6 months	

Table 28.3: Procurement Schedule

Resource Management: The resource management technique can be used by event coordinators to determine the manpower requirements. Typically, during the preparation phase, the manpower requirement is low (even working part-time), but the numbers dramatically increase during the set-up and running of the event.

Description	Mon 1	Tue 2	Wed 3	Thu 4	Fri 5	Sat 6	Sun 7	Mon 8	Tue 9	Wed 10	Thu 11
Preparation	2	2	2								
Set Up Event				20	20						
Wedding						50					
Dismantal Event							10				
Restore								5	5		
Total	2	2	2	20	20	50	10	5	5		

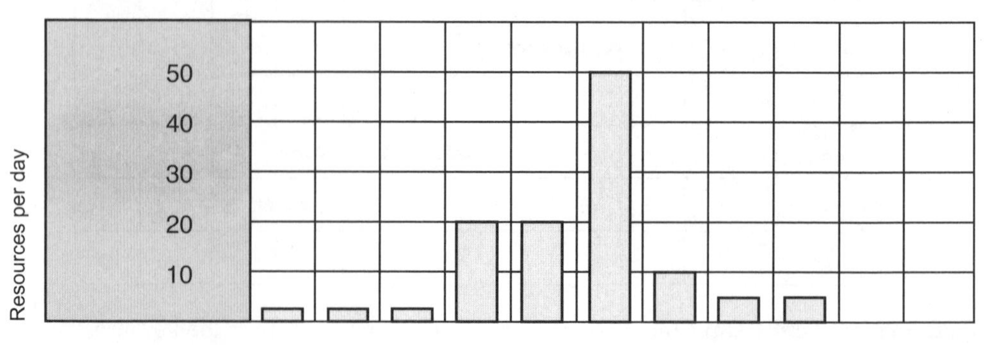

Figure 28.5: Event Resource Histogram – shows how an increase in resources is needed to set up and run the event

Risk Management: The risk management technique can be used by event coordinators to identify the risks that will stop them from achieving their objectives, and develop responses to remove or reduce the impact.

Scope	Objective	Identify Risk	Quantify	Respond
Ceremony	3.00 pm	Bride late – get me to the church on time	Massive impact	Maid of honour to help bride
Ceremony	As per programme	Lose wedding ring	High impact	Give the ring to best man
Venue (garden)	Sociable and comfortable	Rain	High impact	Marquee

Table 28.4: Event Risk Management Table – shows how risks can be identified and managed

Exercises:

1. Discuss how Event Coordinators can use project management techniques.
2. Discuss the different types of events and the special characteristics of events.
3. Events are usually time-limited occasions – you have to get it right on the day. Discuss the importance of planning and how parts of the event can be practised before the event so you hit the ground running.

'Get me to the church on time!'

Appendix 1

This appendix contains the solutions to exercises in the book.

The solutions are presented in chapter sequence.

Chapter 2: History of Project Management

Exercise 2 Solution: PERT Calculation

Using the PERT equation

Expected Time = (o + 4m + p) / 6

o = 12 days

m = 15 days

p = 24 days

What is T (expected)?

$$\text{Expected Time} = \frac{12 + (4 \times 15) + 24}{6}$$

= 16 days

Chapter 7: Project Lifecycle

Exercise 1 Solution: Draw the 'S' Curve

Level of Effort	1	2	3	4	5	6	7	8	9	10	11	12
Feasibility 1	1	2	3									
2		2	2									
3			2									
Design 1				9	9	9						
2					3	4						
3						4						
Execution 1							30	33	24			
2								20	9			
3									6			
Commission 1										12	8	3
2											2	2
3												1
Daily Total	1	4	7	9	12	17	30	53	39	12	10	6
Accumulated		5	12	21	33	50	80	133	172	184	194	200

Figure 7.12: Level of Effort - exercise 1

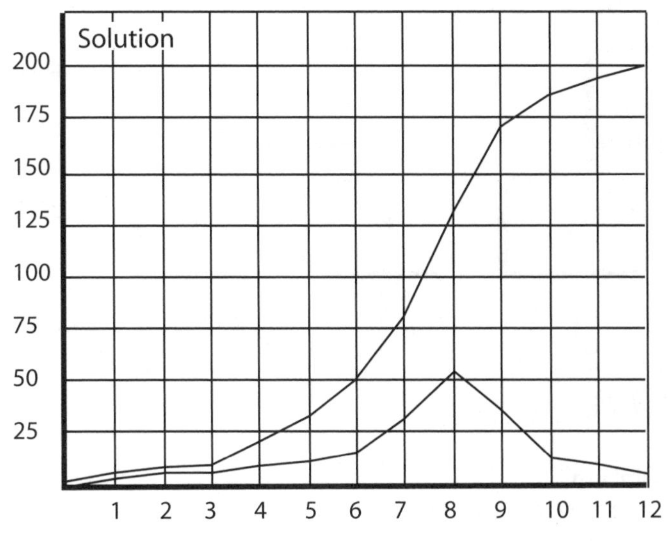

Figure 7.13: Level of Effort 'S' Curve - exercise 1

Chapter 9: Scope Management

Exercise 5 Solution: Checklist - Develop a Scope Management checklist

	Scope of Work Checklist
1.	Have the scope requirements been established by the project sponsor and written into the project charter?
2.	Has the scope of work been adequately defined? Does the scope of work outline the complete project and describe what is to be made (deliverables)?
3.	Has the scope of work outlined what is not included?
4.	Is the scope of work subdivided into manageable WBS work packages?
5.	Do the work packages define unique elements of work; by contractor, by department or by location?
6.	Has the scope verification process been agreed with the client?
7.	Has a scope change control system been designed and implemented?
8.	Has a build-method been developed and agreed with the stakeholders?
9.	Has the configuration management system been designed, and technical experts nominated?
10.	Has a communication system been implemented to gather information and inform stakeholders of the scope of work issues?

Chapter 10: Work Breakdown Structure (WBS)

Exercise 5 Solution: WBS work package numbers

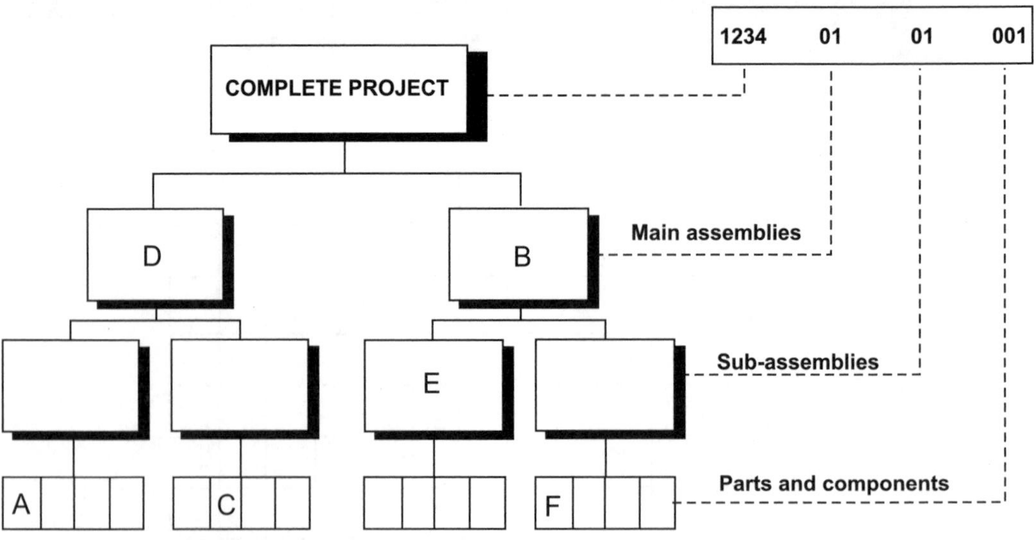

Figure 10.16: WBS – exercise 5

Exercise 4 Solutions:

D = 1234 01 00 000
E = 1234 02 01 000
F = 1234 02 02 001

Chapter 12: Critical Path Method (CPM)

Exercise 2 Solution: Critical Path Method (CPM)

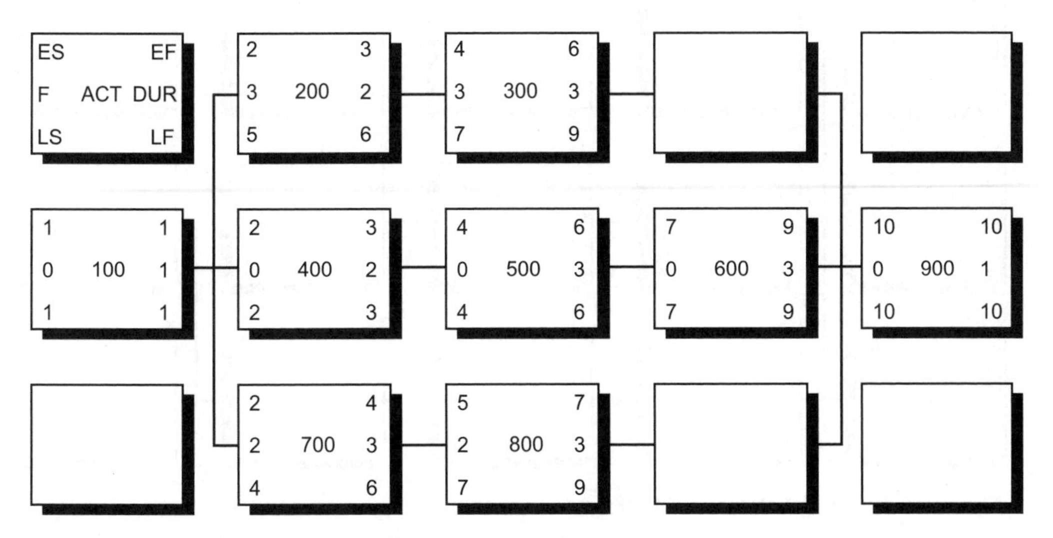

Figure 12.22: CPM 2 Network Diagram – exercise 2

Activity	Duration	Early Start	Early Finish	Late Start	Late Finish	Float
100	1	1	1	1	1	0
200	2	2	3	5	6	3
300	3	4	6	7	9	3
400	2	2	3	2	3	0
500	3	4	6	4	6	0
600	3	7	9	7	9	0
700	3	2	4	4	6	2
800	3	5	7	7	9	2
900	1	10	10	10	10	0

Table 12.9: CPM 2 Tabular Report – exercise 2

Chapter 12: Critical Path Method (CPM)

Exercise 3 Solution: Critical Path Method (CPM)

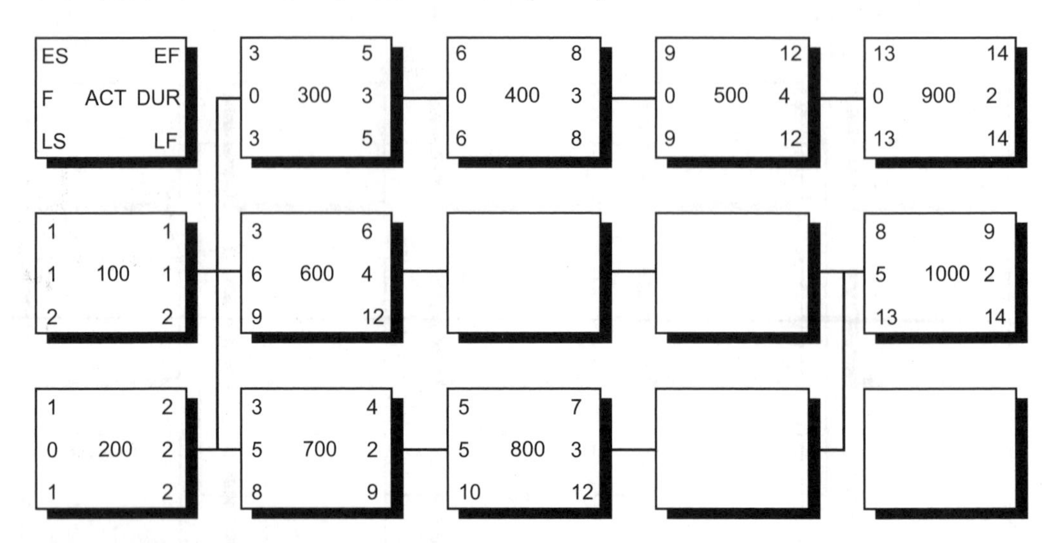

Figure 12.23: CPM 3 Network Diagram – exercise 3

Activity	Duration	Early Start	Early Finish	Late Start	Late Finish	Float
100	1	1	1	2	2	1
200	2	1	2	1	2	0
300	3	3	5	3	5	0
400	3	6	8	6	8	0
500	4	9	12	9	12	0
600	4	3	6	9	12	6
700	2	3	4	8	9	5
800	3	5	7	10	12	5
900	2	13	14	13	14	0
1000	2	8	9	13	14	5

Table 12.10: CPM 3 Tabular Report – exercise 3

Chapter 13: Gantt Chart

Exercise 2 Solution: Given figure 13.11, draw the Gantt chart

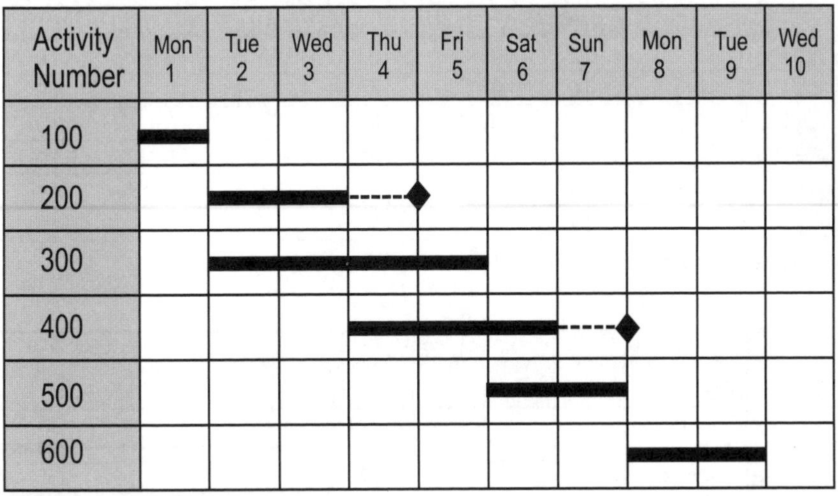

Activity Number	Mon 1	Tue 2	Wed 3	Thu 4	Fri 5	Sat 6	Sun 7	Mon 8	Tue 9	Wed 10
100										
200										
300										
400										
500										
600										

Figure 13.13: Gantt Chart - exercise 2

Exercise 3 Solution: Given figure 13.12, draw the Gantt chart

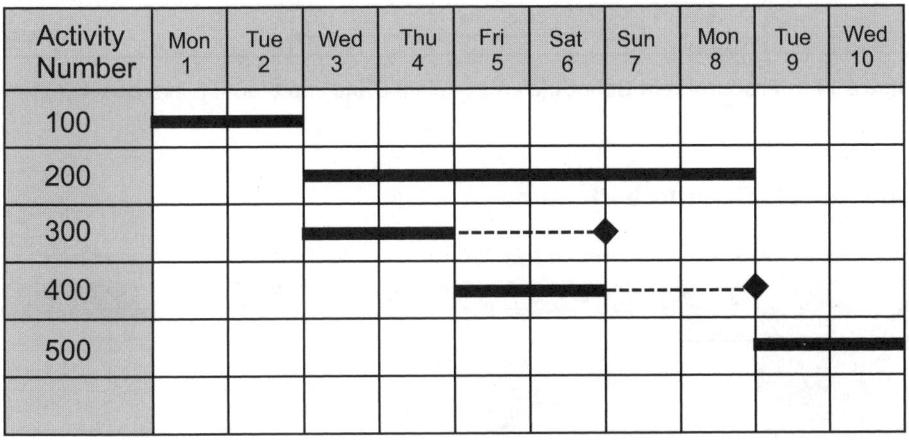

Activity Number	Mon 1	Tue 2	Wed 3	Thu 4	Fri 5	Sat 6	Sun 7	Mon 8	Tue 9	Wed 10
100										
200										
300										
400										
500										

Figure 13.14: Gantt Chart - exercise 3

Chapter 14: Procurement Schedule

Exercise 4 Solution: Procurement Schedule

1. WBS	2. PO	3. Lead Time (LT)	4. JIT	5. Early Start (ES)	6. Order-by-Date (ES - LT - JIT - 1)	7. Delivery Date (DD)	8. Variance (ES - DD)
100	PO 1	3	2	9	3	8	0
200	PO 2	4	0	9	4	8	0
300	PO 3	3	3	9	2	8	0
400	PO 4	4	2	9	2	8	0
500	PO 5	4	4	9	1	9	-1
600	PO 6	2	2	9	1	5	3

Table 14.19: Procurement - exercise 4

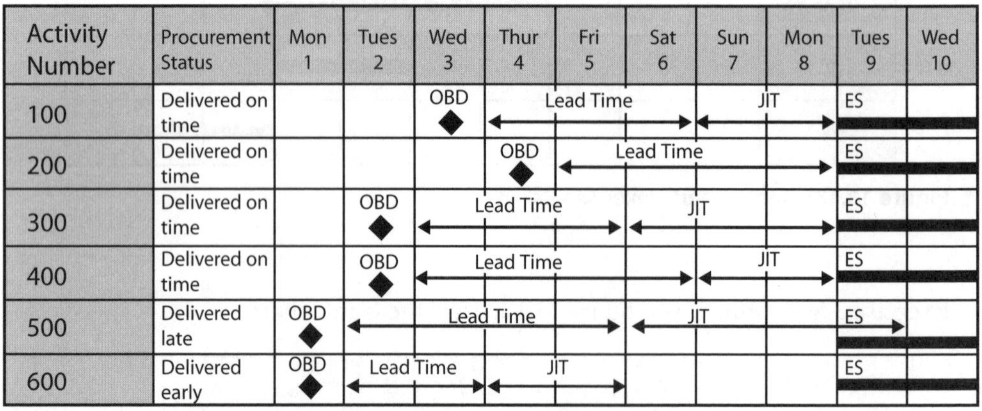

Activity Number	Procurement Status	Mon 1	Tues 2	Wed 3	Thur 4	Fri 5	Sat 6	Sun 7	Mon 8	Tues 9	Wed 10
100	Delivered on time			OBD ◆		Lead Time			JIT	ES	
200	Delivered on time				OBD ◆		Lead Time			ES	
300	Delivered on time		OBD ◆		Lead Time			JIT		ES	
400	Delivered on time		OBD ◆		Lead Time				JIT	ES	
500	Delivered late	OBD ◆			Lead Time			JIT		ES	
600	Delivered early	OBD ◆		Lead Time	JIT					ES	

Figure 14.3: Procurement Schedule - shown in a Gantt chart format, exercise 4

Chapter 15: Resource Planning

Exercise 1 Solution: Resource Smoothing

	1	2	3	4	5	6	7	8	9	10	
100	7										
200		4	4	****	****	****	****	****			
300		****	****	4	4	****	****				
400		3	3	3							
500					3	3					
600				****	****	4	4	****			
700				****	****	****	****	4	4		
800							3	3	3		
900										7	
Totals	7	7	7	7	7	7	7	7	7	7	70

Figure 15.8: Gantt Chart - exercise 1

Chapter 15: Resource Planning

Exercise 2 Solution: Resource Smoothing

	1	2	3	4	5	6	7	8	9	10	
100	5										
200				5	5	****	****	5	5		
300		4	4	****	****	****	****				
400		3	3	3	3	****	****	****	****		
500						5	5				
600						3	3				
700								3	3		
800										5	
900										3	
Totals	5	7	7	8	8	8	8	8	8	8	75
Smoothed	5	7	7	8	8	8	8	8	8	8	75

Figure 15.9: Gantt Chart - exercise 2

Exercise 3 Solution: Resource Smoothing

	Activity Number	Mon 1	Tue 2	Wed 3	Thu 4	Fri 5	Sat 6	Sun 7	Mon 8	Tue 9	Wed 10	Thu 11
	100	5	5									
	200			5	5	2	2					
	300							5	5	1	1	
Move >>>	400					3	3					
Move >>>	500									4	4	
	600											6
	Total	5	5	5	5	5	5	5	5	5	5	6

Figure 15.10: Resource Smoothing - exercise 3

Chapter 17: Project Cashflow

Exercise 1 Solution: Cashflow Statement

	Cashflow	Jan	Feb	Mar	Apr	May	Jun
B/F		1,000	0	-3,000	-2,000	-1,000	1,000
Income	1 month credit	0	0	5,000	5,000	5,000	5,000
Total Income		**1,000**	**0**	**2,000**	**3,000**	**4,000**	**6,000**
Equipment	1 month upfront	1,000	1,000	1,000	1,000		
Labour	Same month		2,000	2,000	2,000	2,000	
Material	1 month credit			1,000	1,000	1,000	1,000
Total Expenses		**1,000**	**3,000**	**4,000**	**4,000**	**3,000**	**1,000**
Closing		0	-3,000	-2,000	-1,000	1,000	5,000

Table 17.5: Cashflow Statement - exercise 1

Exercise 2 Solution: Cashflow Statement

	Cashflow	Jan	Feb	Mar	Apr	May	Jun
B/F		2,000	1,000	-2,000	0	2,000	5,000
Income	1 month credit	0	0	6,000	6,000	6,000	6,000
Total Income		**2,000**	**1,000**	**4,000**	**6,000**	**8,000**	**11,000**
Equipment	1 month upfront	1,000	1,000	0	0		
Labour	Same month		2,000	2,000	2,000	2,000	
Material	1 month credit			2,000	2,000	1,000	0
Total Expenses		**1,000**	**3,000**	**4,000**	**4,000**	**3,000**	**0**
Closing		1,000	-2,000	0	2,000	5,000	11,000

Table 17.6: Cashflow Statement - exercise 2

Exercise 3 Solution: Gantt Chart - 'S' Curve

Activity/Date	$	1	2	3	4	5	6	7	8	9	10
100	100	50	50								
200	150		50	50	50						
300	100			50	50						
400	100				50	50					
500	150					50	50	50			
600	100						50	50			
700	100							50	50		
800	100								50	50	
900	50									50	
1000	50										50
Daily Total		**50**	**100**	**100**	**150**	**100**	**100**	**150**	**100**	**100**	**50**
Accumulated		50	150	250	400	500	600	750	850	950	1000

Figure 17.6: Gantt Chart Solution 'S' Curve - exercise 3

Chapter 17: Project Cashflow

Exercise 3 Solution: Gantt Chart - 'S' Curve

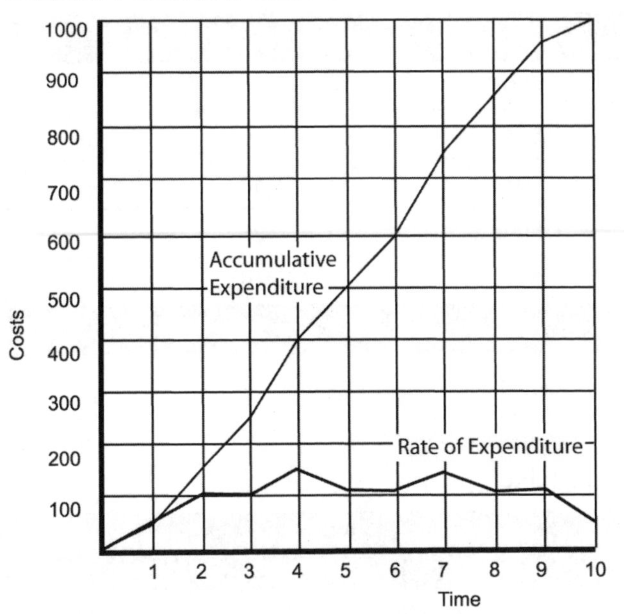

Figure 17.7: 'S' Curve - exercise 3

Chapter 18: Execution, Monitoring and Control

Exercise 4 Solution: Data Capture Table

WBS	Planned Hours	Percentage Complete	Earned Hours
100	100	100%	100
200	120	80%	96
300	80	50%	40
400	50	50%	25
500	60	25%	15
Totals	410	67%	276

Table 18.7: Data Capture Table - exercise 4

Exercise 5 Solution: Data Capture Table

Drawing Numbers	Design 20%	Drawing 70%	Check 10%	Percentage Complete	Planned Hours	Earned Hours	Actual Hours
100	100%	0%	0%	(Measured PC)	100 hrs		20 hrs
	20%			**20%** (Earned PC)		**20** hrs	
200	100%	100%	0%	(Measured PC)	200 hrs		160 hrs
	20%	**70%**		**90%** (Earned PC)		**180** hrs	
300	100%	100%	100%	(Measured PC)	80 hrs		100 hrs
	20%	**70%**	**10%**	**100%** (Earned PC)		**80** hrs	
400	50%	50%	0%	(Measured PC)	120 hrs		60 hrs
	10%	**35%**		**45%** (Earned PC)		**54** hrs	
500	100%	50%	10%	(Measured PC)	100 hrs		40 hrs
	20%	**35%**	**1%**	**56%** (Earned PC)		**56** hrs	
Totals				65%	600 hrs	390 hrs	380 hrs

Table 18.8: Data Capture Table - exercise 5

Chapter 19: Earned Value

Exercise 1 Solution: Continue with the painting example and produce the earned value table similar to table 19.6 for timenow 2 and timenow 3.

WBS Activity	BAC Budget	PV	PC	EV Earned	AV Actual	SV	CV	EAC
100	16 hrs	16 hrs	100%	16 hrs	16 hrs	0	0	16 hrs
200	16 hrs	16 hrs	50%	8 hrs	20 hrs	(8 hrs)	(12 hrs)	40 hrs
300	16 hrs	0						16 hrs
400	16 hrs	0						16 hrs
500	16 hrs	0						16 hrs
Totals	**80 hrs**	**32 hrs 40%**	**30%**	**24 hrs**	**36 hrs**	**(8 hrs)**	**(12 hrs)**	**104 hrs**

Table 19.7: Earned Value Table at timenow 2 (end of Tuesday) - late, over man hours

WBS Activity	BAC Budget	PV	PC	EV	AV Actual	SV	CV	EAC
100	16 hrs	16 hrs	100%	16 hrs	16 hrs	0	0	16 hrs
200	16 hrs	16 hrs	100%	16 hrs	24 hrs	0	(8 hrs)	24 hrs
300	16 hrs	16 hrs	100%	16 hrs	8 hrs	0	8 hrs	8 hrs
400	16 hrs	0	50%	8 hrs	4 hrs	8 hrs	4 hrs	8 hrs
500	16 hrs	0						16 hrs
Totals	**80 hrs**	**48 hrs 60%**	**70 %**	**56 hrs**	**52 hrs**	**8 hrs**	**4 hrs**	**72 hrs**

Table 19.8: Earned Value Table at timenow 3 (end of Wednesday) - ahead, under budget

Booklist

Anantatmula, Vittal, and **Parviz,** F (2005) *Project Planning Techniques*
Andersen, Erling, and **Grude,** Kristoffer, **Haug,** Tor (2009) *Goal Directed Project Management: Effective Techniques and Strategies,* Kogan Page
APM Planning Specific Interest Group (2008) *Introduction to Project Planning,* APM
Association of Project Managers (APM), *Body of Knowledge (BOK) 5ed*

Baca, Claudia (2007) *Project Management for Mere Mortals: The Tools, Techniques, Teaming, and Politics of Project Management,* Pearson Education
Belbin, M (1996) *Management Teams,* Butterworth-Heinemann
Benedetto, R. *Matrix Management Theory in Practice*, Kendall Hunt
Bentley, C (1993) *Configuration Management Within Prince*
BS 5750 (1979) *Quality Management*
Burke, Rory (2006) *Entrepreneurs Toolkit,* Burke Publishing
Burke, Rory (2007) *Project Management Techniques,* Knowledgezone.net
Burke, Rory (2006) *Small Business Entrepreneur,* Burke Publishing
Burke, Rory, **Barron,** Steve (2007) *Project Management Leadership,* Knowledgezone.net
Burleson, C., *Effective Meetings*, Wiley

Chapman, C., and **Ward**, Stephen (1996) *Project Risk Management: Processes, Techniques and Insights*, Wiley
Charland, T (1990) *Advanced Project Management Techniques Handbook*
Clark, C. *Brainstorming*, Wilshire
Cleland, D (1997) *Project Management Field Guide*, Van Nostrand Reinhold
Cobb, Anthony (2006) *Leading Project Teams: An Introduction to the Basics of Project Management and Project Team Leadership,* SAGE
Crosby, P.B (1987) *Quality is Free*, McGraw-Hill
Crosby, P.B (1995) *Quality Without Tears*, McGraw-Hill

Davidson Frame, J (1998) *The Project Office (Best Management Practices)*, Crisp
Dinsmore, P., *Human Factors in Project Management*, Amacom

Fewings, Peter (2005) *Construction Project Management,* Taylor & Francis
Field, M., and **Keller,** L., *Project Management*, Thomson
Frank, M., *How to Run a Successful Meeting in Half the Time*, Corgi
Frigenti, Enzo, and **Comninos**, D (2002) *The Practice of Project Management: a guide to the business - focused approach,* Kogan Page

Gido, J., and **Clements**, J (1999) *Successful Project Management,* South Western College Pub
Gray, C., and **Larson**, E., *Project Management (the Managerial Process),* McGraw-Hill Irwin

Haynes, Marion (1997) *Project Management (Fifty Minute Book)*, Elaine Fritz

Jones, Peter (1997) *Handbook of Team Design: A Practitioner's Guide to Team Systems Development*, McGraw-Hill

Kerzner, H (1997) *Project Management A Systems Approach to Planning, Scheduling and Controlling*, Van Nostrand Reinhold

Lewis, James (1995) *Project Planning, Scheduling & Control: A Hands-On Guide to Bringing Projects in on Time and on Budget*, McGraw-Hill
Lock, Dennis (2007) *Essentials of Project Management*, Gower
Lockyer, Keith (1992) *Production Management*, Financial Times Pitman
Lockyer, Keith, and **Gordon**, James (1996) *Project Management and Project Network Techniques*, Financial Times Pitman

Maylor, Harvey (2005) *Project Management: with MS Project*, Prentice Hall
Milosevic, Dragan (2003) *Project Management ToolBox: Tools and Techniques for the Practicing Project Manager*, Wiley
Morris, Peter (1994) *The Management of Projects*, Thomas Telford

O'Connell, Fergus (1996) *How to Run Successful Projects II: The Silver Bullet*, Prentice Hall
Oosthuizen, Pieter (1998) *Goodbye MBA*, International Thomson

Pennypacker, James (1997) *Project Management Forms*, Cambridge University Press
Pokras, Sandy (1995) *Rapid Team Deployment: Building High-Performance Project Teams (Fifty-Minute Series)*
Project Management Institute (PMI) *A Guide to the Project Management Body of Knowledge (PMBOK 4ed)*

Raftery, John (1993) *Risk Analysis in Project Management*, Routledge
Rosenau, M (1998) *Successful Project Management*, Van Nostrand Reinhold

Schwalbe, Kathy (2008) *Introduction to Project Management*, Second Edition
Stuckenbruck, L., *The Implementation of Project Management: The Professional Handbook*, PMI

Turner, R (1993) *Handbook of Project-Based Management*, McGraw-Hill

Walker, Anthony (1996) *Project Management in Construction*, Blackwell Science
Westney, Richard, *Computerized Management of Multiple Small Projects*, Marcel Dekker

Van Der Waldt, Andre (1998) *Project Management For Strategic Change and Upliftment*, International Thomson

Glossary

Activity: Item of work, task or job. A list of activities is required for the CPM calculation.

AIPM: Australian Institute of Project Managers.

APM: Association of Project Managers (UK).

Audit: An investigation to compare actual performance with planned work, or compare actual management systems with planned management systems.

Backward Pass: After completing the CPM's forward pass, the backward pass calculates the late start and late finish dates.

Barchart: (See Gantt chart).

Baseline Plan: (See Project Management Plan).

Benchmarking: Compares actual and planned project practices to those used on similar projects to identify best practice, generate ideas for improvement, and provide a basis for measuring performance.

Body of Knowledge (BoK): The body of knowledge (of a profession) identifies and describes the generally accepted practices for which there is widespread consensus of the value and usefulness, and also establishes a common lexicon of terms and expressions used within the profession.

Bottom-Up Estimating: An estimating technique based on estimating at the WBS work package level and rolling-up to give the total project value.

Brainstorming: A group method of interaction and a cross-flow of ideas used to generate a flood of creative ideas and novel solutions.

Breakeven Point: The number of products the company needs to sell to cover the set up costs – after this point the company starts to make a profit.

Brief: The project brief is a statement of the situation. It should outline what product or facility might be required, or what problem is to be solved. (See Statement of Requirements).

Budget: Planned allocation of funds to perform a fixed amount of work.

Build-Method: Outlines how to make the product.

Business Case: The business case is owned by the project sponsor. It provides justification to senior management for undertaking the project in terms of evaluating the benefits and alternative options and how they will fulfil the statement of requirements.

Calendar: Outlines the days on which activity work can be scheduled.

Cashflow: The flow of cash in and out of the project's account – typically presented as a monthly snap shot. The project cashflow is determined by integrating the project schedule with the project budget.

Change Control: (See Scope Change Control).

Charter: (See Project Charter and Team Charter).

Client: The client is the key customer who initiates the project, accepts the project and will pay for the project. (See Project Sponsor.)

Closeout Report: The closeout report progressively signs off the completed work, (as-built drawings) and identifies lessons learnt - what went right and what went wrong, useful information for future projects.

Commissioning: The confirming by testing or observation that a project has achieved the required condition as set out in the project management plan.

Communications: A process through which information is exchanged between project stakeholders.

Configuration Management: The process of ensuring the project will achieve the purpose for which it was initiated.

Contingency: An allowance or plan of action to cover unforeseen problems.

Contract: Legal agreement between two parties.

Control: The process of determining progress, comparing planned with actual, and making changes to keep the project on track.

Cost Breakdown Structure (CBS): The hierarchical breakdown of project costs.

Constraints: The identification of boundaries the project has to work within. These could be internal project constraints, internal corporate constraints, and external constraints.

Critical Path Method (CPM): The CPM calculates the start and finish dates of all the activities to calculate the float and identify the critical path.

Data Capture: The process of gathering information to determine the project's progress.

Document Control: The process of managing the movement of project documents to the people identified in the communication plan. This process might also include the storing of documents and signing of transmittals.

Duration: Total time to complete an activity from start to finish.

Earned Value: The integration of cost and time (or man hours and time) to determine the project's progress.

Entrepreneur: A person who spots an opportunity and co-ordinates resources to make-it-happen.

Estimate: Prediction of what will happen in the future. (See Bottom-Up Estimating and Top-Down Estimating.)

Event: A point in time with no duration, for example, the award of contract.

Event Management: Coordinates the preparation, the setup, the running and the restoring of an event or occasion, such as, an exhibition, conference, or sports match.

Execution Strategy: The buy-or-make decision.

Expediting: Is the project support function to anticipate problems before they arise and to offer solutions before delays are encountered. This includes periodic visits to vendors' premises and frequent telephone calls to check on vendors' progress control (materials, resources, workload, progress, quality control, manuals, and delivery).

Fast Track: To start the following activity before the current activity has completely finished, or change the logic to run activities in parallel to reduce the duration of the project.

Feasibility Study: A process usually conducted in the concept and initiation phase to assess if the project can be performed (build-method), meet company requirements (configuration), and make the best use of company resources (business case).

Float: Also referred to as slack, is the amount of time an activity can be delayed without delaying the total project.

Forward Pass: To calculate all the activities' early start and early finish dates.

Gantt Chart: A scheduling tool where the time of each activity is represented as a horizontal bar. The length of the bar is proportional to the duration of the activity.

Hammock: A group of related activities that are shown as one aggregated activity and reported at a summary level.

Impact Statement: Quantifies the impact proposed changes will have on the scope. The nominated experts might quantify the impact with the 'Build-Method' and the 'Configuration Management'.

Integration Management: Project integration management combines and unifies all aspects of the project. Integration techniques include; the project management process, the project management plan, and the project lifecycle.

Job Description: Outlines a person's duties, responsibilities and authority.

Key Date: (See milestone).

Knowledge: Project management knowledge implies that the project manager knows the relevant facts and information, and has an awareness or familiarity gained by experience working on projects. The term *knowledge* is also used to mean the confident use of project management with the ability to use it to achieve specific objectives.

Leadership: The ability to establish vision and direction, to influence and align the team members towards common goals and objectives, and to empower and inspire people to perform.

Level of Effort: A measure of the amount of work planned or performed. This can be presented as a rate of effort curve or an accumulated level of effort.

Management-By-Projects: Packaging a company's work into many small projects to improve focus and accountability.

Matrix OBS: An organization structure where the project manager shares responsibility with the functional managers who supply the resources.

Meetings: Project meetings are held with the relevant stakeholders to manage the project.

Milestone: A significant event that acts as a progress marker of achievement.

Monitoring: Data capture, determining what has happened on the project and its status.

Network Diagram: A graphic presentation of the logical sequence of activities. Drawing the network diagram is part of the CPM analysis.

Organization Breakdown Structure (OBS): A hierarchical breakdown of the organization into management levels or groups for the purpose of planning and control.

Outsourcing: The execution procurement philosophy of contracting-out work, or buying-in facilities or work (as opposed to using in-house resources).

Parallel Activities: Activities that can be carried out at the same time - this will shorten the overall time to perform the work.

Payback Period: The time the income takes to payback the original investment.

Percentage Complete: A measure of the work done.

Phase Review: A review that takes place at the end of each phase, it confirms the work has been carried out as per the project charter.

Plan: An intended future course of action (APM BoK 5ed).

PMBOK: Project Management Body of Knowledge (USA).

PMI: Project Management Institute (USA).

Portfolio: A collection of projects or programs and other work that are grouped together to facilitate effective management of that work to meet strategic business objectives. The project and programs of the portfolio might not necessarily be interdependent or directly related. (PMBOK 4ed).

Portfolio Management: The grouping of projects, programmes and other activities carried out under the sponsorship of an organization. (APM BoK 5ed).

PRINCE2: A project management methodology created for government projects. It is an acronym standing for **PR**ojects **IN** Controlled Environments (second edition).

Procedures: Methods, practices, instructions and policies that explain how work should be carried out.

Process: A linear sequence of steps that is carried out to achieve defined objectives.

Processes: Project management is accomplished through processes which it defines as a set of interrelated actions and activities that are performed to achieve a pre-specified set of objectives, products, results or services (PMBOK 4ed).

Procurement: The buying into the company or project of goods and services. (See Outsourcing.)

Product Breakdown Structure (PBS): A hierarchy of deliverables that combine to form the project.

Program: A group of related projects managed in a coordinated way to obtain benefits and control not available from managing them individually. (PMBOK 4ed).

Programme Management: A group of related projects, which may include business-as-usual activities, that together achieve a beneficial change of a strategic nature for an organization (APM BoK 5ed).

Project (1): A temporary endeavour undertaken to create a unique product, service or result (PMBOK 4ed).

Project (2): A unique, transient endeavour undertaken to achieve a desired outcome (APM BoK 5ed).

Project Charter (1): A document that officially initiates the project or project phase. The project charter can be a simple document outlining, in a few words, what is required, or it can be a much larger document defining precisely what is required and how it should be carried out. The project charter is owned by the project sponsor and is also used to assign responsibility and authority to the project manager.

Project Charter (2): The project charter officially initiates the project by formally adding the project to the company's register of projects. The project is given an identity with a project name, a number and a purpose. The project charter (owned by the project sponsor) sets out the why, what, who, when, where, how to and how much the project aims to achieve to implement corporate strategy.

Project Lifecycle: Shows how a project can be considered as a sequence of distinct phases that provides the structure for progressively delivering the projects. Typical phases include: strategy, feasibility, design, execution, commissioning, operation, maintenance, upgrade and disposal.

Project Management (1): The management of a project using the project management principles, and the special planning and control tools and techniques.

Project Management (2): The application of knowledge, skills, tools, and techniques to project activities in order to meet project requirements.

Project Management (3): The process by which projects are defined, planned, monitored, controlled and delivered such that agreed benefits are realised (APM BoK 5ed).

Project Management Plan (1): The process of documenting the actions necessary to define, prepare, integrate, and co-ordinate all the supporting plans. The project management plan then becomes the main source of information for the project management process (initiation, planning, execution, and closing) (PMBOK 4ed).

Project Management Plan (2): A plan which brings together all the plans for a project. The purpose of the project management plan is to document the outcomes of the planning process and to provide the reference document for managing the project. The project management plan is owned by the project manager. (APM BoK 5ed).

Project Management Plan (3): Is the original plan developed to implement the project, or latest updated plan which progress is tracked against.

Project Manager (1): The person appointed by the project sponsor with single point responsibility to manage and achieve the project's objectives (time, cost and quality).

Project Manager (2): The project manager is the owner of the project management plan and, therefore, responsible for the delivery of the project on time, within budget, and to the agreed quality. The role of the project manager should be outlined in the project charter together with how the project should be managed.

Project Management Process: Managing a project can be subdivided into four process groups; initiating process, planning process, execution, monitoring and control process, closing process. Collectively these process groups are required to make the project. They have internal dependencies and must be performed in the above sequence.

Project Office (PO): Also referred to as the project management office (PMO) is the home of the project team members who are responsible for supporting the information and administration needs of the project manager.

Project Sponsor: The person who initiates the project (through the project charter) and is responsible for acquiring the deliverable benefits from the project to implement corporate strategy.

Resource: The machine or person who performs the work.

Resource Management: Forecasting manpower loading and smoothing resources to match supply and demand.

Responsibility Assignment Matrix (RAM): Assigns work to the responsible person by integrating the WBS with the OBS.

Risk Management: A risk is any event which might prevent the project achieving its objectives. Risk management is, therefore, the process of identifying risk, quantifying risk, responding to risk, and controlling the risk management.

Rolling Wave Horizon: Focuses on the activities happening in the immediate future, perhaps one or two weeks ahead.

S-Curve: Graphic display of cumulative costs, labour hours or other quantities, plotted against time. The curve is flat at the beginning and end, and steep in the middle, as it follows an S shape. It generally describes a project that starts slowly, accelerates and then tapers off.

Schedule: A schedule is a timetable for the project. It shows how project activities and milestones are planned. It is usually presented in a Gantt chart format.

Scope Change Control: The administration of scope changes through the approved scope change control system. This includes logging, monitoring, evaluating and approving the changes (by the designated people) before the changes are incorporated into the baseline plan. This will ensure that the baseline plan always reflects the current status of the project and limits scope creep.

Scope Creep: Unnecessary expansion of the scope of work.

Scope Management: Concerned with identifying what is included and what is not included in the project scope to achieve the stated objectives.

Scope of Work (SOW): A mechanism for subdividing the scope of work into work packages.

Scope Verification: Is the process through which the client and the stakeholders formally accept the completed project deliverables. Verifying the completed work is the process of checking the work has been completed to the approved design and specification.

Single Point of Responsibility: The one person who takes ownership of, and is responsible for, the completion of the project (the project manager).

Skills: A skill (also referred to as a talent) may be defined as an ability or aptitude to perform something well (with the minimum amount of effort and performed efficiently).

Sponsor: (See Project Sponsor).

Stakeholders: An individual or group whose interest in the project must be recognised if the project is to be successful, in particular, those who might be positively or negatively impacted upon during or after the project.

Statement of Requirements: A statement of the needs that the project has to satisfy. The means is justified in the business case.

Supplier: A contractor, consultant or any organization that supplies resources to the project.

Task: (See Activity).

Team Building: Activities designed to increase the performance of the team.

Team Charter: A document that sets out the working relationships and agreed behaviours within a project team.

Team Member: A person who is accountable to, and has work assigned to them by the project manager to be performed either by themselves or by others in a working group (APM BoK 5ed).

Teams: Project teams are formed to carry out a specific task.

Timenow: The date up to which the progress is measured.

Top-Down Estimating: A high level estimate for the project based on limited data, for example, estimating the cost of building a house based on its square meterage, or the cost of building a ship based on its steel weight.

Work Breakdown Structure (WBS): A subdivision of the work into work packages and checklists which can be more easily planned and controlled, and responsibility assigned.

Index

accounting,
- financial, 222
- management, 223
- project, 223

activity,
- duration, 153
- float, 169
- in parallel, 160
- in series, 160
- on-Arrow (AOA), 44
- on-Node (AON), 45

actual value (AV), 251
APM bok, 49
Apple, 42
audits, 59, 274
authority, 313-314

barchart, (see Gantt charts)
baseline plan, 80, 239-242
benchmarking, 275
benefits of project management, 35
bill of materials (BOM), 153
body of knowledge (bok), 47
brainstorming, 304
BS6079, 86
budget, 219-220, 313
budget-at-completion (BAC), 251
build-method, 56, 79, 109
business case, 56, 79

calendar, 163
cashflow statement, (see project cashflow)
centre of excellence, 331
certification (PMP), 50
change control, (see scope change control)
change request (template), 122
charter (see Project Charter)
checklists, 131
client's needs, 98-113
closeout report (see project closeout report)
closing process, 71
CND, 43, 111
code of ethics, 50
cognitive persuasion, 314
commissioning phase, 84-97
communications, 80, 123, 239, 242, 280-295
computing, 33, 44
configuration management, 56, 79, 102, 120-126
constraints, 107-111
contingencies, 258-269
control cycle, 234-249
control sheet, 144
cost accounting, 223
cost control, 221, 241
costs,
- direct, 214
- fixed, 214
- indirect, 214
- labour, 215-216
- procurement, 217
- variable, 214

critical path method (CPM), 78, 156-173

data capture, 243
deadline, (see milestones)
decision-making, 239
decommissioning phase, 258-269
design and development phase, 258-269
direct costs, 213
document,
- control, 290-292
- storage, 193

earned value (EV), 80, 250-257
Eastonian Process, 66
entrepreneur projects, 342
environment, 33

estimating
- bottom up, 213
- costs, 79, 210-221
- data base, 239
- time, 78, 146-155
- top down, 138, 213
ethics, 55
events, 152, 181
event management, 27, 346-353
exception reports, 294
execution, monitoring and control, 70
execution strategy, 78
expediting, 192, 197-198, 239

failure (project), 265
fast tracking, 94
Fayol's management process, 65
feasibility study, 56, 68, 79, 96-113
financial accounting, 222
fixed costs, 214
float, 169
functional OBS, 308-309

Gantt, Henry (chart), 36, 37, 174-185
general management, 32

half-life refit, 258-269
hammocks, 180
handover,
- documents, 292
- meeting, 301
Heathrow T5, 115
history of project management, 38-45
human resource management, 48, 242, 316-327

impact statement, 124
implementation phase, 258-269
indirect costs, 214
industrial relations, 110
information power, 314
initiation process, 68
instructions, 238
issues management, 242
IPMA, 46-51

job description, 152, 238

keydates (see milestones)
Kipling (Rudyard), 77
knowledge, 29, 50

labour costs, 215-216
lessons learnt, 59
level of effort, 90
lines of communication, 280-295

management (types of), 31
management accounting, 223
management-by-exception (MBE), 246, 255
management-by-milestones, 40, 152
management-by-projects, 336
material costs, 217
matrix (OBS), 310-312, 330
meetings, 296-305
methodology (input-process-output), 57,95
milestones, 40, 181
minutes, 294, 300
mobile project office, 256, 336
model testing, 109
monitoring progress, 243-245
monthly reports, 294

NASA, 39, 131
needs analyses, 98-113
network diagram, 159
non conformance report (NCR), 274, 277

OBS, 80
OBS/WBS, 139

PBS, 134
PERT, 40-41
plan (what is?), 76
planned value (PV), 250-257
planning and control cycle, 74-83
planning and control spiral, 236-237
planning checklist, 78
planning process, 69

planning software, 238-239
PMBOK, 48
PMI, 47
PMO, 19, 328-337
PMP, 47, 50
portfolio management, 32
power (authority), 313-314
precedence diagram method (PDM), 159
PRINCE2, 72
problem solving, 239
process management, 62-73
procurement,
- costs, 217
- process, 189, 241
- schedule, 81, 186-199
product lifecycle, 96-97, 342
production line work, 23
production management, 23-25
programme management, 31
progress,
- chaser, 197
- meeting, 303
- report, 293
project
- accounts, 222-233
- cashflow, 80, 222-233
- charter, 54-56, 66, 67, 68, 79
- closeout report, 59-61, 69, 239
- communication, 48, 280-295
- control, (see execution, monitoring and control)
- control cycle, (see execution, monitoring and control)
- cost management, 48, 210-221
- definition, 21
- environment, 33, 43
- estimating, 146, 210
- initiation, 84-97
- integration, 48, 50-59
- lifecycle, 84-97
- management (definition), 29
- management environment, 33
- Management Institute (PMI), 46-51
- management integration, 52-61
- management office (PMO), 328-337
- management plan, 74-83
- management process, 60, 62-73
- Management Professional (PMP), 50
- management software, 33
- management standards, 46-51
- management techniques (history), 39
- management triangle, 43
- methodology, 55, 95
- office (PO), 104, 122, 328-337
- organization structures (matrix), 42, 306-315
- planning document, 79-83
- procurement management, 48, 186-199
- quality management, 48, 270-279
- quality plan, 270-279
- review, 59-61
- risk management, 48, 258-269
- scope management, 48, 114-127
- sponsor, 54, 114
- teams, 316-327
- time management, 48, 146-155
- trigger, 79
- types, 26-28
- work vs production line work, 23
purchase order, 186-199

quality
- assurance, 276
- audit, 274
- control, 277-279
- control plan (QCP), 80, 278-279
- management, 270-279, 242

RACI chart, 139, 288
RAM (responsibility assignment matrix), 139, 288
regulations, 111
reporting, 293-295
- exception, 294
- frequency, 247
- monthly, 294
- status, 293
- trends, 294
- variance, 293

resource,
- histogram, 79, 203-209
- planning, 200-209, 241
- smoothing, 206
responsibility,
- assignment matrix (RAM), 139, 268
- authority gap, 313
revised Gantt chart, 183
reward power, 314
risk,
- continuum, 25
- contracting, 267
- control, 268
- failure, 265
- identification, 263
- management, 80, 258-269
- mitigating, 266
- response, 266
role of the project manager, 34
rolling horizon Gantt chart, 182, 245

S Curve (how to draw), 90, 230-231
sales and marketing continuum, 24
schedule barchart, 174-185
schedule variance (SV), 250-257
scope,
- change control, 120, 239, 241
- creep, 102
- definition, 117
- management, 77, 114-129
- verification, 119
site office, 329
skills, 29, 50
small business management, 344
small projects, 338-345
software, 33
Solent University (model testing), 109
sponsor (see project sponsor)
spreadsheets, 144
stakeholders, 68, 98-113, 287
standards, 46-61
statement of requirements, 29, 56, 79
status reports, 293
storage (documents), 61

target date, 151
task, (see activity)
team,
- building, 316-327
- charter, 316-327
teams, 316-327
technical management, 33, 240
templates (WBS), 143
time,
- cost, quality triangle, 43
- management, 146-155, 240
timenow, 174-185, 251
top-down estimating, 213
total quality management (TQM), 274
Toyota, 274
tracking progress, 243
training (computer skills), 255
transmittal note, 291
transport, 192
trend reports, 294
triangle of forces (time, cost, quality), 43

unit rates, 218
unit standards, 50

variable costs, 214
variance reports, 293

warehouse, 193
WBS, 128-145
- numbering system, 141
- templates, 143
work
- authorisation, 238
- package (see WBS), 137
- pattern, (see calendar)
workshop (brainstorming), 304

This bluewater trilogy by Rory and Sandra Burke includes a preparation guide, a travelogue and a checklist. Bluewater cruising has all the features of a complex project, requiring effective budgeting, procurement, scope management and time planning. Most importantly it requires effective risk management and disaster recovery for the safety of the crew and integrity of the yacht.

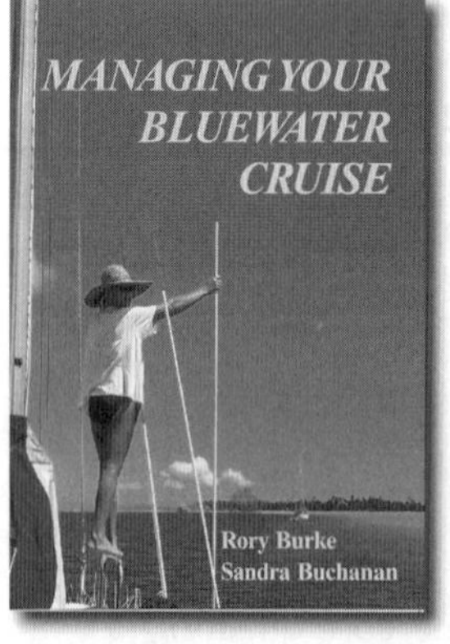

Managing Your Bluewater Cruise

ISBN: 0-473-03822-6
352 pages, 200+ photographs

This preparation guide discusses a range of pertinent issues from establishing budgets and buying equipment to preventative maintenance and heavy weather sailing. The text works closely with the ORC category 1 requirements and includes many comments from other cruisers who are *'out there doing it'*. This book also outlines what training courses to attend before leaving, what gear to take, provisioning strategy and, equally important, how to stow it all. If you wish to bridge the gap between fantasy and reality then your bluewater cruise must be effectively managed.

Greenwich to the Dateline

ISBN: 0-620-16557-x
352 pages, 200+ photographs

This is a travelogue of our bluewater cruising adventure from the Greenwich Meridian to the International Dateline – sit back with a sundowner and be inspired to cruise to the Caribbean and Pacific islands. In this catalogue of rewarding experiences we describe how we converted our travelling dreams into a bluewater cruising reality.

Bluewater Checklist

ISBN: 0-9582 391-0-x
96 pages

Checklists provide an effective management tool to confirm everything is on board, and all tasks are completed. Why try to remember everything in your head when checklists never forget!!! This book provides a comprehensive portfolio of checklists covering every aspect of bluewater cruising. To ensure your bluewater cruise will be successful, it must be effectively managed. Checklists provide an excellent tool for this purpose - even NASA uses them!!!

www.fashionbooks.info

This *Fashion Design Series* by *Sandra Burke* promotes fashion design skills and techniques which can be effectively applied in the world of fashion. In a competitive market it is important to produce designs that are not only stylish and pleasing to the eye, but also commercially viable.

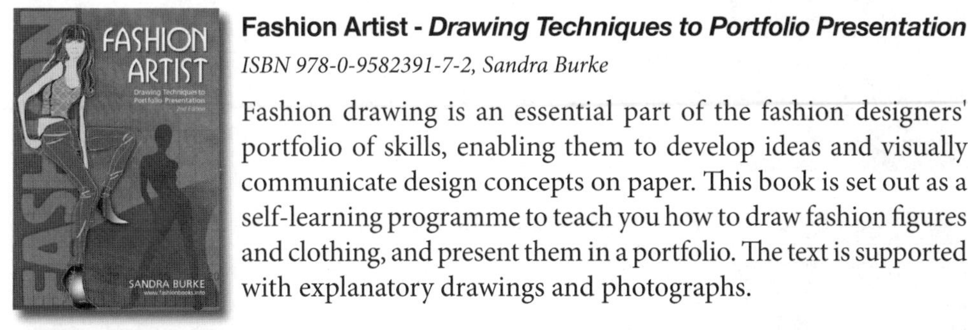

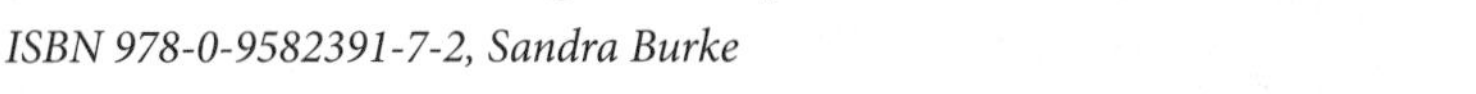

Fashion Artist - *Drawing Techniques to Portfolio Presentation*

ISBN 978-0-9582391-7-2, Sandra Burke

Fashion drawing is an essential part of the fashion designers' portfolio of skills, enabling them to develop ideas and visually communicate design concepts on paper. This book is set out as a self-learning programme to teach you how to draw fashion figures and clothing, and present them in a portfolio. The text is supported with explanatory drawings and photographs.

Fashion Computing – *Design Techniques and CAD*

ISBN 978-0-9582391-3-4, Sandra Burke

This book introduces you to the computer drawing and design skills used by the fashion industry. Through visuals and easy steps, you learn creative fashion computing design techniques. It includes, flats/working drawings, illustrations, fabrics, presentation and the digital fashion portfolio. Specific software includes: Photoshop, Illustrator, CorelDRAW, PowerPoint, Gerber and Lectra Systems.

Fashion Designer – *Design Techniques, Catwalk to Street*

ISBN 978-0-9582391-2-7, Sandra Burke

This book will help you develop your portfolio of fashion design skills while guiding you through the fashion design process in today's fashion industry. It explains how to analyse and forecast fashion trends, interpret a design brief, choose fabrics and colour ways, develop designs, create design presentations and develop collections for specific target markets.

Fashion Entrepreneur

ISBN 978-0-9582733-0-5, Sandra Burke

With your head buzzing with innovative and creative ideas – welcome to the Fashion Entrepreneurs' world of glamour, style and wealth. This book outlines the traits and techniques fashion designers use to spot opportunities and set up small businesses. These topics include: writing business plans, raising finance, sales and marketing, and the small business management skills required to run a fashion companies on a day-to-day basis.

Project Management Series

Fundamentals of Project Management

Rory Burke

ISBN 978-0-9582733-6-7

384 pages

This book is a broad based introduction to the field of Project Management which explains all the special planning and control techniques needed to manage projects successfully. This book is ideal for managers entering project management and team members in the project management office (PMO).

Project Management Techniques *(5ed)*

Rory Burke

ISBN 978-0-9582733-4-3

384 pages

PM 5ed presents the latest planning and control techniques, particularly those used by the project management software and the body of knowledge (APM BoK and PMI's PMBOK). This book has established itself internationally as the standard text for Project Management programs.

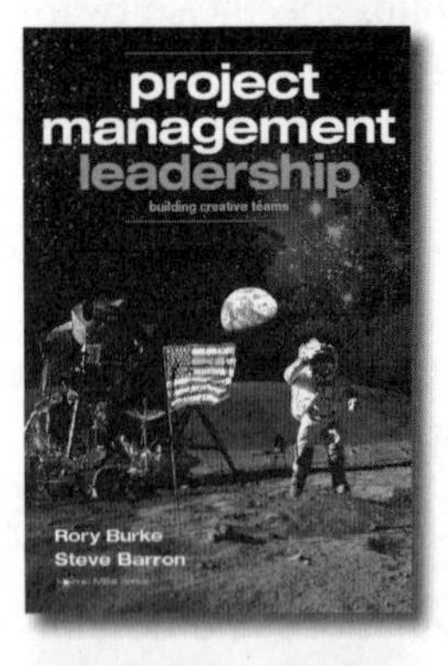

Project Management Leadership - *Building Creative Teams*

Rory Burke and Steve Baron

ISBN 978-0-9582733-5-0

384 pages

This book is a comprehensive guide outlining the essential leadership skills to manage the human side of managing projects. Key topics include: leadership styles, delegation, motivation, negotiation, conflict resolution, and team building.

Advanced Project Management - *Implementing Corporate Strategy*

Rory Burke

384 pages

Advanced project management focuses on the project methodologies companies can use to implement corporate strategy to achieve competitive advantage. This book is ideal for project sponsors responsible for developing the statement of requirements and the business case to acquire the benefits from configuration management, project finance and the project management process.

Project Management Series

German translation of:

Project Management - *Planning and Control Techniques*

Rory Burke

ISBN: 3-8266-1443-7

Project management techniques are perfect for a country renowned for its precise time keeping and production quality.

Greek translation of:

Project Management - *Planning and Control Techniques*

Rory Burke

ISBN: 960-218-289-X

The 2004 Olympic Games clearly showed how project management techniques can be used to control an event to meet a fixed end date.

Chinese translation of:

Project Management - *Planning and Control Techniques*

Rory Burke

ISBN: 0-471-98762-X

With the boom in demand for Chinese manufacturing, so there is an associated boom in Chinese infrastructure projects and the need for project management techniques.

Project Management Series

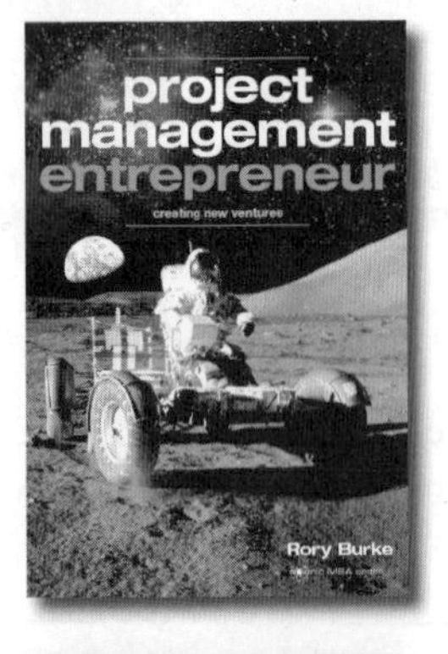

Project Management Entrepreneur - *Creating New Ventures*

Rory Burke

ISBN 978-0-9582733-2-9

384 pages

This is the first book to integrate the three management skills of Entrepreneurship, Project Management, and Small Business Management. Entrepreneur skills are required to spot opportunities, project management skills are required to implement the new venture and small business management skills are required to run the business on a day-to-day basis.

Entrepreneurs Toolkit

Rory Burke

ISBN: 978-0-9582391-4-1

160 pages

Entrepreneurs Toolkit is a comprehensive guide outlining the essential entrepreneurial skills to spot a marketable opportunity, the essential business skills to start a new venture and the essential management skills to make-it-happen.

Small Business Entrepreneur

Rory Burke

ISBN: 978-0-9582391-6-5

160 pages

Small Business Entrepreneur is a comprehensive guide outlining the essential management skills to run a small business on a day-to-day basis. This includes developing a business plan and sources of finance.

BBC interview

Rory Burke was educated at Wicklow and Oswestry. He has an MSc in Project Management (Henley) and degrees in Naval Architecture (Southampton) and Computer Aided Engineering (Coventry). After working internationally on marine and offshore projects, Rory set-up a publishing business to strike a work-life balance between sailing and writing. Rory is a visiting lecturer to universities in Britain, America, Canada, HK, New Zealand, Australia and South Africa.